LOW-COST HOUSING TECHNOLOGY

An East-West Perspective

Other Pergamon Titles of Interest

PAMA	Low Income Housing: Technology and Policy
HOWELL	Your Solar Energy Home
EGGERS-LURA	Solar Energy in Developing Countries
	Solar Energy for Domestic Heating and Cooling
FAREED	Industrial Housing Systems — An Evaluation
RAD	IAHS International Symposium on Housing Problems 1976

Related Journals Published by Pergamon

Mazingira

Building and Environment

International Journal of Housing Science and its Applications

Underground Space

Cement and Concrete Research

Disasters

LOW-COST HOUSING TECHNOLOGY

An East-West Perspective

Edited by

L. J. GOODMAN
R. P. PAMA
E. G. TABUJARA
R. RAZANI
F. J. BURIAN

East-West Center, Hawaii, USA

PERGAMON PRESS

OXFORD · NEW YORK · TORONTO · SYDNEY · PARIS · FRANKFURT

U.K.	Pergamon Press Ltd., Headington Hill Hall, Oxford OX3 0BW, England
U.S.A.	Pergamon Press Inc., Maxwell House, Fairview Park, Elmsford, New York 10523, U.S.A.
CANADA	Pergamon of Canada, Suite 104, 150 Consumers Road, Willowdale, Ontario M2J 1P9, Canada
AUSTRALIA	Pergamon Press (Aust.) Pty. Ltd., P.O. Box 544, Potts Point, N.S.W. 2011, Australia
FRANCE	Pergamon Press SARL, 24 rue des Ecoles, 75240 Paris, Cedex 05, France
FEDERAL REPUBLIC OF GERMANY	Pergamon Press GmbH, 6242 Kronberg-Taunus, Pferdstrasse 1, Federal Republic of Germany

First edition 1979

British Library Cataloguing in Publication Data

Low-cost housing technology.
1. Underdeveloped areas - Housing 2. Underdeveloped areas - Poor
I. Goodman, L J II. East-West Center.
Resource Systems Institute
301.5'4 HD7391 79-40197

ISBN 0-08-023250-7 Hardcover
ISBN 0-08-023249-3 Flexicover

In order to make this volume available as economically and as rapidly as possible the authors' typescripts have been reproduced in their original forms. This method has its typographical limitations but it is hoped that they in no way distract the reader.

Printed in the United States of America

Contents

Acknowledgement

The East-West Center gratefully extends special acknowledgement to the East Asia Regional Housing and Urban Development Office, U.S. Agency for International Development, Seoul, Korea, for the grant to support the publication of this book. The grant also made it possible to purchase 2,000 copies of a soft cover edition for distribution to cooperating institutions by the Agency for International Development and the East-West Center.

Preface

The former Technology and Development Institute (TDI) now merged into the Resource Systems Institute of the East-West Center is pleased to collaborate with its partners in Asia and the United States in this two-part study of low-cost housing technology.

This book represents a timely and most unusual contribution to the state of the art in supplying low-cost housing units to low-income families in a variety of socio-economic settings in the Philippines, Indonesia, Thailand, Korea, and the state of Hawaii. The contributing authors are recognized experts in fields of direct concern to providing adequate housing, including (1) innovative low-cost building materials and design concepts, (2) safe water supply systems and sanitation, (3) relevant information exchange systems, and (4) policy links within each country to implement the results of research.

The collection of chapters represents a concerted effort by project directors and their colleagues in the TDI network of cooperating institutions in the housing problem areas, the history of which will be described in the Introduction. This network was created in July, 1973 as a direct result of an international seminar on low-cost construction materials, followed by a major workshop to establish specific priority problem areas common to nations, East and West. It was agreed that low-cost housing for low-income families was needed in *all* countries in the world. It was further agreed that a cooperative problem-solving approach to low-cost housing technology was necessary, stressing the need for multinational and multidisciplinary research and development on a cooperative and coordinated basis. An important ingredient in this cooperative approach was the professional expertise of four senior fellows at TDI who followed up their research activities as contributing authors to cover specialized areas of materials, housing design that takes into account natural hazards such as earthquakes, strategies for in-country linkages and curriculum development. These topics are covered in Part II.

The book is in two parts, with Part I covering Country Chapters, and Part II containing Special Topic Chapters. There are eighteen contributing authors, twelve from Asia and six from the United States. It is truly an East-West book, cooperatively produced, providing many insights for a vital global problem area.

The general focus of the text is on technological innovations in the low-cost housing field with appropriate consideration to a number of economic and sociological factors, such as:

1. Reduction of the cost of an acceptable housing unit, considering building materials, design concepts and land use.

2. Creation of a significant number of new jobs through the net effect of technological innovations on the combined building material and construction industries.

3. Consideration of the attitudes of the people concerned, so that the final product is both useful and usable.

4. Provision of low-cost water supply and waste disposal systems for public health control measures.

As a result of discussions by the majority of the contributing authors in Honolulu in October, 1975, it was agreed that each of the country chapters would focus on the following four points:

1. The book should be envisioned as a resource, which implies that each contributing author should deal with the various aspects of the housing field with which he is most familiar.

2. Each author should emphasize what is unique to his country, or to his project, or to any combination thereof.

3. Particular attention should be given to those aspects of country-specific project areas which can be applied directly, or adapted to other countries.

4. Attention should be devoted to problems and activities, particularly in the context of common problem areas and kinds of solutions being generated to solve these problem areas. For example, each country chapter should give some attention to the necessary in-country linkages between the R&D institution and policy-making/implementing groups such as national housing authorities and construction industries.

The chapter by each of the senior fellows who have been intimately involved with the project focuses on his particular area of knowledge and reflects his contribution to the project. These areas are itemized in the Table of Contents.

The role of TDI was one of catalyst, coordinator, and researcher. This included (1) convening miniconferences and roving workshops to provide a forum for on-the-spot effective exchange of results, problems and ideas, (2) providing necessary input regarding new research ideas on low-cost construction materials and design concepts, and a strategy for developing in-country linkages with various groups from public and private sectors concerned with the housing problem, (3) initiating cooperative evaluation of Research and development activities at network institutions, and (4) implementing plans for relevant information exchange within the network and with other organizations involved with similar activities in different parts of the world.

This volume represents one outcome of the project.

The coordination center for the entire project was shifted to the Asian Institute of Technology, Bangkok, Thailand in October, 1977.

The general editors wish to extend warmest thanks to their many colleagues at the cooperating institutions in Asia and the United States who have performed the necessary research and development activities to make this manuscript a reality. Special acknowledgment is accorded colleagues for their role in coauthoring a paper from which excerpts were utilized both in the Introduction and Chapter 1, Louis J. Goodman, Albert G. H. Dietz, Hasan Perbo and Fredrich Burian, "Problems and Issues of Low-Cost Housing", East-West Center Technology and Development Working Paper, September, 1974. The project book is consistent with the East-West Center's mandate to foster better relations and understanding through cooperative study, research and training. It is also consistent with the general goals of all Center programs, with particular attention to enhancing the quality of life among the peoples of Asia, the Pacific and the United States. Indeed, this book symbolizes a true partnership approach to solving a critical problem area, East and West.

Louis J. Goodman
Ricardo Pama
Ernesto G. Tabujara
Reza Razani
Fredrich J. Burian

List of Contributors

A. Country Chapters

1. Indonesia	Prof. Hasan Poerbo	Senior Lecturer Department of Architecture Institute Technology Bandung Bandung, Indonesia
	Ir. Albert Kartahardja	Director, Directorate of Building Research Bandung, Indonesia
2. Korea	Dr. Sung Do Jang	Principal Investigator and Head Ceramic Materials Laboratory Korea Institute of Science and Technology Seoul, Korea
	Mr. Hang Koo Cho	Manager, Technical Division Korea National Housing Corpo- ration Seoul, Korea
3. Philippines	Dr. Ernesto G. Tabujara	Professor of Civil Engineering and Executive Officer Building Research Service University of the Philippines Diliman, Quezon City, Philippines
	Prof. Geronimo V. Manahan	Associate Professor of Architecture and Planning University of the Philippines Diliman, Quezon City, Philippines
4. Thailand	Dr. Tongchat Hongladaromp	General Manager Expressway & Rapid Transit Auth. of Thailand Bangkok, Thailand

	Dr. Ricardo Pama	Associate Professor of Structural Engineering Asian Institute of Technology Bangkok, Thailand
	Dr. Seng-Lip Lee	Professor and Head, Department of Civil Engineering University of Singapore Singapore
5. United States State of Hawaii	Mr. David Firth	Former Director Hawaii Community Design Center Honolulu, Hawaii

B. Special Topic Chapters

1. Iran	Dr. Reza Razani	Professor of Civil Engineering Pahlavi University Shiraz, Iran
2. Philippines	Mr. Eduardo Canela	Research Associate Institute for Small-Scale Industry University of the Philippines Diliman, Quezon City Philippines
3. U.S.	Prof. Louis Goodman	Assistant Director Resource Systems Institute East-West Center Honolulu, Hawaii
	Dr. Albert Dietz	Professor of Building Engineering Massachusetts Institute of Technology Cambridge, Massachusetts
	Dr. Floyd O. Slate	Professor of Engineering Materials Cornell University Ithaca, New York
	Mr. Fredrich J. Burian	Research Associate Resource Systems Institute East-West Center Honolulu, Hawaii

Introduction

HISTORY OF THE EAST-WEST CENTER TDI LOW-COST HOUSING PROJECT

I.1. BACKGROUND

Today, low-cost housing represents a serious national problem in both developing and developed countries. The acuteness and magnitude of the problem are obviously more pronounced in developing societies, but increasingly the issue of low-cost housing cuts across economic, social, technological and political issues. National governments must adopt and implement appropriate policies and priorities for housing, and multinational cooperative research and development programs must be initiated and coordinated to search for solutions to the multifaceted problems of providing decent housing for the untold millions of people who are unable to own or rent minimal essential shelter. The argument for multinational and (interregional) cooperation rests on the premise that many unknown or little-known efforts are being made in this field globally with too little and, in many instances, no interchange of problems and potential solutions.

Given this background, the East-West Technology and Development Institute (TDI) stimulated the establishment of a network of cooperating institutions in Asia and the United States to undertake an action-oriented project concerning low-cost construction materials and design concepts for low-income family housing units in a variety of socioeconomic settings. This project included activities comprising research, professional development, and graduate degree study. The objective of this cooperative project was the provision of better housing for low-income family groups. The cost of housing to be studied was based on a general "rule of thumb", adopted by a number of agencies concerned with this problem, which states that the cost should not exceed 2 — 2.5 times the annual income of a family.

I.2. NETWORK OF COOPERATING INSTITUTIONS

The following institutions constituted the cooperative network for the action-oriented project discussed in this book. Those shown with an asterisk (*) represent the embryonic network in 1973 which consisted of five institutions. The number increased to eleven, with the University of Singapore joining the network in October, 1975, and Pahlavi University in August, 1976. In some cases, the institutional project team was multidisciplinary. For example, the team at the University of the Philippines was represented by the Building Research Service, comprised of the Colleges of Architecture and Engineering and the Institute of Planning.

The first important meeting was the conference on "Relevance in Engineering Education for Asian Needs with Workshop on Nature and Concept of Intermediate Technology[1]* held in February, 1972. This seminar-workshop involved a panel of eminent technologists from key Asian institutions of technology. The main result was a summary of defining characteristics of appropriate technology which covered some examples of projects of common concern to both East and West, including low-cost housing. The second conference was a research seminar on "Low-Cost Construction Materials[2] held in November, 1972.

Participants were invited from the Philippines (University of the Philippines and Mindanao State University), Thailand (Asian Institute of Technology), Korea (Korea Institute of Science and Technology), Indonesia (Institute of Technology, Bandung), India (Central Building Research Institute, Roorkee), and the United States (University of Hawaii, Cornell University, Georgia Institute of Technology's Industrial Development Division, Agency for International Development, National Bureau of Standards, Department of Housing and Urban Development, and Technology and Development Institute, East-West Center).

The participants were administrators, engineers, and architects who had relevant practical and research experience with low-cost construction materials and designs in their respective countries.

The seminar was conducted using case studies on both practical and research experience, presentation of country problems, open discussion on topics presented, and a group report that considered future research and development needs as well as institutional cooperation in the problem areas. The focus was on the following specific objectives:

1. To exchange experiences and to generate innovative concepts in low-cost construction materials and designs for housing and a variety of public works projects.

2. To explore cooperative endeavours in these research and development areas.

3. To exchange information on country peripheral problems related to low-income housing and low-cost public works projects.

4. To provide specific guidelines for a later workshop on low-cost housing materials and design concepts.

As a result of participant interest and interaction in the November seminar, a workshop was planned and designed for July, 1973 with the theme "Feasibility, Design and Implementation of Low-Cost Housing for Low-Income Families."[3] There were 30 participants from 11 countries in Asia, the Pacific and the United States. The July workshop had as its objectives:

1. To discuss a number of prepared case studies regarding low-income housing projects from inception to implementation, including project feasibility studies, related policy and decision-making, cost data, etc. Both successes and failures were to be treated in this regard.

2. To exchange experiences and generate new ideas regarding low-cost building materials and design concepts for low-income housing to include the following:

---

*Superscript numbers refer to notes and references at the end of each chapter.

*1. Asian Institute of Technology, Bangkok, Thailand

*2. Institute of Technology, Bandung, Indonesia

*3. Korea Institute of Science and Technology

*4. University of the Philippines, Manila, Philippines

*5. University of Hawaii, Honolulu, Hawaii

6. Cornell University, Ithaca, New York

7. Directorate of Building Research/United Nations Regional Housing Center, Bandung, Indonesia

8. Hawaii Community Design Center, Honolulu, Hawaii

9. Korea National Housing Corporation

10. University of Singapore, Singapore

11. Pahlavi University, Iran

I.3. THE EAST-WEST TECHNOLOGY AND DEVELOPMENT INSTITUTE PROGRAM

A. Evolution of Program

Three research conferences and one workshop held at TDI were basic to the evolution of the concept of a cooperative action-oriented research and development program in low-cost housing materials, design and construction. In addition to these formal activities, a series of in-house discussions between TDI staff and senior fellows took place during the months of January and February, 1972 to explore in depth such concepts as technology transfer, intermediate technology, adaptive technology, and appropriate technology.

Earlier, one of the general editors had prepared a paper proposing a Technology Adaptation Research Program (TARP) in concert with the University of Hawaii College of Engineering. The primary objective of the TARP proposal was to establish an R&D interdisciplinary program oriented toward the goal of improving living standards in low-income countries. This was proposed in November, 1971 and became part of the College of Engineering Six-Year Plan. TARP would initially focus on applying technology within an interdisciplinary framework to meet the needs for low-income housing in Asia and the Pacific. Structural, public health (water supply and waste disposal), and electrification requirements were to be an integral part of the research program. Attendant social and cultural considerations would ultimately be included in the studies. Interdisciplinary interaction among architects, economists, engineers, and social scientists would provide the necessary framework to view the totality of the numerous problems inherent in applying low-cost housing technology to different socioeconomic environments. Linkages would be formed with key institutions in Asia and the Pacific region to provide input for relevant research and training in the utilization of indigenous materials and skills for low-cost housing.

(a) Selection of materials;

(b) Adequate structural design considering natural hazards;

(c) Efficient and economical construction methods;

(d) Functional roofing systems;

(e) Sanitary considerations.

3. To explore further cooperative endeavors in the above-mentioned research and development areas.

Among the priority items identified by the workshop were (1) the need for a well-coordinated cooperative research and development program in low-income housing, (2) the need for realistic structural and sanitary standards, (3) innovations in roofing systems, and (4) the establishment of an information exchange network among the institutions involved in low-cost housing research and development.

Then, Dr. Albert Dietz of the Massachusetts Institute of Technology and Prof. Hasan Poerbo of the Institute of Technology, Bandung, Indonesia joined TDI as senior fellows in September, 1973 to assist in the preparation of a research proposal seeking external funding for a cooperative research and development effort initially linking five institutions. These institutions were the University of the Philippines in Manila, the Bandung Institute of Technology in Indonesia, the Asian Institute of Technology in Bangkok, the Korea Institute of Science and Technology in Seoul, and the University of Hawaii.

A mini-conference[4] was held in May, 1974 to complete, on a cooperative basis with the project directors from each of the aforementioned cooperating institutions, the preliminary housing proposal prepared by Dr. Dietz, Professor Poerbo, and TDI staff. Another objective of the miniconference was to initiate planning for an initial Roving Workshop scheduled for January, 1975. The roving workshop concept was an important component of a continuing series of programmatic activities in the low-cost housing technology area designed to strengthen further the embryonic network of cooperating institutions.

In summary, the TDI program concerning Low-Cost Housing Technology resulted from two and a half years of careful planning and searching for relevant program orientation in the context of establishing a network of cooperating institutions in both East and West to perform the necessary research and development on a partnership basis. It is noteworthy that the May 1974 miniconference resulted in complete agreement on the final form of the housing proposal to be submitted to potential external funding agencies.[5,6]

B. Objective of Project

This cooperative action-oriented project was concerned with adapting, innovating, and diffusing low-cost construction materials and design concepts for low-income family housing units in different socioeconomic environments. The major objective was to enhance the capabilities of participating institutions to develop their R&D programs more effectively through collaborative study and exchange of common problems and solutions than by working alone.

The vehicle to accomplish overall project objectives was the establishment of a network of cooperating institutions in Asia and the United States to work as partners on the necessary research and development. The project dealt primarily, but not exclusively, with the production aspect of housing as part of the formative processes in the development of human settlements. Clearly, the system of influences and constraints outlined previously strongly affects the production process. The prime objectives were (1) the provision of better housing and (2) the reduction of costs, through action-oriented research and development by research institutions in cooperation with other bodies associated with housing. Utilization of local materials and skills, innovative designs, and application of the most appropriate technologies were central problems in the project. In-country linkages with national housing authorities, developers, construction industries and financial institutions were considered essential.

C. *Methods, Procedures, and Agenda of Activities*

The project strategy was formulated to encourage country participants to decide for themselves which area of research they wished to pursue within the general framework of the cooperative research plan. It was felt that participating institutions would have their own strengths and weaknesses, and conditions would be different from country to country. Cooperation among the institutions was viewed as a continuous and systematic exchange of experience and R&D information, and, whenever possible, also an exchange of personnel and mutual assistance. The collective increase in knowledge would be broadened to include a systematic gathering and examination of experience in other parts of the world. The TDI project team, including both staff and senior fellows, planned to follow through with cooperative research endeavors with the institutions in the network to provide the necessary input regarding new ideas on low-cost housing materials, design, and construction; this includes implementation of plans for the recommended information exchange network. The following activities occurred in the fiscal years 1975 and 1976:

1. Research

A continuing review, evaluation and dissemination of available information in the area of low-cost construction materials and innovative design concepts by each of the cooperating institutions, concentrating on the following topics, which received special attention in the past seminar and workshop series: (a) selection of materials, (b) adequate structural design considering natural hazards, (c) efficient and economical construction methods, (d) functional roofing systems, and (e) sanitary considerations.

Additional related research areas studied the role of national housing authorities and other in-country linkages such as construction industries to create an awareness and understanding of the functions and activities of each group concerned with the housing problem.

2. Roving Workshops

The concept of a roving workshop was one of the recommendations TDI received from the July, 1973 workshop participants. The primary intent of the roving workshop was to provide on-the-spot observations of actual research and development activities underway at each of the cooperating institutions, and to examine the capability and potential of each to participate in an information exchange network.

Roving Workshop I convened in January, 1975 with a core group of two representatives each from the five institutions involved in the embryonic cooperative network. This core group of ten persons along with two TEI staff, one senior fellow and a participant from Cornell University spent five days each in Indonesia, Philippines and Hawaii. Within each country, the core group was augmented by an additional

30-40 persons coming from a representative cross section of university, government and private sectors with demonstrated interest and competence in, concern for, and power to implement housing for low-income families.[7] Roving Workshop II was held in October, 1975, with a similar sequence of events in Thailand, Korea and Hawaii.[8]

These two workshops were planned and organized in such a way that the representatives from each of the cooperating institutions would have an opportunity to (a) mutually explore and share the broad aspects of the numerous problems inherent in adapting, innovating and diffusing low-cost construction materials and design concepts for low-income family units in different socioeconomic environments, (b) delineate and document through video tape and slides the kinds of research and development presently being carried on, both in the laboratory and in the field, (c) discuss the kinds of information needed to implement the cooperative R&D in a meaningful way, and (d) establish an effective method of coordinating efforts, including the exchange of relevant information.

The participants agreed that Roving Workshop I was highly successful in satisfying the basic objectives of the Workshop. Much of the credit for this success was due to the administrative officials and project directors from the cooperating institutions in the countries involved with this first Roving Workshop. The Workshop was also highly effective in pulling together the various institutions in the network so that the cooperative project represented a block-building process in the ultimate establishment of a world-wide network that must be organized for problem solving in low-cost housing technology for developing countries. Related to this most important matter was the twofold need for developing relevant in-country linkages, and expanding the embryonic network of cooperating institutions. A direct result of the Roving Workshop was the Directorate of Building Research/UN Regional Housing Centre in Bandung and the Korea National Housing Corporation in Seoul became the first significant in-country linkages in the network. Cornell University and the Hawaii Community Design Center also joined in the network, representing additional U.S. institutions to work as partners with our colleagues in Asia.

This experience of traveling, living, and working together for a 15-day period in three countries was viewed as a most rewarding experience from both professional and personal points of view. Indeed, this type of activity symbolizes the mandated goal of the East-West Center "to promote better understanding and relations among the peoples of East and West through cooperative study, research and training."

Roving Workshop II took place in October, 1975 with scheduled visitation and seminars in Thailand, Korea, and Honolulu. Again, this activity was evaluated by the participants as highly successful in providing on-the-spot exchange of R&D results; further strengthening the solidarity of the network; pulling together the various groups from public and private sectors to assure implementation of relevant R&D results; confirming the need for establishing an effective information exchange system; and creating an awareness of the need for attention to sociological factors in supplying low-cost housing for low-income families.

3. Documentation and Training

A systematic visual documentation strategy was employed for the duration of this project. Selected features of the emerging cooperative R&D research conducted at network institutions were documented, using a suitable range of audio-visual materials and methods. Audio and visual information collected periodically by appropriate network personnel together with comprehensive documentation during the Roving Workshops were used to track project development. In 1976, professional development interns from the network institutions spent four months at TDI developing documentation skills, examining available low-cost housing documentation and planning for collection of additional documentation, producing media presentations from available materials, and developing a cooperative framework for inter-institution information exchange.

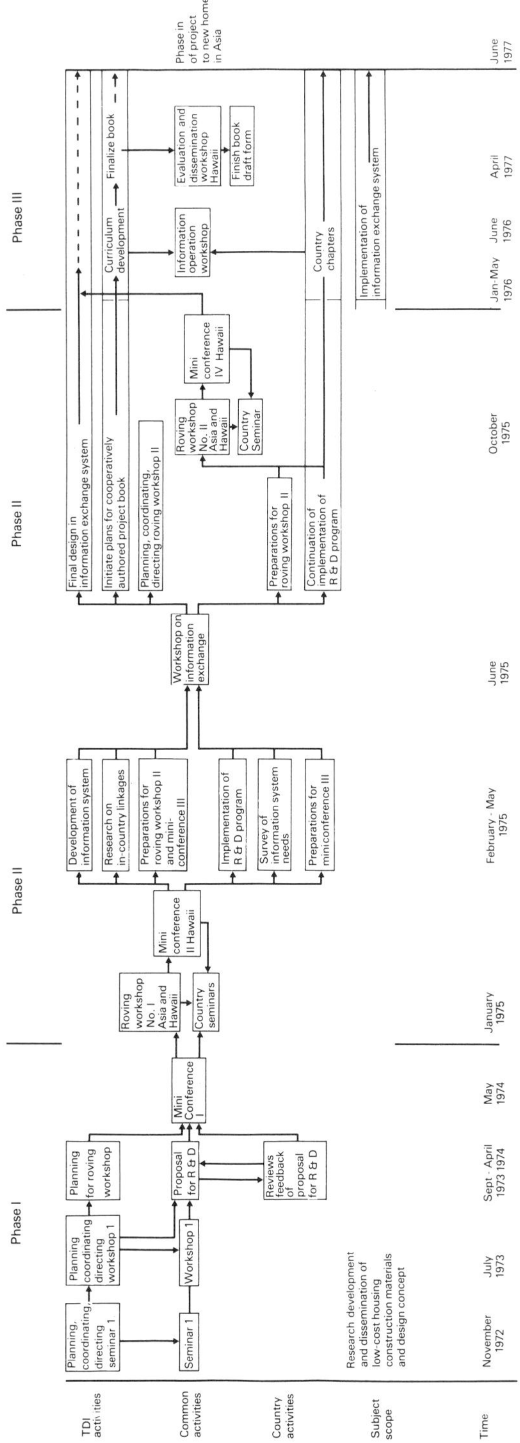

Fig. I. 1. Cooperatively designed plan for flow of activities, 1973-1977

D. *Project Cycle*

Figure I.1 is a diagrammatic flow sheet that summarizes the various phases of the project from its inception, including future plans, and shows the relationships of the various participants. TDI activities are at the top of the chart, activities carried out by the countries and Research Institutes are at the middle, and common activities are between. The scope of the activities is shown below the flow diagram and the approximate times are at the bottom. The block at the left indicates the activities already carried out and supported by TDI.

E. *Project Outputs*

The project was able to extend the impact of the initial Low-Cost Housing Network of cooperating institutions through (1) site visits to network countries to evaluate on-going R&D activities and meetings with a variety of individuals and groups in each country involved in low-cost housing, and (2) the establishment of a practical information exchange system to support the diffusion of low-cost housing technology information within and among Network countries. These mechanisms were designed to enhance the viability of in-country and international linkages among relevant groups such as national housing authorities and construction industries. The overall result has been a pulling together of necessary forces from private and public sectors to implement the results of the R&D institutions in a more effective manner than has generally been the case.

I.4. PROGRAMS AT COOPERATING RESEARCH AND DEVELOPMENT INSTITUTES

In the Asian countries initially involved in this program, the research and development institutes were in a position to play a key role in the development and implementation of cooperative programs. They were unique in these areas because they had the capability of doing the necessary research, taking promising research results through the development stage, and bringing them to public and private agencies in a position to implement them. Indeed, the research institutes could be actively engaged in the implementation process.

To carry out these functions successfully, the research institutes needed to stretch their own scarce resources, which were limited legislatively primarily to technology and design. They needed to draw on the resources of other agencies working on the broad aspects of public policy, the social aspects of housing, labor, population, economics, and finance. And they needed further to reach out beyond their own particular regions to explore common efforts carried on in other regional institutes, and to explore common requirements bearing on housing in those regions. Even beyond this, however, they needed to be able to recognize similar problems in other parts of the world, such as Latin America and Africa, where solutions which were applicable to Asian areas might already have been sought and found. Finally, they needed to draw upon the experience of areas where industrialization and technology of housing are considerably advanced, such as Europe, North America, and Japan, to determine not only to what extent the technology is applicable, but what the social, political, and economic constraints have been, how the market has been aggregated, and how the whole process has been organized and managed.

Various efforts in international cooperation in housing have already been made. As matters now stand, however, the individual research institutes, although they occupy potentially key positions in the housing picture in their respective countries, tend to operate in isolation from each other and from the activities of the rest of the world. They lack good mechanisms to bring about closer liaison among themselves, and to obtain access to developments elsewhere. They are, therefore, less effective than they could be in this critically important area.

The aim of the TDI program was to encourage the development of such links among the research institutes and to extend these links to other institutes through collaborative action-oriented research and development and the development of a common information system.

The next two sections summarize current and planned research and development activities at the cooperating institutions, along with illustrative examples.

I.5. CURRENT AND PLANNED RESEARCH – NETWORK INSTITUTIONS

Although the network has only been in existence for a relatively short time, member institutions have already begun to distinguish themselves in the vital area of research and development. The Roving Workshop concept in particular has provided an excellent forum for exchange of common problems and generation of solutions in both innovative building materials and design concepts for appropriate low-cost housing in a variety of socioeconomic settings. The general editors stressed the need for continuing collaborative research in areas of innovative building materials, design concepts, public health factors, in-country linkages and the necessary foundations for an effective information exchange system in order to maximize the expertise available from both East and West.

The following is a summary of a representative cross-section of research conducted at each institution. The intent of this summary and the following section on examples of low-cost housing in each of the network countries is to provide the reader with a brief view of what each cooperating institution is doing in this vital field. The details of country research and development will be covered in each of the country chapters in the first volume.

Thailand – Asian Institute of Technology (AIT)

1. Mechanical properties of bamboo-reinforced slabs;

2. Design and evaluation criteria for low-cost housing;

3. Design of asbestos-cement houses for low-income families;

4. Prefabrication of low-cost housing;

5. Mechanical properties of wood-cement composites;

6. Evaluation of a corrugated asbestos cement roof panel;

7. Performance evaluation and design of asbestos cement low-cost houses.

Korea – Korea Institute of Science and Technology

1. Utilization of coal wastes for building materials:
   Ceramic Materials Lab;

2. Utilization of zeolite in waste treatment:
   Environmental Engineering Lab;

3. Foamed polystyrene (styrofoam concrete for building components: Plastic Lab;

4. Study on thermal insulation and fuel economy in building units: Chemical Process Development Lab;

5. Design and development of lime-sand brick manufacturing plant.

Indonesia – Institute of Technology, Bandung (ITB)

Current Research Areas

1. Linkages in wood-based industries as context for identifying what kind of wood-based components for housing can be developed;

2. Product development: wood-based components for housing.

Related Programs

1. Cooperating with regional housing center in Bandung;

2. Survey on linkages in wood-based industries;

3. Survey on housing markets and construction methods;

4. Formulation of performance specification for wood-based components;

5. Development and testing of wood-based components;

6. Prospective field lab: PN Jatilukur;

7. Promoting components for production and marketing.

Indonesia – Directorate of Building Research/UN Regional Housing Centre

1. Binding materials:

   (a) Urea formaldehyde adhesives;

   (b) Latex as binding agent for building board manufacturing.

2. Building element/component:

   (a) Wood-wool board production;

   (b) Development of prefab concrete building elements and components.

3. Housing units:

   (a) Prefabrication of houses;

   (b) Prototype houses.

4. Timber construction:

   (a) Development of prefabricated timber structures.

5. Waste materials:

   (a) Development of particle board from forestry waste material.

Philippines — University of the Philippines

Current Research

1. Information classification and building typology for typical house types in the Philippines;

2. The Barangay as an operational system in the disposal of solid waste in urban areas;

3. Application of methane gas/septic tank system in housing design and layouts;

4. User requirement study of a prototype urban workers' condominium in Metropolitan Manila.

Research Proposals

1. Asphalt impregnated building material study;

2. Investigation of industrial waste materials for building purposes;

3. Identification and locational analysis of inorganic building material sources in the Philippines;

4. Design of building and finishing hardware for low-income housing;

5. Evaluation of industrially produced bamboo laminates for building purposes;

   (a) Long-term physical properties of KAWOOD

   (b) Effectiveness of component designs of KAWMAT

6. Adaptation of traditional multiuse concepts to the development of multifamily housing projects in urbanized areas of the Philippines.

United States — University of Hawaii

1. EW/TEI graduate student research in areas related to low-cost housing;

2. Development of graduate courses concerned with construction materials and construction management and water supply/waste disposal;

3. Cooperating with TDI in developing an information exchange system for the network;

4. Research program: materials (development and testing) and design concepts by graduate assistants, preferably EWC grantees.

United States — Cornell University

1. Study of ways to organize, teach, and do research in a multidisciplinary mode for low-cost housing;

2. Design of adobe and rammed earth houses to resist earthquakes;

3. Modification of existing adobe and rammed earth houses to resist earthquakes;

4. Use of coconut hull fibers to make fibrous concrete for houses;

5. Survey of a Limon, Costa Rica squatter colony to be resettled to determine demographic information and many such things as housing preference, ability for self-help, and their suggestions for solving housing problems;

6. Housing delivery systems in Ghana;

7. Rural and urban basic systems in Ghana.

This list does not include the large number of one-semester projects (some are research) being done by teams of advanced students as part of their course work.

United States — Hawaii Community Design Center

1. *Waianae Self-Help Housing Proposal*
   A self-help housing funding and operational proposal for Model Cities Housing Task Force to construct 180 houses in Waianae.

2. *Waimanalo Church Roof*
   Drawings of roof structure and kitchen space to assist church in fund raising and completion of existing addition.

3. *Kokokahi YWCA Cottages Renovation*
   Using Community Quest Students help in repairing existing cottages, steps, hand-rails, lanai and windows.

4. *Kahuku Housing Corporation Plan*
   General planning assistance to aid residents in maintaining life-style community and housing, for eventual construction of 244 self-help houses.

5. *Solar Energy Proposal*
   Research and evaluation of different solar water heating systems for a CAP proposal for 240 houses on Molokai.

United States — Additional Project Initiatives

A tripartite relationship involving TDI, the University of Hawaii Department of Civil Engineering (UHCE) and the Hawaii Community Design Center (HCDC) has resulted from a series of meetings catalyzed by Roving Workshop I. Plans are under way to design and construct a prototype soil-cement house on State property utilizing a self-help approach. This house will cost approximately $4,000 in an area where low-cost housing starts at $25,000. This represents an excellent example of West learning from East (soil-cement housing has, of course, been developed extensively in the Philippines).

In fact, it was through contact with the Philippines delegation at Roving Workshop I that HCDC first conceived the idea of soil-cement housing for the State of Hawaii.

TDI, UHCE, and HCDC feel that soil-cement housing has great potential in the State of Hawaii for the development of low-cost housing. In addition to providing a *different* type of housing, it will conserve land and energy (this type of house can be clustered, thus it is ideal for an urban setting), make use of cheaper building materials (and eliminate the cost of transporting building materials), maintain current lifestyles (design of the house is extremely versatile), and will make use of ecological utility systems (in addition to providing excellent insulation, thus no energy consumption for cooling, *and* elimination of the normal upkeep on a wood house).

The labor force which will construct the houses will be composed of individuals from low-income community groups. Commitments have been received from low-income groups and individuals to do the actual construction work. In addition, the Project participants hope that low-income people from their respective communities come out and take part in the program.

NOTES AND REFERENCES

1. *Report on Conference on Relevance in Engineering Education for Asian Needs with Workshop on Nature and Concept of Intermediate Technology*, February 23-26, 1972 (Honolulu: East-West Center Technology and Development Institute).

2. *Papers Presented at Seminar on Low-Cost Construction Materials*, November 27-29, 1972 (Honolulu: East-West Center Technology and Development Institute).

3. *Proceedings: Adaptive Technology Workshop II, Feasibility, Design and Implementation of Low-Cost Housing for Low-Income Families*, July 9-20, 1973, Vols. I and II (Honolulu: East-West Center Technology and Development Institute).

4. *Final Report: Mini-Conference on Cooperative Action-Oriented Research and Development in Low-Cost Housing Materials, Design and Construction*, May 16-18, 1974 (Honolulu: East-West Center Technology and Development Institute).

5. *Project Proposal: Cooperative Action-Oriented Research and Development in Low-Cost Housing Materials, Design and Construction*, July 1974 (Honolulu: East-West Center Technology and Development Institute).

6. *Cooperative Action-Oriented Research and Development in Low-Cost Housing Materials, Design and Construction: Supplemental Report Series No. 1*, August 1974 (Honolulu: East-West Center Technology and Development Institute).

7. *Proceedings: Low-Cost Housing Technology Project Roving Workshop*, January 15-28, 1975 (Honolulu: East-West Center Technology and Development Institute).

8. *Proceedings: Low-Cost Housing Technology Project Roving Workshop II*, October 6-20, 1975 (Honolulu: East-West Center Technology and Development Institute).

1

LOW-COST HOUSING: GUIDELINES AND ISSUES

Louis J. Goodman

1.1 THE NEED FOR LOW-COST HOUSING

Housing is a basic human need. Secure shelter therefore represents a basic human right. On October 24, 1970, the United Nations General Assembly adopted the following Declaration on the International Development Strategy for the Second United Nations Development Decade:

> Developing countries will take steps to provide improved housing and related community facilities in both urban and rural areas, especially for low-income groups. They will also seek to remedy the ills of unplanned urbanization and to undertake necessary town planning. Particular effort will be made to expand low-cost housing in both public and private programmes on a self-help basis, and also through cooperatives, utilizing as much as possible local raw materials and labour-intensive techniques. Appropriate international assistance will be provided for this purpose.

Low-cost housing is clearly a high-priority program for all countries in the world. Population growth, urbanization, and the attendant growth of slums and squatter colonies all contribute to the urgency of the problem. The need for new housing alone is so staggering that, according to estimates by the United Nations Center for Housing, Building and Planning, the demand for construction of all kinds during the last half of the twentieth-century will exceed the total volume of building throughout the whole of human history. Their 1973 report on world housing conditions reaffirms that conditions continue to deteriorate at an alarming rate.[1]

If the housing problem is to be solved in a practical way, programs must be formulated that will utilize as many local materials and skills as possible. Only essential materials that will not interfere with primary economic development should be imported. Indeed, housing programs should complement such development wherever possible. Finally, program planners must recognize that the problem cannot be solved by quick recipe but instead requires a multifaceted approach considering (1) planned utilization of land, (2) increased indigenous technical knowledge and skills, (3) increased use of indigenous building materials, (4) promotion of labor-intensive construction technologies, (5) the development of suitable public policy and finance mechanism. The solutions will require interdisciplinary interaction among architects, economists, engineers, and social scientists who will provide the necessary framework to view the totality of the numerous problems inherent in applying low-cost housing technology to different socioeconomic environments. Ultimately, the total spectrum of the problems of human settlements must be addressed, ranging from research and development on building materials and designs to methodology of physical planning, social and functional aspects of low-cost housing, and development of a viable indigenous building industry base.

1.2 SOME PROBLEMS AND ISSUES

There is a vital need for developing and developed societies to conduct cooperative research and development in the housing field on a partnership basis (see Introduction for an example of such a multinational partnership program among research and development institutions). Furthermore, technological adaptations and/or innovations must satisfy a number of economic and sociological factors. Among these are: (1) reducing the cost of an acceptable housing unit, (2) creating a significant number of new jobs through the net effect of technological innovations on the combined building material and construction industries, (3) considering the attitudes of the people concerned, so that the final product is both useful and usable, (4) providing low-cost water supply and waste disposal systems for public health control measures.

Among the many problems and issues that must consequently be taken into account, the following deserve special consideration. They cannot be considered as separate from each other. Indeed, they form a system of influences and constraints that must be viewed as a whole.

Materials, Transportation and Production

1. To the fullest possible extent, materials should be indigenous to avoid contributing to the depletion of often-scarce foreign exchange. Where imported materials can *clearly* contribute to lowered costs and efficient use of indigenous materials, however, they should be considered.

2. A two-pronged approach to the most efficient use of materials, transportation, and products seems to be indicated. Drawing on indigenous materials and keeping in mind the limitation of transport, and utilizing local production facilities, planners can devise a system of housing components that can be assembled by low-skilled or unskilled labor, with minimum supervision and training. The two prongs, therefore, are (1) the application of labor-intensive but efficient field assembly methods to the production of housing, with maximum use of indigenous resources of materials and labor, (2) the shop production of simple, easy-to-assemble components.

3. Materials should be combined composites and assemblages to the fullest possible extent.

(a) To make the most efficient use of materials and to obtain composite properties not attainable with the individual materials acting alone;

(b) To reduce the number of components and parts that must be transported and assembled in the field. If, for example, roof and ceiling can be combined into one composite, the amount of material can often be reduced.

Aggregating the Market

If any building system based on components produced in a shop for assembly at the site is to succeed, the housing market must be aggregated. Demand must be reasonably predictable if programmed production in the shop is to continue at a sufficiently sustained level to obtain the potential cost reduction; sporadic operations can cost more than conventional construction. Some means must be found to assemble enough dwellings to be built as a group.

Labor

In many developing areas, labor falls into two more or less distinct categories. The first consists of trained workers associated with established and growing industries. These people, secure in their jobs and reasonably well paid, tend to form a more or less exclusive group. The second, usually larger, labor category comprises the unskilled or low-skilled underemployed workers, casual laborers with no secure employment doing whatever they can, and receiving uncertain income. Often, they badly need housing but do not have the skills or organization to provide it by conventional building methods. On the other hand, their potential need is extremely large, and they constitute an equally large potential labor pool to provide that housing if the means to utilize their abilities can be organized. A case study of a successful housing subdivision in the Philippines demonstrates how such labor can be organized and encouraged to provide its own housing on a partial self-help basis.[2]

Customs, Preferences and Prejudices

Perhaps in no other phase of existence people have deeper emotions than those related to their homes. They have spent their lives in a home, have become attached and accustomed to it, and are not easily diverted from the way of life associated with it. They are generally strongly conservative in their views respecting what constitutes a home, especially if they own it. Thus, custom, preference, and prejudice constitute strong constraints. They are not, however, insurmountable.

Perhaps no other aspect of housing design is so difficult and important as this. Research is badly needed to determine what constitutes socially acceptable housing. This entails searching for answers and guidance in explorations that are presently under way and it demands that new experiments be undertaken if these answers are not forthcoming.

Climate and Geography

If the cost of housing is to be kept as low as possible, undue reliance cannot be placed on mechanical means of controlling living conditions, but the housing must be designed to consider carefully the climate and weather, taking advantage of those features that promote comfort, and minimizing those that do not. The sun, prevailing winds, temperature and humidity variations, and natural shading or favorable exposure must all be considered. Design must turn these factors to advantage to the fullest possible extent or, where appropriate, minimize their effects. Similarly, geographical features such as terrain, soil, ground cover, earthquakes, and possibility of flooding must be considered more closely than ever, and the simplest and least costly methods of combating, avoiding, or taking advantage of them must be sought.

Much scattered and fragmentary information exists, but that information and methods which have already been devised need to be gathered together and analyzed. If necessary, further research and development must be undertaken to find solutions.

Public Policy

Public policy is of crucial importance. It can encourage housing or it can raise formidable and even insurmountable barriers through public action or inaction. Public policy can encourage or frustrate the development of the level of housing production needed for combined industrialization and rationalized field operation by action or inaction in such vital areas as availability of housing finances, interest rates, establishment of housing banks and savings and loan associations, encouragement of cooperative action, codes, planning and zoning, and provision of sites and services.

Most important, if the political agencies, both nationally and locally, support a steady policy, housing can proceed on an efficient cost-reducing basis. If not, housing becomes erratic and costs rise.

Many national, regional, and local housing authorities have been set up with variable success. These, and housing policies generally, need to be studied and the elements of successful ones analyzed for possible adoption.

Information Acquisition and Dissemination

The need for effective methods of information acquisition, processing, and dissemination is manifest in any organized research and development effort. To a large measure, the internal viability of an R&D organization rests with its ability to scan appropriate information environments, have access to information relevant to its developmental objectives, and identify gaps in information preliminary to the conduct of original research. To just as great an extent, the external viability of an R&D organization depends upon its ability to communicate its findings to appropriate audiences and receive feedback from those audiences.

Access to the most complete information possible is essential to the efficient and successful conduct of research and development. This is particularly true of the complex field of housing. Relying on existing information acquisition and dissemination capabilities of institutions currently engaged in low-cost housing R&D, several recently developed communication strategies (microforms, magnetic tape displays, film projections, among others), may be employed to enhance information exchange among R&D counterparts located at the far-flung corners of the world.

Project Management

There is need to improve the capabilities and skills of managers of a variety of public works projects, including housing, in both East and West. Indeed, this is such a vital issue in many countries that the East-West Technology and Development Institute planned and designed a separate project area concerned with the management of development projects. One of the objectives of this new programmatic activity was to develop a cadre of effective indigenous project managers in a number of countries with the necessary skills to implement successfully a specific project.

1.3 EXAMPLES OF LOW-COST HOUSING RESEARCH AND DEVELOPMENT

There are a number of examples of research and development efforts resulting in low-cost housing units for low-income families in both Asia and the United States.[3]

A project team at the Asian Institute of Technology (AIT) in Bangkok, Thailand is investigating the use of locally produced asbestos-cement sheets for both exterior walls and roofing systems. The results of this team effort are illustrated in Fig. 1.1. The cost does not include land, which is on the AIT campus. The cost of labor is approximately 20% of the total cost.

Fig. 1.1. One of two prototype duplex houses built on the campus of the Asian Institute of Technology, Thailand; utilizing corrugated asbestos cement sheets for both roof and longitudinal walls. It provides approximately 50 m^2 in floor area for each unit with a total construction cost in early 1974 of US $1,250 per unit, or $25 per m^2.

A research team at the University of the Philippines cutting across the Colleges of Engineering and Architecture, and the Institute of Planning, is conducting research on "development of design criteria and methodology for the low-cost, low-rise buildings to better resist external winds". Figure 1.2 shows one of two experimental low-cost houses designed and constructed as a part of this research study. In this case, as in the AIT Prototype, the cost does not include land.

A low-cost housing unit in Korea, including a breakdown of the cost analysis (excluding land) is shown in Fig. 1.3. The exterior walls are constructed of stabilized soil building blocks, and the roofing system consists of asbestos-cement tiles.

Indonesia has long been active in the area of low-cost housing development. A prototype low-cost house built in the early 1970s is illustrated in Fig. 1.4.

In one of the highest cost-of-living areas in the United States, the island of Oahu in the state of Hawaii, the Hawaii Community Design Center, staffed by three VISTA volunteer architects, has designed and built a prototype "Minimum House" intended for self-help construction (see Fig. 1.5). The construction cost in 1974 was $5,500 including all materials but excluding the cost of labor and land. This is indeed a remarkable feat in an area where low-cost housing for low-income families generally falls in the $25-35,000 range.

Fig. 1.2. The round sensor discs on the roof and walls of this experimental house at the Science Garden, University of the Philippines, monitor the wind pressure on the house to test its resistance. The double walls and ceilings are made of locally-produced modularized chipboard panels and the roof is made of galvanized iron sheets. The direct construction costs of this 46 m^2 house in 1973 were US $1,100.

An outstanding illustration of a pilot project adapting local soil and cement as a low-cost building material is found at Mindanao State University, Marawi City, Philippines. A Cinva-ram was used to produce the soil-cement building blocks and floor tiles. The total cost of the house shown in Fig. 1.6 was $600 (excluding land) in 1969-70, for a floor area of 45 square meters. Labor accounts for approximately 30% of this total cost. Research is continuing, investigating soil-cement hollow blocks, soil cement roofing panel systems and bamboo reinforcement.

1.4 SUMMARY

Both the Introduction and Chapter 1 have shown that low-cost housing for low-income families is a problem of prime and immediate importance. We have briefly summarized the many interrelated problems and issues, stressing the need for multinational and multidisciplinary research and development on a cooperative and coordinated basis. The stage is now set for the following chapters in this unique two-part work.

NOTES

1. Report of the Secretary-General, United Nations Economic and Social Council (Committee on Housing, Building and Planning), "World Housing Survey", September 1973 (New York: United Nations).

2. "A Housing Cooperative for Industrial Workers in Iligan City", in *Leadership Strategy in Development Projects: Public Leadership Workshop II, Part II,* October 15-26, 1973 (Honolulu: East-West Center Technology and Development Institute).

Fig. 1.3. This prototype house was built by the Bureau of Housing and Urban Planning, Ministry of Construction, Government, Republic of Korea utilizing stabilized soil blocks and asbestos-cement roofing tiles. With a total floor area of 40 m^2, the cost of construction was US $747 in 1970-71, which can be reduced to $600 with self-help labor, and less than $500 with self-production of blocks. The breakdown of costs is as follows:

Item	Cost	Percent
Foundation and footing	$ 46.7	6.3
Blocks and Block-works	211.5	28.3
Wood works	156.8	21.0
Roofing	80.7	10.8
Mortar finishing	62.0	8.3
Door and windows	66.0	8.8
Water closet	35.5	4.7
Miscellany (Painting, Wall papering, Electricity, Heating-floors)	88.2	11.8
Total	$747.4	100.0%
Self-help labor	122.0	
Self-made blocks	150.0	
Estimated Net	$475.4	

3. *Cooperative Action-Oriented Research and Development in Low-Cost Housing Materials, Design and Construction: Supplemental Report Series No. 1*, August 1974 (Honolulu: East-West Center Technology and Development Institute).

Fig. 1.4. This is one unit of a prototype housing project in Bandung, Indonesia. The project provides housing for low- and middle-income families. The unit shown is one of the low-cost designs being evaluated at the project site. Designed and built by the Regional Housing Center, Bandung, the house is 45 m^2 in floor area, with walls of woven bamboo mats, burned clay tile roof and a total construction cost of $550 in 1970; excluding land.

Fig. 1.5. The Hawaii Community Design Center staffed by three Vista volunteer architects have built a prototype "Minimum House" designed for self-help construction. The house is 71 m^2 constructed of wood with asphalt sheets over plywood for the roof. The cost of the house in 1974 was $5,500 inclusive of kitchen and bathroom fixtures but excluding labor.

Fig. 1.6. This house utilizing soil-cement building block materials for both exterior walls and floor tile has been constructed at Mindanao State University, Philippines. The floor area is 45 m^2 and total cost of construction in late 1969 was somewhat lower than US $600, excluding land. The galvanized iron sheeting roof is sub-functional and research is continuing to develop more functional and lower cost roofing system.

Part I

LOW-COST HOUSING: BY COUNTRY

2

LOW-COST HOUSING IN THE PHILIPPINES

E. G. Tabujara and G. V. Manahan

2.1 HOUSING AND HOUSE-FORMS

The Basic House

For ages the primitives of the Philippines used for their shelters rock caves very similar to those of the ancient Tabon Man of Palawan. At times they would roam the rain-forests seeking shelter in huge trees. In their further desire to protect their abodes from intruders and the evil spirits, broad-leaf plants were ceremoniously placed in the entrances. They later realized that these indigenous materials as well as fibrous plants had some practical purposes in protecting their abodes from the elements. It was simple enough for them to improvise the building of lean-to shacks when they ventured into the plains and fertile flats by the banks of rivers along the coasts of the country-side. During the lulls away from the hunt, they would improve their shelters by devising contraptions by weaving long-leaved mattings. Whenever they made use of these mattings as their shelter from the harsh sun, they would prop the mattings, hence the early beginnings of the lean-to type of construction. This led to the emergence of the nipa hut.

The Bahay Kubo

The nipa hut, or *bahay kubo* was and still is the typical house type in agricultural areas of the country. The materials used vary according to the materials that abound in the region. For example in regions where cogon, bamboo, nipa or rattan grow, these are used to advantage.

Housing of the Filipino mountain dwellers vary from the cave dwellings of the Tasaday tribe, the four-trunked house of the Ifugaos and the tree-houses of the Higaonons. The Ifugao house consist of wall panels, floors, and framework of hardwood. Layers of grass are used for thick thatchings. The area of the elevated house is from 16 to 20 m^2.

In the fishing village of Southern Philippines, away from the devastating typhoons, the typical rural house is on stilts right above the water. Materials used are grass reeds for walling and a thatch roof. Logs having a diameter of 10-15 cm are used as framework.

Bi-sectoral View of Housing

The *bahay kubo* was standard house form for the masses even in old Manila. It was only in the 1920s when the traditional building methods for housing the masses drastically changed. Two strong development waves can be attributed to this phenomenon:

1. The adoptation of the Burnham Plan for Manila during the turn of the century, and its consequent implementation in the 1920s when the country experienced a monetary windfall from the coco excise tax. The Burnham Plan provided the momentum for the modernization of the city on a broader geographic scale. In effect the projects and improvements initiated in Manila became a model to be followed by the smaller cities of the country.

2. The promulgation of the building and zoning code of Manila in the 1920s prevented the wide use of the nipa hut and thus inhibited the use of traditional materials which were very fireproof. The districts of Tondo, Santa Cruz and Sampaloc which had dense communities of low-income families had to use masonry and wood for their house instead.

These two techniques adopted by government of coping up with growing urbanization in the private city resulted in viewing housing for the masses as a two-pronged situation. House forms and materials for the urban sector has since then been viewed differently from the greater majority of the people's housing situation.

Fig. 2.1. Typical Rural Nipa Hut (Bahay Kubo) in Luzon Island

Fig. 2.2. Typical Timber House of the 1950s Leyte Island.

2.2 POLICY FORMULATION FOR HOUSING

Policy-making with the respect to housing in the Philippines has been adopted and pursued by government as far back as 1934. But for 40 years funding support has been intermittent, for housing has been bereft of a comprehensive program in carrying out the incipient state philosophy for housing the greatest number of the lowly as implied in the development of Barrio Obrero. Furthermore, the growing strength gained by socially conscious housing policy-makers was disrupted and delayed by the ravages of World War II. This was further compounded by the country's loss in housing inventory as a result of the war. This was particularly true in the urban areas of the country. The development effort was thus geared to economic and social reconstruction. Consequently, the economy at that time was not in a position to support a viable housing program. These problems were complicated by the disparate distribution of population which conglomerated towards the principal sources of employment. By the mid-sixties the national housing need was established at 470,000 dwellings or 12 new houses per 1,000 population. Despite the absence of regional disaggregates of the housing figures it has been set that 140,000 units are needed in the urban areas and 330,000 units in the rural areas of the country.

An analysis of the census data of 1970, shows that there are some 6,308,632 households in the country which gives an average of 5.8 persons per family. On that census year, there existed 5,186,873 dwelling units. There is an implication, therefore, that a housing deficit of 1,121,759 units exists. This figure is assumed to be a high estimate because of the typical Filipino characteristic of extended families and doubling up. Shortages therefore, are underestimated.

Housing Shortage

Pursuing the assumptions made by the Social Service Unit of NEDA, with regards to population, the urban-rural distribution that is based on actual household enumeration were then correlated with the available housing inventory, to wit:

	Rural	Urban
Households	4,300,794	2,007,838
Dwellings	4,097,545	1,089,328
Apparent shortage	203,249	918,510
Percent apparent shortage	4.6%	40.7%

Government has also pin-pointed that the housing shortage is more apparent in Metro Manila where it is estimated that two-thirds of the country's shortage of urban dwelling units are most needed.

It must, however, be recognized that the following phenomena must have to be included to the annual housing deficits:

1. The creation of some 200,000 new families every year;

2. The existence of 389,873 "barong-barongs". Even though these structures are habitable, these would have to be environmentally upgraded;

3. The reduction of the housing stock due to deterioration, calamities, and poor habitability. The number can be anywhere close to 250,000 units.

Housing Supply

Housing supply in the Philippines is dependent on effective demand. The slow pace of housing development is however attributed to the population's low-effective demand. This is greatly due to the lack of financial resources of those who are in need of housing as well as the lack of relevant institutions and systems that can help translate their housing problems in physical form. For the past decade, the rate of housing supply is merely two dwelling units per 1,000 population. Most of these are generated by the private sector. It is estimated that this can amount to 80-90% of the supply. The housing market which various financial institutions of the country services are categorized in the following:

1. Private banks and non-bank financial institutions cater to the upper income bracket;

2. Social Security System and GSIS-financially assisted projects are often grabbed by middle-income families;

3. Government Social Housing schemes which are supposedly for the low-income, most often go to the lower-middle income families.

Thus far, the lower-income classes of the nation either rent or squat (if not caught) in urban areas, or on the basis of the "bayanihan" system, with the help of their friends and relations erect a traditional "kubo" in the rural areas. Government has protected the renters with a rent-control law for tenants paying less than $40 a month. On the other hand, rural families have been granted by the Development Bank of the Philippines the opportunity to get loans up to $670 for their rural houses.

Concept of Affordability

As a result of the insurmountable problems faced in financing the infrastructure of housing for the country, the concept of affordability of the household has cropped up in the policy-making levels of the government. In terms of existing building quality and standards accepted by financial institutions, urban families must have an annual income of $10,000 to afford open-market housing. Housing analysts have estimated that these compose only 12% of the urban families of the country. The next bracket belong to the $533-$1,332, which number about 23% of the urban families. They can afford home ownership if they are extended long-term financial assistance at reasonable interest rates and with a minimum of land development and construction costs.

The residual consists of 65% or urban families who cannot afford adequate shelter even at reduced rates. Government aims to provide for them social housing under subsidized or partially-subsidized schemes. Within given specific housing development areas, cross-subsidies shall be tried as an alternative. In the rural areas where over 70% of the Filipinos have annual incomes of $133 per family, housing costs are not as high as in the urban areas. What is urgently required are sufficient provisions for utilities and services as well as appropriate settlement patterns to extend to the populace the benefits of modern service facilities. The use of indigenous materials will require better technology and intensive studies.

Fig. 2.3. PHHC (Philippine Homes & Housing Corporation) Houses in Quezoncry.

Table 2.1 Summary Table on Affordability of Housing Benefits and Minimum Payment Terms
National Housing Authority June 1976

Annual income class	Distribution of families[a]		Monthly ave. income[a]	Expenditure for housing[b]	Weighted expenditure	Interest rate	Housing benefit	Amount 100 hhlds.	% Fund allocation
Indigent	(h_i) [c]		(y_i)		$(q_i y_i)$		(c_i)	$(h_i c_i)$	
3,000 below	35.30		156.38	15.64	15.64	3% 25 years	3298.11	116423.28	6.68%
Social									
3000-3999	15.1	35%	287.75	57.55	20.14	6% 25 years	8932.16		
4000-4999	9.8	22	371.83	74.37	16.36	"	11542.73		
5000-5999	7.7	18	453.50	90.70	16.33	"	14077.26		
6000-7999	11.0	25	592.58	114.52	28.63	"	17774.29		
	43.60	100%		Average	81.46	6% 25 years	12643.14	551240.90	31.16%
									38.29%
Economic									
8000-9999	7.1	39%	742.75	185.69	72.42	10% 25 years	20434.67		
10000-14999	8.4	46	1007.17	251.79	115.82	12% 25 years	23906.59		
15000-19999	2.7	15	1435.17	358.79	53.82	14% 25 years	29805.75		
	18.20	100%		Average	242.06	11.7% 25 yrs.	23437.42	426561.04	24.46%
						Sub-Total		1094225.23	
Open Market (may be for commercial/industrial development aside from residential housing) if for residential housing:									
20000 over	2.9		2786.75	696.69		18% 25 years	45912.50	649534.71[c]	37.25%
	100.0% population			Average interest Earnings		12% 25 years[c]		1743759.94	100.00% Funds

NOTES: [a]Distribution of Families taken from Table 5 BCS '71 Survey Hhlds Series 34; Average Monthly Income from Table 2. Note, however, that if widely accepted '76 estimates were available, their use would be more appropriate.

[b]Computed at 10% of Monthly Ave. Income for INDIGENT, 20% for SOCIAL, 25% for ECONOMIC; these are the q_i referred to in note [c] 20% is guide used by NEDA Housing Sector Study, January '75.

[c]Reference is made to FINANCIAL FRAMEWORK AND MODEL FOR FINANCIAL PLANNING which expresses the financial situation of the Authority as the simultaneous occurrence of two equations: $\Sigma h_i c_i \neq a = A$ and $\Sigma h_i q_i y_i \neq fa = f'A$ where f is the annuity factor appropriate to the open market; a, the amount of open market funds, f' the annuity factor for the average interest earnings of Authority funds and A the total amount of funds. In this case $\Sigma h_i c_i = 1094225.23$ $\Sigma h_i q_i y_i = 8509.24$; f = .015174 or 18% 25 years; and f' = .010532 or 12% 25 years. Solving the two equations simultaneously yields A = 1743759.94 and small a = 649534.71.

Fig. 2.4.

Fig. 2.5.

Fig. 2.6. Rural Houses, Llyte Island, Nipa roof, bamboo walls, timber stilts.

2.3 MILESTONE LEGISLATION IN HOUSING

Government efforts in housing have proliferated among various agencies until President Marcos on 31 July 1975, issued Presidential Decree 757 creating the National Housing Authority and dissolving all existing housing agencies. The urgency to concentrate all government housing efforts, resources, functions and activities was thus realized.

The Housing Program

Under Presidential Decree 757, a comprehensive and integrated housing program shall be prepared covering among others, housing development and resettlement, financing schemes, delineation of government and private participation. Also, the program called for specifying the priorities and targets in accordance with the integrated National Human Settlements Plan prepared by the Human Settlements Commission. The factors to be considered in the preparation of the program are the following:

1. The management of urban development to promote the economic and social well being, stabilize the physical mobility of the people, and facilitate industrial growth and dispersal;

2. The conservation of land for housing development as well as the regulation of land use to achieve optimum urbanization patterns;

3. The organization of public and private resources into financial intermediaries to meet the demand for housing, including provisions for incentives and facilities to broaden the private sector's participation in housing investments;

4. The extensive use of building systems, which shall maximize the use of indigenous materials and reduce building costs without sacrificing sound engineering and environmental standards.

Objectives of the Housing Program

The National Housing Authority was also given the task to develop and implement the housing program for the country. It was organized as a government corporation directly under the Office of the President having the following purposes and objectives:

1. To provide and maintain adequate housing for the greatest possible number of people;

2. To undertake housing development, resettlement or other activities as would enhance the provision of housing to the people;

3. To harness and promote private participation in housing ventures in terms of capital expenditures, land, expertise, financing and other facilities for the sustained growth of the housing industry.

Basic Principles of the Housing Program

To give flesh to the objectives of the NHA, the agency has adopted the following basic principles:

1. Housing is a process of development. The approach should be group-oriented. Emphasis must be given to environmental hygiene;

2. Housing is a concern of everybody. Slums are not the problems of government alone;

3. The thrust of the housing effort is the very low-income group;

4. Land use must be maximized in urban areas;

5. Housing must be affordable;

6. All agencies of the government must contribute to the housing effort;

7. The private sector must be tapped for housing; and their resources coordinated.

The Authority was given an authorized capital of $500 million which shall be subscribed by the Republic of the Philippines. Upon the approval of the decree, $50 million has been appropriated, $5 million of which was released upon the organization of the Authority and the balance when needed. Thereafter, the NHA shall receive $50 million for every subsequent fiscal year for a period of 9 years which shall be included in the General Appropriations Act.

Housing Agencies Abolished by the Decree

The decree abolished the following agencies and their functions, assets, and records were transferred to NHA:

1. The People's Homesite and Housing Corporation (PHHC);
2. The Presidential Assistant on Housing and Resettlement Agency (PAHRA);
3. Tondo Foreshore Development Authority (TFDA);
4. Central Institute for the Training and Relocation of Urban Squatter (CITRUS);
5. Presidential Committee on Housing and Urban Resettlement (PRECHUR);
6. Sapang Palay Development Committee;
7. Inter-Agency Task Force to Undertake the Relocation of Families in Barrio Nabacaan, Villanueva, Misamis Oriental;
8. All other agencies: task forces and *ad hoc* committees involved in housing.

All urban estates of the government, the Department of Agrarian Reform and the PHHC were also transferred to the NHA.

Powers and Functions of the Authority

The Authority was also given the following powers and functions to be exercised by the Board according to the Human Settlements Plan of the Human Settlements Commission:

1. Develop and implement the comprehensive and integrated housing program;
2. Formulate and enforce general and specific policies for housing;
3. Prescribe guidelines and standards for the reservations, conservation and utilization of public lands identified for housing and resettlement;
4. Exercise the right of eminent domain or acquire by purchase privately owned lands for the purposes of housing development, resettlement and related services and facilities;
5. Develop and undertake housing development through joint ventures or other arrangements with public and private entities;
6. Issue bonds, or contract loans, domestic or foreign for the implementation of its housing programs;
7. Discharge all responsibilities of government with respect to commitments on housing and resettlement;
8. Promote housing development by providing technical assistance;

9. Prescribe and enforce guidelines, standards and rules designed to protect the house and lot owners through the regulation of the real estate trade and business;

10. Regulate the relationship between owners and lessees of residential properties.

Under the decree, the Board of Directors of the National Housing Authority are:

1. Secretary of Public Works, Transportation and Communication;

2. Director General of the National Economic and Development Authority;

3. Secretary of Finance;

4. Secretary of Labor;

5. Secretary of Industry;

6. General Manager of the Authority.

2.4 URBAN LAND POLICY

Urban Land Problems

In urban areas of the country, land as a housing resource has become very limited. This has been brought about by:

1. Increasing demand for intensive land uses;

2. Private speculation which has forced useful land into idleness;

3. Absence of a land management policy for the country.*

Because cheap land within urban areas is not available in adequate sizes, the low-income sectors of the community are priced out of the residential market. As a case in point; within the inner core of Metropolitan Manila, residential land ranges from \$53 to \$106 per m^2. This progressively goes down to \$10.67 per m^2 some 20 km out. These are still expensive for low-income people. Furthermore, an added burden in commuting from place-of-residence to place-of-work is shouldered by the working members of the household.

For the past two years, government has moved towards the realization of a meaningful urban land policy which can halt unjustifiably high land prices. Several existing laws and recently promulgated decrees directed towards new concepts for land ownership in the country are now being enforced. However, it may take a generation before these root in Philippine society. Under the Philippine Constitution, public lands are controlled by the government. The "Regalian Doctrine" is followed. The Public Land Act of 1936 is presently enforced. The law states the conditions and procedures how a Filipino citizen can acquire lands from the public domain.

---

*The Human Settlements Commission, as early as 1975, had initiated studies towards the promulgation of Presidential Decrees on Land Resource Management Systems.

Land Policy for the Future

A milestone in land policy formulation was the Presidential decree promulgated in June 12, 1974. President Marcos during the Independence Day celebration stated that henceforth all alienable and disposable land belonging to the public domain are to be owned by all. It shall no longer be sold or transferred to private persons but shall only be subject to lease or similar disposition which will result in greater benefits for the citizens of the country. Due to this declaration, government agencies affected by this policy have undertaken studies to quantify the effectiveness of the policy directions on real property taxes and lease prices, as well as the repercussions of policy on socioeconomic development and land resource management. Transitory provisions have also been included in the policy so that owners of agricultural lands who are in the process of acquiring the lands they cultivate can have up to a given date to file applications for the confirmation of their title to the land. After that date, it will be considered that the actual occupants hold the lands under leasehold.

The policy of 1974 still recognizes the concept of private land ownership. However, all land is treated as an instrument of national development. No one can do what he wants on his land. The occupant cannot alter the original purpose for which the land has been classified unless a permit is granted by a national government authority. Similarly, a new decree is in course of preparation which will control and regulate the change of land use. This will all be embodied under a land resource management decree being studied by the Human Settlements Commission.

Under the Philippine Constitution government is authorized to expropriate private lands upon payment of just compensation. Such lands can be subdivided into small parcels and conveyed at cost to the people. Local government are also authorized to do likewise. This provision is generally not resorted to by government agencies because of the "just compensation" clause. Oftentimes the fair market value is very difficult to be agreed upon. This has resulted in the promulgation of Presidential Decree No. 76 which attempts to solve the predicament. It required all landowners to declare what they consider as fair market value of their landholdings so that the same shall be used as basis from pricing land; either for real property tax purposes for payment of compensation if the lands are to be bought by government.

The idea of taxing idle urban lands has long been discussed in academic circles. Currently, a number of policy studies spearheaded by the Bureau of Lands and the University of the Philippines Law Center have made recommendations with regards the controversial idea. The policy studies have added still another dimension on land taxation. This is the idea of added tax assessment on lands which have risen in value due to improvement made by government or utility companies. The concept has assumed that the incremental value added due to public improvements will necessarily control speculation.

Innovative Land Ownership Concepts

A new concept on land ownership now being experimented upon, although in a limited scale, by private land developers is the concept of making land an expendable resource and not as an ordinary commodity of trade. This concept, inhibits one to abuse his property and be mindful of the ecological consequence of ownership. The idea embodies certain rights to property like regulations on development rights as well as transfer of rights to a large organization to manage and maintain land for the greater good of the community. Trends indicate that Philippine society is receptive to this revolutionary concept of land ownership as has been applied in the Caliraya Leisure Community at Lagura. This can change development patterns for low-income housing. Normally, the time-frame for such ideas to take root is about 20 years. However, the Bagong Nayan concept of the National Housing Authority may be able to shorten their time horizon.

2.5 HOUSING FINANCE

Financial Requirements

Currently government is encouraging the private sector to participate in formulating the housing finance programs of the country. Issues are expected to be resolved in this manner so that the appropriate policy programs can be immediately implemented. This is a clear recognition of the role the private sector plays in the housing industry of the country. As an example, in 1973, investment in housing is about 2.3% of the GNP. This is only about $116 million. But 80-90% of this has been contributed by private investors, mostly the upper-income and middle-income levels of Filipino society. Still UN housing advisers to the Philippines have stated that the country has to increase housing investments to about 4% of the GNP if a genuine effort is to be made in dealing seriously with the housing problem of the nation.

Just as in any country of the Third World, the estimated financial requirements for the housing program in the Philippines are even below the "4% of GNP" dictum of the United Nations. It can fall anywhere in the region of $133 million per annum. This estimate has been made by NHA financial planners and is based on a yearly incremental demand analysis. Currently, the NHA is reevaluating the housing standards from which funding requirements were based so as to match its rationality with national housing policies. Similarly, funding sources which can sustain the strategies on a long term are being identified.

Due to the lack of one unifying concept on how to reduce risks attendant to the beneficiaries of indirect and social housing, the private financial sector has a very minimal impact on this area. Private banking institutions and insurance companies concentrate their resources for housing on the upper-income levels which are considered less financially risky. This situation gives greater responsibilities to the NHA in preparing the financial programs for housing. This is significant because the indigent and social sectors of the country can range from 70% to 75% of the urban areas. Concerns that the NHA is considering in the design of its first financial program are: the affordability by both the government and the people in sustaining the housing program in the long term; the high credit risks involved if financing concepts are to be based on traditional banking systems and institutions; and the administrative costs involved in innovative methods are not introduced.

Cross Subsidy Approach

The National Housing Authority is strongly inclined to introduce the "cross-subsidy" approach in its programs of financing a major segment of its operations. Instead of merely developing the housing areas, the strategy that the NHA is developing is that it shall get involved in complete developments of the community where industrial and commercial land uses are considered essential elements of the development. On this basis, the costs for development of the indigent and social housing segment of the community are to be assumed by the earnings realized from the sale of commercial and industrial developments in the community.

The "cross-subsidy" approach has required a deeper evaluation of the existing financial climate of the country so that breakthroughs can be identified, tested for viability, and implemented as a total system.

Areas of Financial Concern

The areas of concern, as identified by the financial consultants of the NHA are:

1. For the low-income sector to afford the money for housing, it should be on a long-term basis ranging from 20 to 25 years and at low-interest rates of 3-6% per annum;

2. More innovative systems should be devised to replace traditional administrative means of granting loans and collecting amortizations if these expenses are to be reduced. These expenses are expectedly high if the NHA carries out the administrative work attendant to collection of amortizations as practiced presently by the Authority;

3. It has been strongly recommended to the NHA that a system of indexing of amortizations on housing be made regularly. The advantage of the method is that over the long period, the NHAs capability to implement its program can be assured. Furthermore, the beneficiaries of the NHAs program can adjust their future payments and consider inflationary increases in costs. This will allow the NHA to generate more funds in order to accommodate higher replacement costs.

4. The NHA has now recognized that sites and services is a viable approach worth considering in effectively spreading the benefits of housing to the lower-income levels of the country. Inherent in this approach is the legitimization of certain squatter communities in areas that are not inimical to overall national development. In Metro Manila alone, this concept can effectively benefit 20% of the area's population. The advantage of sites and services has long been validated by social surveys as reflected in the following case:

In a survey conducted in a slum area of Manila, specifically in Tondo, a question regarding preference for living in a single-family detached house, an accessoria, or a multi-family walk-up unit showed an overwhelming choice for the single-detached house. Some 85% chose the single-detached house. The accessoria comes second for half of the respondents. A follow-up query which shows values for building form as against tenure status was then administered. Fully 74% opted for the apartment that can be bought rather than a single-detached house that can be rented. The main reason given is that ownership would be more of a priority than housing where one would be worrying about payments every month. A further query as to relationship of owning properties and place-of-livelihood showed clearly that the respondents immediately preferred the latter.

In essence the advantage of the sites and services scheme are:

(a) The approach will avoid causing disruptive dislocation to many families who are on the subsistence level of living in urban areas;

(b) The existing stock of housing can be prolonged by upgrading. Despite this, caution should be expressed in matters of pricing the upgraded housing stock so that there is a reasonable correlation of the amortization period with the actual life of the structure. Otherwise the replacement costs may be beyond the capacity of the beneficiaries.

5. Even with subsidized financing terms, the full cost of housing benefits cannot be shouldered by the low-income groups. Of particular importance is the cost of developed land. A more effective low-income housing program should assume costs of land and its development, either fully or partially. A consequence of this is the requirement for government to land-bank potential urban development areas and that long-term arrangements for leasehold which appeal to the majority be analyzed.

Because of the paucity of financial sources, managerial talents, and technical expertise that the NHA can master in a short time, the strategy is for the Authority to work jointly with the private sector. Today, money in housing is expensive. There is not enough of it being channeled to the housing sector. No existing formal system can effectively channel sources and other means of funding toward low-income housing. Part of this lack is the absence of a mechanism for rediscounting housing mortgage documents.

Recommendations for Financing Housing

The issues that NHA financial advisers have recommended for further evaluation are:

1. Because of the limited market for 20-year term bonds in the country, the NHA cannot rely on bond sales as a continuous source of funds. The 20-year maturity of NHA Bonds is ideal because it matches the amortization term of the loan. Unless a secondary mortgage market is developed the ability of the NHA to use bond issues will be severely reduced.

2. Another approach that the NHA is considering is the concept mentioned earlier of cross-subsidies where earning and cash flows from commercial and/or industrial developments are expected to support housing projects for the low-income. However, the NHA realizes that it cannot rely fully on earnings from developments that are incidental to its basic objectives. It cannot commit a substantial part of its resources to commercial and industrial developments.

2.6 RURAL HOUSING

One of the largest yet long unrecognized groups of home-builders in the country is the informal rural sector. This is the unquantified area of housing in the country. Most of the building approaches for this sector are traditional, generic and generative. It will require informal support mostly geared to improve their traditional concepts for building without disrupting their cultural beliefs about what is a home. There is need to provide this sector with better techniques and methods to make their abodes more hygienic, constructively stable and appropriately located for more effective use of utilities. Government likewise will have to provide infrastructure support, rationalization of house sites to employment sources in order to make new settlements usable. On the whole what this informal sector will require is substantial information for a self-reliant approach to housing: a how-to-do it approach to rural building. This is imperative as most of the skills needed in rural house building have been dissipated by the influx of construction workers in urbanizing areas of the country and the export of construction labor to the Middle East and Africa.

In rural Philippines, a house can mean a lowland farm house made of bamboo, timber and thatched roof, a timber house in the mountain regions, or stilt houses along the foreshore or marshlands or a boat house by the sea as that of the Badjaos. The early Filipino's response to the need for shelter is simple in origin. His basic requirement was to be assured of protection from depredations of the elements and surprise attacks from animals.

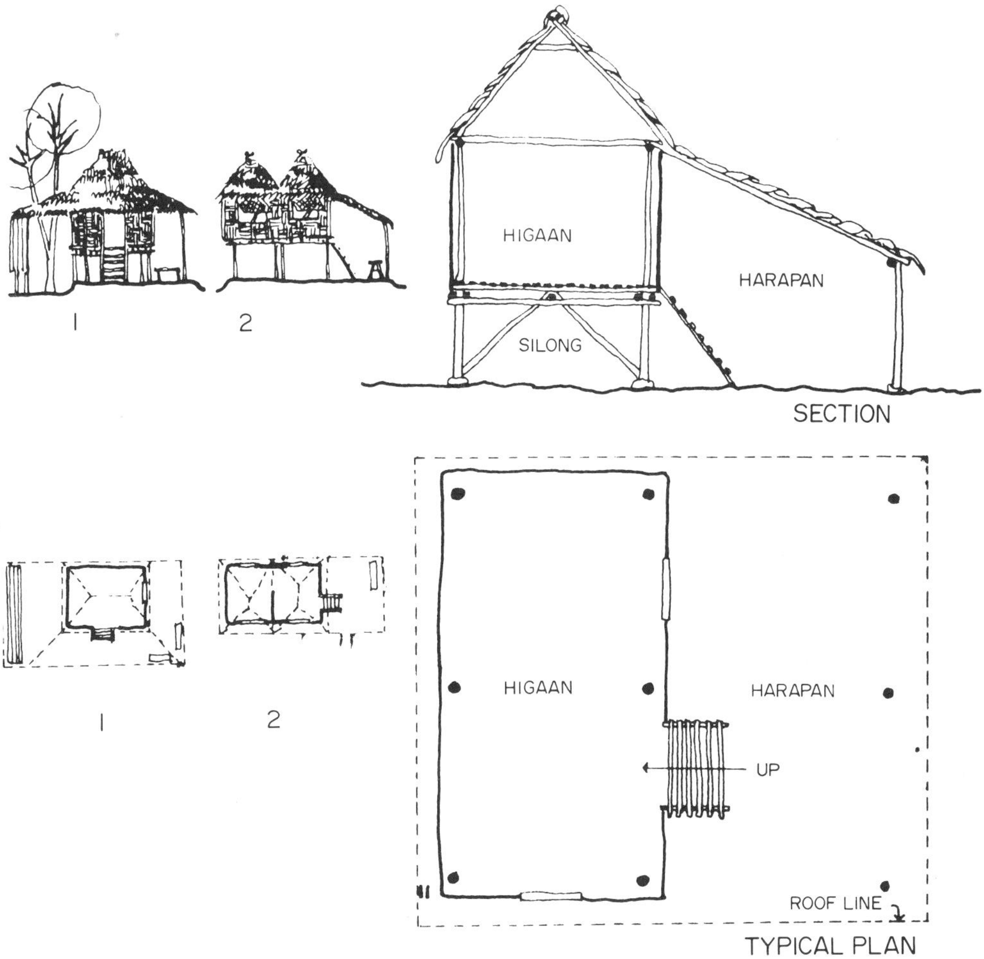

Fig. 2.7. A simple "Bahay Kubo"

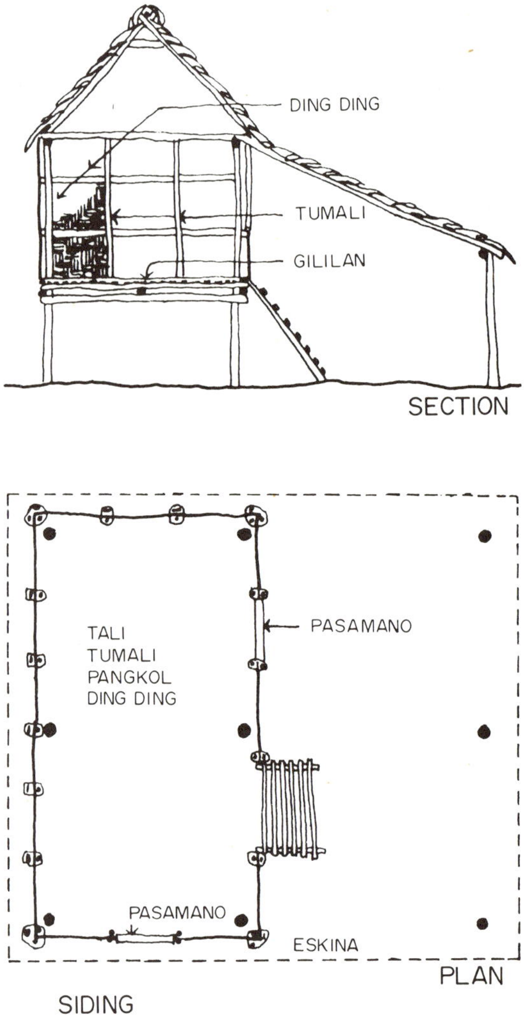

Fig. 2.8.

Cultural Beliefs

Superstitious beliefs enter in the selection of construction materials and in orienting the hut. For example, in some places in the Southern Tagalog Region, the steps leading to the ground must face the Southern Cross so that the house will always be "occupied". In the traditional Muslim areas, the slopes of house sites must be towards Mecca. For good luck, the number of steps must follow numerological beliefs.

A simple "bahay kubo" is mainly the "higaan" — an enclosed and elevated space that functions as sleeping area and storage space. The "silong", the space underneath is where animals and farm implements are kept. The roof extension in front, called the "harapan", serves as social and entertainment space. Occupants of the "kubo" are supposed to sleep with their head towards the east.

Roof Layouts

Space expansion of the "kubo" means extending the roof which is called "pahulog". This increases the covered ground area which is sufficient to protect daily activities that take place around the hut. Should an additional room be needed for the increasing number of the family, it is built adjacent to the existing room. The new room will have to be under its own "balangkas" (hip roof) and not a "pahulog". The building mass assumes a different silhouette as two ridges are formed. Being under a "pahulog" is avoided because it would imply that the new family is "nakikisunong" (being carried on the head), or a burden to the first family.

"Balangkas" means framework, but it also refers to a type of roof — the hip roof. Other types are the "pasibi" (lean-to) and the "pakamalig" (gable roof). The roof of the hut is constructed first, before any other frame of the house. The roof frame is assembled on the ground. Roof pitch is high, to facilitate drainage.

Framework

Two ways of setting up the posts are by the "papatong" (placed on) and the "pabaon" (to bury) methods. In the first method, the horizontal members ("sikang", "kahab-an", "yawi" and "patukuran") are attached to the posts ("haligi"); then the assembly is raised so that flat stone slabs which serve as the foundation can be placed underneath each post. In the second method, holes are dug, posts set in, and the horizontal members attached on to the posts.

The people believe that all the posts should be of the same species. If different materials have to be used, posts of the same kind should come in pairs-coupled or "married" ("magkakaasawa"). If there are odd number of posts, the odd one would have to be paired with a "kilo" or a "soleras" of the same species. The posts are made in pairs so that the couples "would not be lonely". Wood species used for posts are the local baraba (*Lagerstroemia speciosa*), malaruhat (*Eugenia calubcob)*, mulawin (*Vitex parviflora*), yakal (*Shorea gisok)* and guyong-guyong (*Decaspermum fruticosum*).

Bamboo Flooring

The floor or "sahig" is made of one-inch wide bamboo slats spaced at about 3 cm on centers. The "gililan" (sole or floor sill) frames the "sahig" and supports the walls. The "soleras" (floor joists) which support the "gililan" and "sahig" are supported by the "patukuran" (girder). The "yawi" on which the "patukuran" rests completes the frame. Should the span between posts be too wide, an intermediate post or "tukod" may support the "patukuran".

Sidings

Placing the walls would present a problem if they had to be attached directly onto the crooked posts. Instead, they are attached to the "tumali" (wall studs) that rest between the relatively aligned "gililan" (floor sill) and beams.

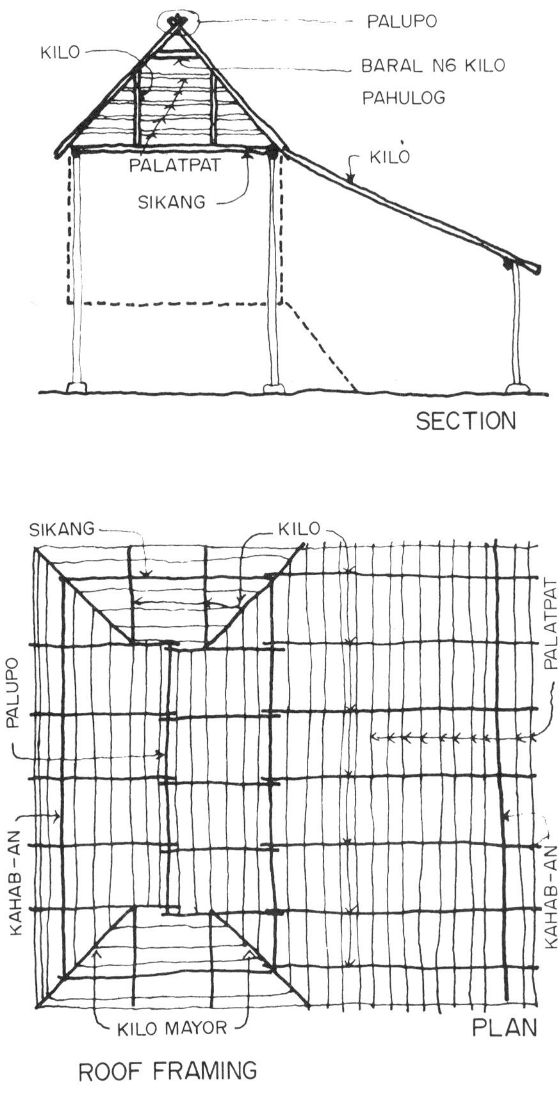
PALUPO
KILO
BARAL N6 KILO
PAHULOG
KILO
PALATPAT
SIKANG
SECTION
SIKANG
KILO
PALATPAT
PALUPO
KAHAB-AN
KAHAB-AN
KILO MAYOR
PLAN
ROOF FRAMING

Fig. 2.9.

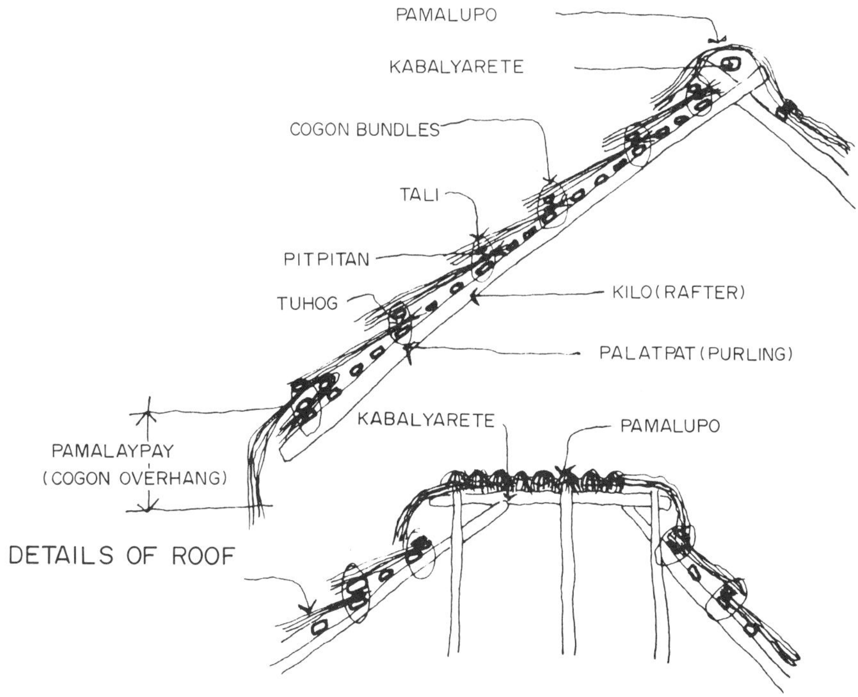

Fig. 2.10.

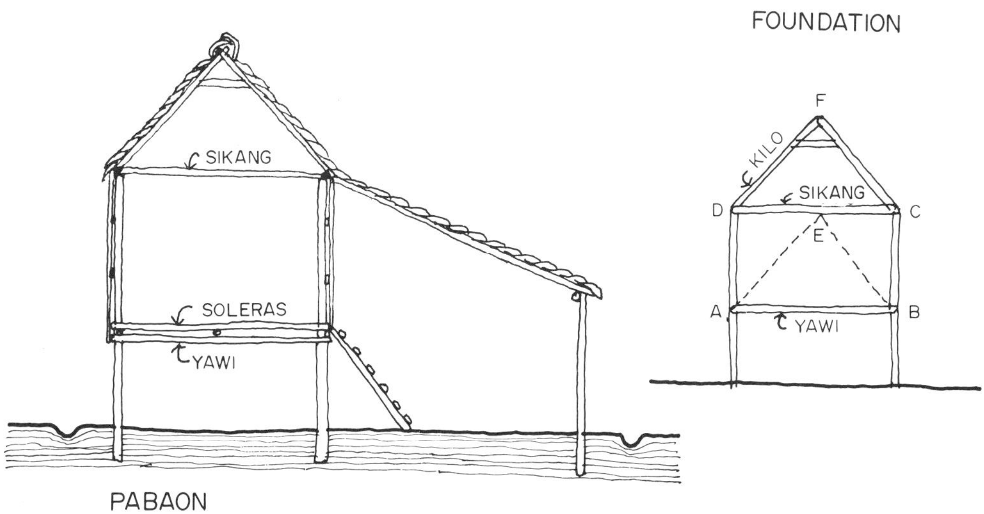

Fig. 2.11. A trench dug around the house keeps the ground area from being flooded during rains. The earthen floor is "ginagstiya" which means treating the shaded ground area with a mixture of carabao manure, dried grass and water.

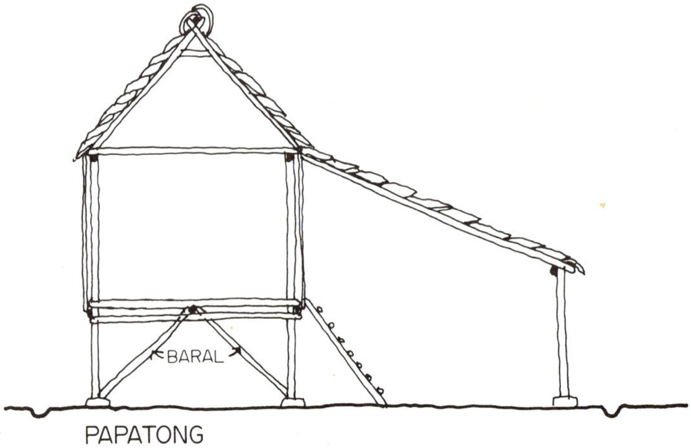

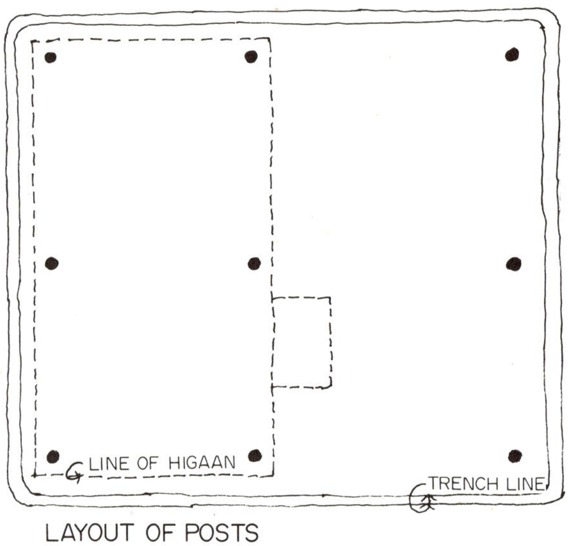

Fig. 2.12.

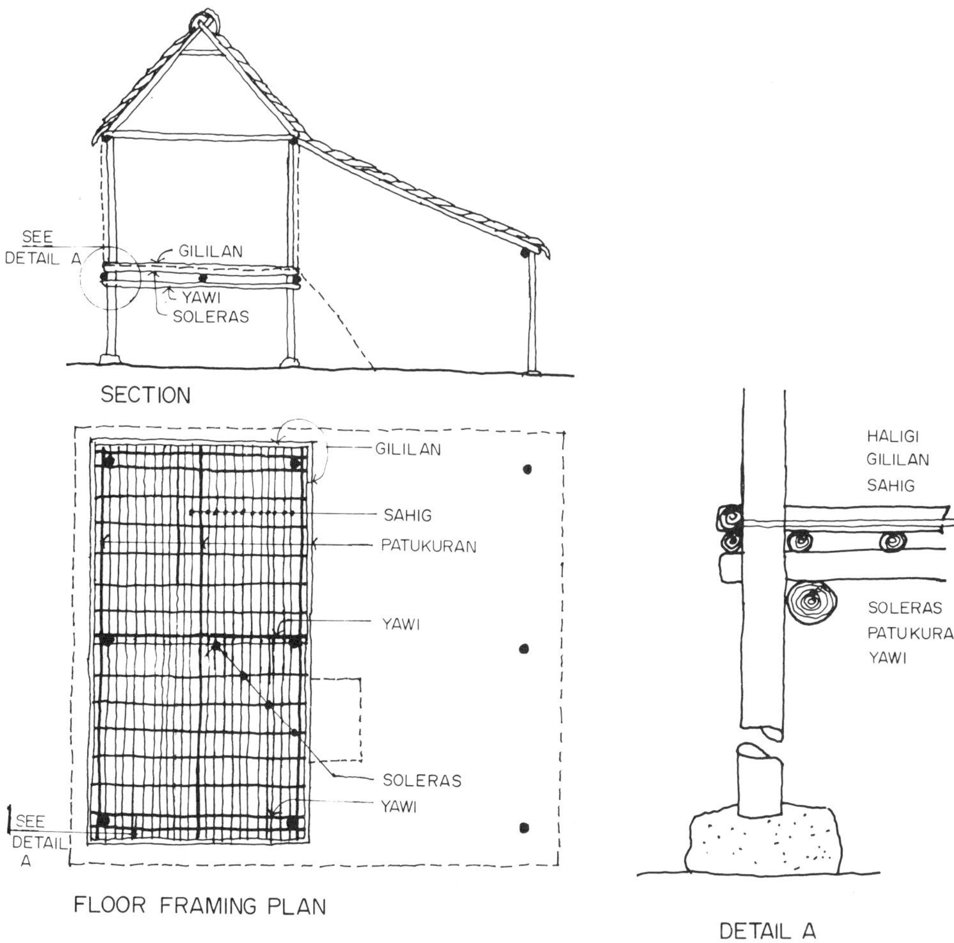

Fig. 2.13.

The most common siding material is "sawali". This is a bamboo mat made by weaving and framing the thin and flattened bamboo strips. The "sawali" wall is light and cool but it burns easily. Some rain may be blown through it during storms. However, if it is not exposed to the elements, it could last up to 20 years. It may be used for ceilings, partitions, doors, and window shutters.

Research for rural housing as promulgated by the National Science and Development Board have the following objectives:

1. Improving the material to make the rural house more durable and resistant to fire, decay and wear;

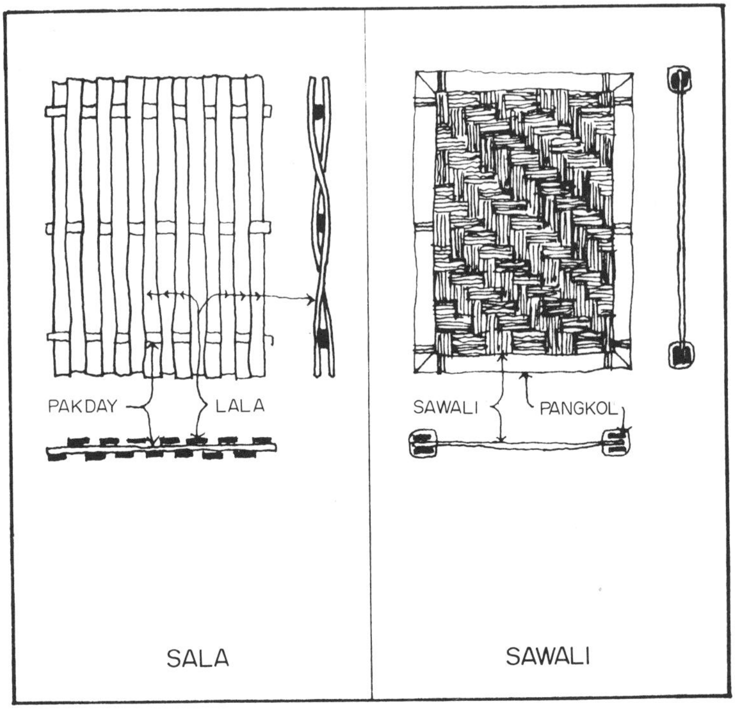

Fig. 2.14a. Sidings

2. Improving the design of the rural house and its components to better resist storms, adapt to increasing industrialization processes, and anticipate the higher quality of life for the rural inhabitants.

Research that shows promise is the use of composite cladding of traditional house materials made of an inner panel of bamboo matting (sawali), as the structural layer; a core of light material, and an outer face of aluminum sheet. All three surfaces act as a sandwich panel which can be used as a wall or a roof. Another sandwich panel using ring segments of bamboo culms to form a hollow core was also tested for service characteristics. Results revealed that applications were not found marketable for the intended low-income market.

2.7 FORMAL HOUSING FOR THE MASSES

Formal approaches to low-income housing now being tried in the Philippines range from those with high technological inputs in both the soft and hard sciences, to those with large amounts of intermediate technology and citizens participation. Some cases in point are the experiments of the Philippine Business for Social Progress and the Pleasant Hills Housing and Service Cooperative, Inc. Both are cases for actual research laboratories to find new alternatives to urban housing for the masses.

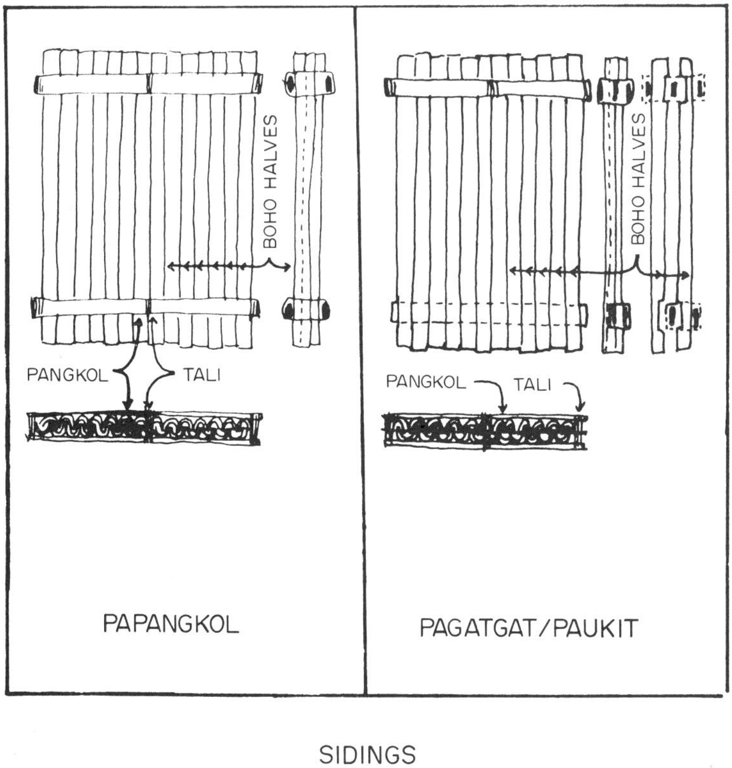

Fig. 2.14b.

As the country is regularly visited by disasters, government has initiated mechanisms for sustaining services when communities are wrecked by disasters. One such procedure is that being tried in Mindanao by the National Housing Authority.

The Sambahayan Concept

The most notable condominium project designed particularly for the low- and middle-income families is the Sambahayan Condominium of the Philippine Business for Social Progress, a non-profit foundation organized by the private sector and the Social Security System.

The PBSP-SSS Sambahayan project is intended to house 353 low-income families whose breadwinners work within one bus ride from the housing complex.

During the initial stages of construction, there was a noticeable resistance from prospective buyers due to the "bad image" attached to condominium living in the Philippines. People then tended to associate condominiums, especially those for low-income households, with loss of privacy. Also, they believed then that since various families are housed in one tall building, they would be easy prey to delinquents and the bad elements of the community.

However, the PBSP has since launched an intensive education campaign that made extensive use of mass media. The PBSP says that the campaign has been successful and it is now on its second phase of monitoring this major social experiment.

Sixto K. Roxas III, who was chairman of the PBSP board of trustees at the time the project was initiated, emphasized that the PBSP, in conducting the Sambahayan project, seeks to demonstrate that the low-cost housing venture for urban workers is one that can be replicated if proper infrastructure services such as drainage, roads, water mains, power lines, and sewerage can be assumed by government in support of the project.

The Sambahayan, he explained, is designed to become a community that can reach a stage where the members can be able to sustain themselves. The Sambahayan, which is located in Mandaluyong, is expected to become a model that other similar undertakings in the future may follow.

The Sambahayan project was started with a ₱960,000 loan from the SSS. However, rising prices and the oil crisis pushed the cost up.

The PBSP had originally expected to recover only about ₱8 million to ₱9 million of the investment and give away the rest. The project is composed of five buildings along Sacrepante Street in Kandaluyong. The project provides living areas ranging from 34 to 56.64 m^2 having nine standard types of residential units. Several units in the upper floors have mezzanines.

The cost of each unit which ranges from a socialized price of $2,000 to a high $5,730 is payable in 30 years through the Social Security System Housing Loan Assistance. A 10% down payment is also required. One-half of this 10% down payment, however, can still be paid in 36 equal monthly installments interest free but guaranteed by the employers where the prospective resident works. The total monthly payments, on the other hand, will be from $20.00 to $57.30 a month. This monthly amortization includes the monthly dues payable to the SSS, Real Estate Tax, Fire Insurance Premium, Mortgage Redemption Insurance (the insurance which automatically pays in full the unpaid balance of the applicant's loan in case he dies), water, light, and contributions to the maintenance and repair of the common facilities in the Sambahayan.

Eco-settlement Concept for Self-help Housing

Recognizing the need for an integrated approach to housing development for the indigent families of Metro Manila, the Samahang Bagong Buhay Foundation, Inc., a Catholic welfare organization undertook a housing project for resettling residents from Mandaluyong. This organization, founded in 1975, is composed of Salesian Fathers, social workers, seminarians from Christ the King Seminary, and volunteers from various professions. The aim of the project was to "build socially and economically self-sustaining communities through multi-purpose cooperatives with the development of the whole man and a Christian community". The model that was adopted for the resettlement community is the so-called "ecology settlements" which aim for a high level of interdependence between human and other environmental resources. The development process was tried for the displaced residents of Mandaluyong.

A pilot project in Antipolo, consisting of 5 hectares, was chosen to settle some 350 squatter families from Mandaluyong. The development process included participative planning as well as surveys of the client families so as to determine needs, aspirations and resources available. The people involved in the process were Fr. Leo Schmitt, SVD, the Director of SBBFI, the squatter families, community development workers, volunteer architects, volunteer engineers, skilled workers, architecture students from U.P. and A. Bruce Etherington, the Director of Asia-Pacific Housing and Development Research Corporation of University of Hawaii.

The planning process included study and research, interviews, staying with the client families to study patterns of living, behavior circuits and spatial needs. A cyclical process to generate design solutions was made so that filtering of ideas, dialogue with the displaced people presenting the proposed design, and developing a technology feasible for the development can be accomplished in the shortest time possible.

Both resettlement and housing construction are expected to be carried out by the settlers themselves. Through the process of generative planning, the settlers will assist in determining the qualitative and physical characteristics of the new settlement. To train settlers in construction methods for housing, the "pyramid" concept of skills training was adopted.

Selected settlers will construct, under professional supervisors, a model house. After mastering techniques of construction, and the procurement of building materials in proper quantities, the settlers with new building skills will then be dispersed to train and supervise predetermined groups of settler households in the construction of their homes. Each group will be located in a physically defined area called a compound and will cooperatively construct all houses for their group. This will permit certain amounts of building specialization within the groups.

When the first compound group has completed their homes a selected number of settlers from these new settled groups, who have mastered the techniques of house construction, will then undertake to train and supervise succeeding compound groups. This process is then repeated until all settlers to be resettled are housed. Construction techniques will include a structurally sound building system which will permit expansion of the core house either horizontally or vertically. Costs of dwellings are expected to be highly competitive with current low-cost housing projects. Each house will also have its own independent sewage system and water supply system. House plots will have a minimum density of 70 families per ha. and will be of sufficient size to accommodate a family of 6-8. A garden sufficient to supply the household's nutritional needs, except for 80% of carbohydrates and 75% of protein requirements is also incorporated in the site. The typical compound designed for 6-20 families will include facilities for compatible cottage industry, children's playground, a landscaped circulation and public area, a place for a sari-sari store, to name a few. An enclosing wall will provide security and control within the community. An integrated automatic irrigation system will provide water to all private and public areas. This is possible because the physical features of the site are so disposed as to ensure gravitational flow of water. The eco-settlement will provide communal fish ponds and reservoirs, within a flood control/drainage system. This provision for communal food production facilities will help reduce the protein and carbohydrate production shortfalls of the private gardens.

The socioeconomic program for the community will include the development of community-based commerce and cottage industry, and manpower training through rehabilitation and skills re-education. At present, the Samahang Bagong Buhay housing project for resettled people, known as an eco-settlement is at its first phase of development. The people are in the process of producing the materials needed for house construction.

Disaster and Emergency Housing

The National Housing Authority has adopted a three-phase policy for providing housing in disaster-stricken areas of the country. Initially, the work is being coordinated by the National Disaster Coordinating Center. The activities are divided into three phases. Phase I is primarily geared to provide emergency housing assistance to hopeless victims of fires, earthquakes, or insurgency. In the August 1976 earthquake that hit southern Mindanao, 11,399 of 11,500 homeless families have been given housing assistance worth $67 each. Coupons worth $53 were used to purchase basic building materials from accredited suppliers in order to build temporary shelters. A cash grant of $13 was also given to the household for transportation and sundry expenses. Dwelling units made of nipa, bamboo, and other locally available materials were used. Construction was done on the spirit of self-help and "bayanihan". Communal toilets, open spaces, pathways and other community facilities were also constructed by the residents with the Authority providing the technical assistance. Phase II begins with the granting of loans to homeowners for the rehabilitation of damaged houses. This is started as soon as the emergency housing of Phase I is completed. Up to $667 loan can be granted per family by the NHA. This is intended to be payable in 10 years.

Phase III consists of a housing development program which will provide permanent settlements for homeless families. Coordination is made directly with the Commissioner of the Region, but under the direct technical supervision and control of the NHA through the General Manager.

This policy has now been used for sometime in the Muslim areas of Mindanao.

2.8 ALTERNATIVE TECHNOLOGIES

Owing to the urgent need for applicable alternative solutions to the energy problems of the country, new sources of power are at present being studied. Geothermal energy is expected to be operational by the 1980s in Bicol, Southern Luzon and Leyte with the power grid entirely linked-up for the whole of the Island of Luzon. Despite apprehensions of risk of pollution, the government had to take the painful decision to establish a nuclear power plant in the growing industrial region of Bataan. Wave power, solar power and tide power have also been studied by several agencies of the government. But the most promising for the low-income families in the rural areas is bio-gas. The Energy Development Board, the National Science Development Board and the Bureau of Animal Industry have coordinated their research and development activities very closely.

Energy, if not fully harnessed for community development will go for naught. The UNEP has in Tondo an on-going applied research which demonstrates what alternative technologies can do in an integrated approach to development.

Bio-gas

Felix D. Maramba, Sr., is the pioneering Filipino in bio-fuels and agriculture recycling. He says that the bio-gas system is a fourfold answer to human survival problems. And he talks from experience as he has built an operational bio-gas plant at the Liberty Flour Mills Maya Farms in Rizal.

Maramba states that by a process of fermentation, the bio-gas system produces a versatile fuel. Then it is a source of disease-free organic fertilizer. Thirdly, it controls pollution by disposing of organic waste such as animal manure, crop waste and garbage. And lastly, it is an effective way of conserving natural resources

The first bio-gas plant at Maya Farms was initially intended for pollution abatement only. However, it was discovered that the sludge was an excellent fertilizer for the farm's vegetables and sweet corn. This saved about 75% on fertilizer cost and increased yields by approximately 50%. Continuous cropping became a possibility because the soil conditioning value of sludge made water absorbity of the soil more efficient so that water pretention is facilitated during the dry season.

Bio-gas was utilized as fuel at Maya Farms in the meat processing, canning, and rendering plants. By the time the Philippines experienced a shortage of LPG, the company had sufficient bio-gas to convert most of their equipment using LPG to bio-gas. Today bio-gas is used as a direct substitute for LPG, gasoline and diesel fuel in the heating, drying, cooling and lighting requirements of the company. Maya Farms now operates a bio-gas plant producing 850 m^3 per day. This saves the company some 20-25% of ordinary power consumption. The cost of constructing and installing small-scale bio-gas plants of different sizes is approximately as follows:

1. A capacity of 1.7-2.0 m^3 of bio-gas/day which is sufficient for lighting and cooling, plus 16 tons of fertilizer per annum will cost $267-$320;

2. A capacity of 4.25-5.10 m^3 of bio-gas/day which is sufficient for lighting, cooling and pumping water, plus 40 tons of fertilizer per annum will cost $400-$480;

3. A capacity of 7-8.5 m^3 of bio-gas/day which is sufficient for lighting, cooking, pumping water and running a gas refrigerator, plus 65 tons of fertilizer per annum will cost $667-$880.

The UNEP Project at Tondo

Another area where alternative technologies are being studied in the country is the UNEPs demonstration in Tondo. The project deals with the improvement of the environment by using an integrated approach on the social, economic, physical and environmental aspects of communities.

The project is a joint undertaking by NHA with UNEP on a 3-year demonstration scheme that commenced last July 1976. The project is concentrating its efforts on squatter areas with the objective of looking at new ways to improve living conditions of the urban poor by selecting concepts that are ready for massive applications but still require to be demonstrated. These are:

1. The utilization of renewable materials for building such as agricultural products and other plant materials;

2. The use of industrial waste as building materials particularly from the oil, mining and refining industries;

3. The use of recycled building materials rather than discarding them at the end of the useful life of the building;

4. The use of other sources of energy such as wind and sun to produce power and to utilize the minimum quantity of water for washing and cleaning by fine-droplet spraying;

5. The use of rainwater and means of extracting water from the air by the phenomenon of condensation;

6. The use of solar power to purify sea-water as well as a prime mover for cooking, heating and cooling;

7. By recycling water and separating it by functions for potable use, cleaning, conveying and cooling or watering plants;

8. The use of practical modular coordination to avoid waste of materials;

9. The production of protein foods through aqua-culture.

Attempts at correction of environmental disruptions are at present concerned with reducing the rural migration to the Tondo area. The land tenure problem is in course of solution. Work opportunities and improved sanitary conditions are being tried. It is expected that better opportunities can then be afforded to the squatter area as environmental conditions are improved.

2.9 BUILDING MATERIALS

Recognizing the need for building materials that are locally available, replenishable and reusable, research on these is aimed at import substitution, technology transfer and the use of agricultural waste products. In addition, the government of the country and its research agencies are showing interest in composite products combining synthetic and natural materials, prolongation of the life of traditional materials, and improvements in constructional techniques.

Cement Products

Concrete hollow blocks are considered the typical low-cost construction materials in urban areas of the country. Several manufacturing concerns have established huge enterprises for the production of approved quality concrete hollow blocks which generally satisfy international standards for testing of materials. It has also become a cottage industry in the countryside when the cimva-ram was introduced sometime in the 1960s. Where sand and gravel abound these have become the aggregates used with cement. However, in inland areas of the country, where sand and gravel are scarce or very expensive, the National Science Development Board has launched a coordinate research project to determine other substitutes for sand and gravel. The hollow blocks produced were aimed at providing a low-cost housing material for single storey detached houses that do not require the same high mechanical properties as those sought when concrete hollow blocks are utilized for multi-storied structures. Below is a corporative table of results of tests made on normally manufactured cement hollow blocks. Based on an acceptable compressive strength of 7.05 kg/cm^2 for non load-bearing blocks it can be easily seen that several substitutes for sand and gravel can be used in the manufacture of low-cost cement blocks.

Ferro-Cement

In 1974 the Patent Office of the Philippines granted a utility patent to a ferro-cement product called HYPAROOF. This product invented by Tolosa & Mejia of Iligan City is now being used in some housing projects in the south. It measures 3.80 x 8.80 m which can be pre-cast or cut on one side. The form follows a hyperbolic paraboloid which weighs 1 ton. Reinforcements used are woven 6.1 wires. No special equipment is necessary except a crane for erecting the structure.

Table 2.2. Results of Tests on Manually-manufactured Hollow Blocks (Prepared by FORPRIDECOM)*.

Mixture proportion by volume	No. of specimens tested	Ave. weight	Compressive Strength Ave.	Range of individual values Min.	Range of individual values Max.
Soil-cement (10x 20 x 40 cm)					
1:1	5	20.4	443	410	466
1:2	5	20.2	259	206	309
1:3	5	19.1	268	177	357
1:3½	5	17.8	167	134	247
1:4	5	17.8	178	82	286
1:4	5	18.3	175	158	201
1:4½	5	17.8	129	99	158
1:5	5	18.8	173	147	221
1:6	5	18.1	127[a]	106	144
Soil-cement (15 x 20 x 40 cm)					
1:3	5	25.2	109[a]	57	198
1:3½	5	24.4	113	76	153
1:4	5	23.6	114	74	160
1:4½	5	24.6	145	119	191
Sand-cement (10 x 20 x 40 cm) Control					
1:3	5	22.7	222	100	352
Sawdust-cement (15 x 20 x 40 cm)					
1:3	5	15.1	75	53	117
Baggage-soil-cement (10 x 20 x 40 cm)					
I II I					
1:3	5	20.9	119	76.8	141
1½ 1½ 1					
1:3	5	19.4	189	104	227
2 2 1					
1:4	5	20.2	232	164	287
2½ 2½ 1					
1:5	5	17.5	142	95.9	193
3 3 1					
1:6	5	17.0	94.2	70.3	145
Coconut Coir-soil-cement					
2 2 1					
1:4	5	11.0	156	-	-

*Sampled from those actually used at the FORPRIDECOM Housing Site.

[a]Acceptable for non-load-bearing hollow blocks is of gross area.
NOTE: 100 psi = 7.05 kg/cm^2; 1 kg = 2.2 pounds.

Fig. 2.15 Hyparoof by Tolosa

Sulphur Concrete

Sources

Elemental sulphur in the Philippines is associated with volcanic rocks, mostly pyrites and marcasite. To date most of the sulphur bearing ores have no commercial value to the chemical industry due to high extraction and processing costs. However, refined sulphur is not necessary in the production of sulphur concrete for the housing industry. Ores with as little as 30% sulphur can be utilized for making sulphur concrete.

Sulphur can be recycled without any expensive equipment. Only 30 k cal/kg of energy is needed to liquify sulphur. The Philippines has some 59.3 million metric tons of sulphur reserves. These are related to the action of inactive sulfataric-fumerolic activity. The location, volume and purity of the known sulphur deposits are as follows:

Table 2.3. Deposits with Over 35,000 Metric Tons.

Location		MT	% Analysis
Camiguin Island		61,100	10-15
Mt. Pinatubo, Botolan, Zambales		35,000	over 10
Mt. Malinao, Tiwi, Albay		4,940,000	0.39-10.45
Mt. Mandalagan, Talisay Negros Oriental		1,209,000	16.19 Ave.
Pamplona, Negros Oriental		52,590,000	17.19 Ave.
Cabiran, Biliran Island, Leyte		490.091	7.98 Ave.
	TOTAL	59,325,191	

Physical Properties

The density of sulphur is 2 tons m^3. It has a compressive strength which can be as high as 464 kg/cm^2 depending on the aggregate use. Tensile strength is 47 kg/cm^2 and this can be increased by the addition of fibrous reinforcements, glass fibers or organic fibers. The melting point of sulphur ranges from 110° to 119° C depending upon environmental conditions. Its boiling point is 440°C and its combustion temperature is 248° C. However, its burning is not sustained. Sulphur fires can be easily detected by smell, and are best put out by dry chemical sprays. Hot sulphur is capable of producing severe burns.

Sulphur has high bonding strength, impermeability to water, good insulation value and mixes with a variety of aggregates like sand, gravel, earth or clay. Sulphur has a tendency to develop static changes thereby presenting a fire and explosion hazard. Unless a fire-retardant to sulphur is used it should be avoided in places where there is a fire hazard. Similarly, despite its non-toxic qualities, the odor of sulphur may be disagreeable to people.

Table 2.4. Deposits with Smaller or Maximum Reserves.

Location	MT	% Analysis
Gonzaga, Cagayan	n.a.	
Bontoc, Mt. Province		
Mercedes, Camarines Norte	n.a.	
Mt. Isarog, Goa, Camarines Sur	ltd.	
Malino, Albay	65	51.22
Bacon, Sorsogon	ltd.	
Sorsogon, Sorsogon	755	40.75 Ave.
Tinambacan, Samar	100	39.92
Burawon, Leyte	2,250	-
Bacon-Dawin, Negros Oriental	2,333	34.34 Ave.
Taya san, Negros Oriental	n.a.	
Valencia, Negros Oriental	n.a.	
Mt. Hibok-Hibok, Camiguin	n.a.	
Mt. Apo, Cotabato	n.a.	
Sarangani Island, Davo Sur	n.a.	
Balut Island Davo Sur	n.a.	

Sulphur Concrete in Tondo

Sulphur concrete is formed by mixing molten sulphur with different types and grade of aggregates. Experiments are being carried out on these substances as a building material for low-cost housing in the Philippines. The procedure being applied in the manufacturing sulphur concrete is as follows:

1. Heat the sand or aggregate to a temperature of 106° C;
2. Simultaneously heat the sulphur up to its melting point;
3. The molten sulphur is added to the hot sand;
4. The mixture is poured into molds.

A prototype housing module is now being implemented in Tondo with blocks weighing 13 kg. A block has 50% gravel, 20% sand and 30% sulphur. For low-cost housing, studies done in the Philippines show that only 105-175 kg/cm^2 is needed to resist most loads in residential construction.

To date, the problem of extraction of sulphur and the transportation expense attendant of it have limited the uses of sulphur for low-cost housing in the Philippines.

Potential Uses of Sulphur in Housing

Remelting sulphur does not diminish its physical properties. It can be used to waterproof bamboo mats, thatch, earth walls and other indigenous materials. The method in applying molten sulphur is by immersion or by painting. If the melting temperature can be maintained, it can also be applied by means of spraying.

The following inherent potentials of sulphur are being studied by the scientists of the country to determine the suitability of the material as:

1. A bonding agent for building blocks;

2. A substitute for portland cement;

3. An agent to increase moisture resistance.

Cement-asbestos Boards

Cement-asbestos materials are not new in the Philippines. They were introduced in the country by Eternit Corporation, when in early 1959, the Philippine Homesite and Housing Corporation undertook a massive housing project in and around Manila. At that time cement-asbestos was used as a corrugated roofing material. Today, one can still see several roofs still covered with the original corrugated cement-asbestos sheets in the housing projects of Quezon City. There was a time that cement-asbestos sheets could not compete in the housing market which was cornered by the cheaper corrugated galvanized iron sheets — this is no longer the case. The higher costs of imported iron sheets made cement-asbestos viable as a low-cost roofing material. Late last year (1976) Eternit Corporation introduced in the market the Placa Ondula type of roofing. Since it is ideal for the climate of the country which experiences adverse weather conditions ranging from strong typhoon winds to long, dry spells, the Placa Ondula has advantage of lightness when compared to ordinary clay tile roofs and is more durable than ordinary G.I. roofing sheets. Each tile is pre-painted and is supplied with its own fittings, composed of metal straps, self-tapping screws and a leak-proof vinyl washer. In between the tile is a strip of polyethylene foam which is compressed when the strap pulls the laps tightly together. The unit cost of one tile 1.07 x 0.813 x 0.0045 m including accessories is $2.19. When in place it can withstand a pressure of 22 kilos/6.5 cm^2 and wind velocities of up to 220 km/hr.

Placa Ondula is primed with various colors in the factory. Each tile is given a tough, non-fading highly adherent and water-resistant color treatment.

An account of the need for the Philippines to import asbestos fibers, attempts are being made to replace them by long-fiber organic materials. Examples of these are bamboo fibers, abaca fibers and ramie. Abaca appears to be a good potential substitute. Today a manufacturing process is available in the Philippines wherein abaca fibers are converted into pulp — principally for disposable linen and for security printing. At this time, research and investigation are concerned with studying the weathering resistance of cement-orbitor brand and the reaction of abaca pulp when mixed with cement.

Water Hyacinth

The Philippines is blessed with innumerable water bodies which are the natural habitat of the water hyacinth (*Eichornia crassipes*). This plant was originally introduced into the Philippines as an ornamental. It has, however, become widely spread throughout the Philippines and grows in abundance. Until recently the water hyacinth, despite its beautiful giant blossoms, has always been regarded as useless and a nuisance. Along lakes and rivers they are considered navigation hazards. Sometimes it enters rice paddies near lakes in such quantities as to interfere with the growing of rice. During the rainy season, these aquatic plants are said to clog the normal flow of flood waters. Children bathing in creeks avoid these plants because of the bodily irritation they cause when touched.

In the fishing regions of the country it is used very extensively in the building of fish traps. The bases of the leaves, particularly under certain conditions form conspicuous rounded bladders. These are used as feed meal or poultry and livestock. The plant has fairly large potash content and can be used for soil conditioner.

It is expected that the water hyacinth which is now entering the handicrafts market will have uses in the low-cost housing field. Today the plant is being used to manufacture bags, mats, slippers, belts and other items which are exported from the Philippines. In 1976, the price of processed stalks was ₱3 per thousand. A patent has been granted to a Filipino for the "process of producing fiber strands from water hyacinth leaves and fiber strands produced thereby".

The process converts the long water lily stems into sturdy and pliant fibrous strands after proper treatment. In general the process is as follows:

1. The dried water hyacinth stalks are subjected to compression to expel air from the hollow cellular structure of the stalks;

2. The pressed stalks are trimmed at the edges of the entire length on both sides to produce the needed strands;

3. Braiding and plaiting or weaving is introduced at this stage;

4. Stalks are finally treated chemically to eliminate the poisonous coating of the fibers that causes body itch and which produces molts.

Rattan

Relatively less popular than bamboo or coconut trunks for housing materials is Rattan. This plant is a climbing vine usually found in thickets or second-growth forests at low and medium altitudes. Rattans are currently used by the furniture and interior design fields. Rattans are now used for beds, riggings for hauling medium weight materials, room partitions, ceiling panels and even suspensions for short-span bridges.

In the Philippines, the rattans are represented by 61 species of two genera.* Canes are harvested on the 15th year when their stems attain an average length of 25-35 m and a diameter of 1.5-3.5 cm. Thereafter, selective cutting of the matured canes may be done at intervals of 3-4 years. Harvesting of the canes is more favorable during the dry season when the poles can be treated with fungicides.

The processing of canes starts when these are scraped to remove the thin silicious coating. Once this is removed the yellowish luster of the material is brought out. This could be done either by rubbing vigorously a wet "sinamay" cloth and ground over the canes or by pulling the half bent canes back and forth around a bamboo stake. To minimize fungal attack the canes are dried to not more than 20% moisture content based on oven dry weight. Chemical treatment is done by soaking the canes overnight in a 5% solution of commercial zinc chloride. Care must, however, be taken at all times because areas treated with this chemical corrode iron easily.

Bamboo

Bamboo belongs to the same family as corn, wheat, and other grasses. There are about 700 known species of bamboo all over the world. Thirty-two of these are found growing in the Philippines. They are distinguished from other members of the Gramineae family in that they are woody perennials. Bamboo varies in height from 15 cm to over 30 m at maturity. More than 30 cm a day is not an unusual rate of height growth. After attaining full growth, the culms may require 2-5 years to harden and mature. If left standing too long, or until they became even-matured, bamboos are reported to lose much of their resiliency.

Today the botanical character of bamboo is still not well established. The reason is that most bamboos produce flowers and fruits at intervals of 30 years or longer. Soon after flowering, many species die.

In the use of bamboo as a reinforcing material in concrete, the elastic strength of bamboo is of prime importance. Kauayantinik (*Bambusa bluneana*), have been used as reinforcements for beams and columns.

Bamboo is used as a framework of a great majority of rural houses in the country. "Sawali", a common bamboo matting in tropical Asia is very attractive as an interior finish. Currently, bamboo is being processed as a laminated product by a private company in the Philippines with the trade name KAWOOD. However, most of its production is now absorbed by the furniture and the building interiors industry. Its potential as a house component is very great. It can be used as a parquet floor tile which has wearing characteristics comparable with those of wood used for the same purpose.

---

*The following species are commercially exploited:

Lambutan (*Calamus halconensis*, Becc.)
Limuran (*Calamus ornatus*, Becc.)
Lokuan (*Calamus reyesiamus*, Becc.)
Padlos (*Calamus microsphaerian*, Becc.)
Palasan (*Calamus maximus*, Becc.)
Panus (*Calamus ramolusus*)
Sika (*Calamus spinifolius*)
Tagiktik (*Calamus filispadix*)
Tumalyn (*Calamus mindoreneis*)
Ditaan (*Darmonorops mollis*)

As a plant material for house construction bamboo is most important because it is strong in relation to its weight; easy to transport due to its lightness; available everywhere in the developing countries of Asia; has good thermal properties and is relatively cheap.

In the Philippines, a plantation made of 3-4 years old erect-bamboo species yields about 5,991-6,900 kg of air-dry bamboo per ha per year. FOREPRIDECOM has developed a kind of bamboo-wood tile, 22 cm^2 consisting of wood-veneer crossband and back, and bamboo-veneer face. It has retained excellent glue bond between the plies after a number of years of exposure to foot traffic and stresses brought about by moisture changes. The wearing characteristics of the bamboo face are comparable to those of wooden tiles of similar construction.

The uses of bamboos are innumerable and on account of their workability, they are employed for almost every purpose as wood. They are commonly used for bridges, houses, forces, rafts, vessels for carrying and storing water, water pipes, cooking utensils, splints for baskets, mats, vehicle shafts, furniture packing and shipping, poles, masts, sporting goods, phonograph needles, nails, handles, toys, and the like. Young shoots of many species are relished for food.*

Coconut Trunk

Coconuts grow extensively in the Philippines and mature at the age of 35 years. The present area planted to it is over 2.1 million ha or about 7% of the total land area of the country. Density would be from 100 to 144 trees to a hectare. To date, several coco plantations are reaching diminishing productivity particularly those planted before World War II. In Quezon Province alone, some 14 million coconut trees are 69-80 years old. These have to be replaced with new growth. Research on how to make use of the potentials of coconut trees that are no longer productive has been conducted by the National Science Development Board. Of relevance to low-cost housing is the utilization of the coconut trunk. Each tree has an average potential trunk volume of 0.5 m^3, but only one-third of the volume, usually the lower portion or the butt, has the coco hard portion suitable for low-cost housing materials.

---

*Erect species of bamboo growing in the Philippines:

Anos (*Schizostachyum lima* (Blanco) Merr.)
Arundinaria (*Arundinaria niitakayamensis* Hayata)
Bayog (*Dendrocalamus merrilianus* Elm.)
Bolo (*Gigantochloa levis* (Blanco) Merr.)
Botong (*Dendrocalamus latiflorus* Munro)
Buho (*Schizostachyum lumampso* (Blanco) Merr.)
Giant bamboo (*Gigantochloa aspera* Kurs.)
Guadua (*Guadua philippinensis* Gamble)
India bamboo (*Bambusa arundinacese* Wild.)
Kalbang (*Schizostachyum textorium* (Blanco) Merr.)
Kauayan-china (*Bambusa multiplex* (Lour.) Raeusch.)
Kauayan-tinik (*Bambusa blumeana* Schultes f.)
Kauayan-kiling (*Bambusa vulgaris* Schrad.)
Kauayan-tinik (*Bambusa blumeana* Schultes f.)
Lopa (*Bambusa cornuta* Munro)
Merrill bamboo (*Bambusa merrillii* Gamble)
Palawan schizostachyum (*Schizostachyum brachyladum* Kurz.)
Pole-vault bamboo (*Phyllostachys nigra* Munro)
Spineless India bamboo (*Bambusa tulda* Roxb.)
Yellow bamboo (*Bambusa vulgaris* Schrad. var. *striata* (Lodd.) Gamble)

Coco trunk is not commercially available in the country. Problems of technology have yet to be solved, particularly the milling of coco trunk into lumber.

Cross-section

The gross physical character of the cross section of a coco trunk in terms of color are different from those of wood. The darker portions (coco hard) are found on the outer portion near the bark. On average, coco hard occupies 47.43% of the cross sectional area. Coco soft (the soft core) is 52.5%. There is also a variation of the coco hard/coco soft ratio from stump height to the upper portion of the trunk. In most cases, coco hard occupies 55% of the cross-sectional area at the stump height to 28% (0.77 m from the ground) at the level 2 m below the origin of the leaves.

Table 2.5. The Strength and Related Properties[a] of Green Coconut Trunks from Tiaong, Quezon.

Property	Unit	Species mean position along cross-section Hard outer layer	Core
Moisture content	Present	121	287
Specific gravity[b]	-	0.530	0.299
Static Bonding			
Stress at proportional limit	kg/cm^2	310	144
Modulus of rupture	kg/cm^2	527	242
Modulus of elasticity	1,000 kg/cm^2	73.6	30.6
Compression Parallel to Grain			
Stress at proportional limit	kg/cm^2	169	73.7
Maximum crushing strength	kg/cm^2	294	123
Modulus of elasticity	1,000 kg/cm^2	109	47.8
Compression Perpendicular to Grain			
Stress at proportional limit	kg/cm^2	38.4	18.6
Shear Parallel to Grain			
Maximum shearing stress	kg/cm^2	53.1	22.5
Hardness[c]			
Side	kg	524	137
End	kg	488	117
Toughness	kg cm/specimen	323	175

[a]All tests conducted followed the procedures of the ASTM Standard Methods of Testing Small Clear Specimens of Timber. STM Designation: D143-52. Part 16, 1970 Annual Book of ASTM Standards, American Society for Testing and Materials, Philadelphia, Pa.

[b]Based on oven-dry weight and volume at test.

[c]Load required to embed a 1.128 cm steel ball to 1/2 its diameter.

Sawing

Coco hard is the most important building material to obtain from coco trunks. Coco hard has higher strength properties and greater decay resistance. Coco soft is susceptible to drying defects like cupping and warping. The appropriate saw milling techniques should, therefore, take the above-mentioned factors into consideration. Sawing with the use of stellate-typed saw blade was found to be the most efficient and practical.

Particle board

Experiments on particle-board-making with use of coco trunk has also been undertaken by FORPRIDECOM. Mixtures of 100% coconut trunk, and a 50:50 ratio of coconut trunk and wood particles, manufactured into particle-board, using urea formaldehyde at 8% and 10% resin-content levels, has been found to be technically feasible.

FORPRIDECOM has arrived at a recommended proportion of 50:50 coconut trunk and wood-particle mixture at 8% resin which appears to be the optimum condition, both in point of strength and economy in production cost.

Other research and development work

(a) Coco trunks are sawn into lumber in the form of stone-cut sidings and as flooring in the form of tongue and groove sections. The soft core of the coconut trunk is being studied for ceilings in the form of T & G boards, as well as for window jalousie blades.

(b) Coconut-trunk sawdust has been studied as an ingredient in hollow blocks having a proportion of 1 part cement to 3 parts sawdust by volume.

(c) Exploratory studies undertaken by the National Science Development Board at the factory of the Aras-Asam Timber Co. indicate the technical feasibility of producing slice-veneer plywood from coconut trunk. The trunk-bolts were steam-heated for 12 hours in order to produce satisfactory slice veneers.

(d) An attempt was made by some wood scientists to produce shingles out of coconut trunk. Prospects for production are bright but studies are still being done on the preservative treatment of coco-shingles.

(e) To attain more utilization of the coconut trunk, a trial test was made to utilize the trunk for parquet flooring. Service tests are still being done so that significant information is not yet available.

(f) An exploratory study has been made on the use of coco lumber for low-cost roofing. The roof was made of a modified T & G profile. A sealing compound was used on the joints of the pieces and then sprayed with a water-proofing solution. An accelerated service test is now being conducted at the FORPRIDECOM Housing Site.

REFERENCES

Casin, R. F., Ordinario, E. B. and Tamayo, Jr. G. Y., *Solar Drying of Apitong Narra,* Red Lauan & Tanguile, FORPRIDECOM, Laguna, 1976.

Tamolong, Francisco, *Studies on the Potential Uses of Coconut Trunk in the Philippines*, FORPRIDECOM, Laguna, 1976.

Tamolong, Francisco, *The Utilization of Coconut Trunk and Other Parts in the Philippines,* FORPRIDECOM, Laguna, 1976.

3

MASS HOUSING IN INDONESIA: IN SEARCH OF NEW SOLUTIONS

Hasan Poerbo and Albert Kartahardja

3.1 MASS HOUSING IN INDONESIA

Housing is a large and unwieldy subject which has been already much explored. Yet new concepts are always emerging as new premises, perceptions and insights replace old ones.

This Chapter concerns itself with one of the many aspects of housing: productions of mass-housing for the low-income group in Indonesia. The paper addresses itself to the problems of the management of building processes, institutional and manpower development and the development of appropriate building technologies within the context of the Indonesian situation.

The ideas set forth are not novel in industrialized countries. However, the transfer of these ideas into an entirely different cultural and industrial setting brings to our attention new dimensions in the problem of mass-housing production. It is our hope that our limited experience may contribute to a better understanding of the prospects and pitfalls of the transfer of mass-housing technologies in different settings.

The Setting

Indonesia is a country with about 130 million population in 1971, spread out over thousands of islands. However, population distribution is very uneven, Java having the largest population of around 76 million. Population densities are therefore also very uneven. Kalimantan for instance has a population density of less than 20 persons per km^2. By contrast Java has a density of more than 700 persons per km^2.

These imbalances govern national development policies in Indonesia, which put heavy emphasis on population redistribution coupled with control of population growth and natural resource development to attain an economic growth of around 7-8% annually, to sustain a growing population which has a rate of annual increase of around 2.5%.

Mass-housing

Mass-housing, in the sense of production of houses in large quantities as an organized effort, has now become more and more a common phenomenon.

Population redistribution has brought with it the need for creating new settlements as part of the transmigration program, in which families from overpopulated islands such as Java and Bali are resettled in newly opened areas in other less populated islands. Transmigration housing is a mass-housing problem with unique characteristics, as the areas in which the new settlements have to be built are usually isolated in remote areas, are rural in character and have to be built with minimum standards of quality and costs.

On the other hand, resource development such as oil, mining, modern agriculture and forestry have induced the development of new settlements in the form of company towns. These are usually settlements with higher standards of quality, compared to transmigration settlements, even for their labor housing.

Then there is a third category of mass-housing which has now become part of national development programs. As the national economy improves, funds have become available, both public and private, to be channeled into housing. Social housing development is now being undertaken by the National Urban Development Corporation (NUDC). Private commercial estate developers are now also active in several large cities. At present, plans are being prepared to stimulate the growth of private non-profit housing organizations, such as housing cooperatives and associations.

This Chapter is intended to deal with mass-housing in urban areas, which comes under the third category.

Mass-housing by NUDC

The NUDC has been established as an executive arm of the National Housing Authority (NHA). The NHA itself is an interdepartmental Board, responsible to the President. It has as chairman the Minister of Public Works and Power. Its members are the Ministers of Social Affairs, Interior, Industry and Finance. Its function is to coordinate housing policies and programs.

The NUDC operates in the regions, where it cooperates with local governments and private non-profit organizations to develop "low-cost housing", next to its other statutory duty to manage the implementation of the Sites and Services and Kampung Improvement Programs.

In these activities the NUDC cooperates with the National Mortgage Bank, whose function is to finance loans to middle-low and low-income families to acquire houses built by the NUDC.

NUDCs mass-housing program for the Second Five-year Plan can be seen in Table 1.

Various types of houses have been developed by the Directorate of Building Research (UN-Regional Housing Center) in Bandung, which are now used as standard by the NUDC. The houses are designed for the lower-middle income group, which corresponds to an income of between Rps. 30,000-Rps. 60,000 per month ($74-$148). There are basically four standard types of approximately 36, 45, 60 and 70 km^2, designed as detached, semi detached and row-houses (one and two story), flats and masonites. Building cost (excluding land and infrastructure) is put at around Rps. 22.000-Rps. 25.000 per km^2 ($53-$60).

Table 3.1. Targets for "Low-cost Housing" in the Second Five-year Plan 1974-1979.

Location	74/75	75/76	76/77	Total 74/79
I. Jabotabek				
(Jakarta and surroundings)	-	1,200	23,800	25,000
1. Klender	-	-	1,000	1,000
2. Cengkareng	-	-	5,000	5,000
3. Depok	-	1,100	3,900	5,000
4. Tanggerang	-	-	5,000	5,000
5. Bekasi	-	-	5,000	5,000
6. Bogor	-	100	3,900	4,000
II. Outside Jabotabek	120	880	24,000	25,000
7. Surabaya	-	-	7,000	7,000
8. Bandung	-	250	4,000	4,000
9. Semarang	-	380	2,500	3,000
10. Cirebon	-	250	1,500	2,000
11. Yogya/Solo	-	-	2,000	2,000
12. Medan	-	-	3,000	3,000
13. Ujung Pandang	-	-	2,000	2,000
14. Padang/Palembang	-	-	2,000	2,000

Source: Saleh Amiruddin, ME. Dipl. DP. *Low-cost Housing in Urban Areas*, Directorate for Building Research (UN Regional Housing Center), 1975.

Private "Low-cost Housing"

The oil boom in the early 1970s has induced investment in housing by real estate developers. Bright economic prospects at that time and an acute housing shortage in the large urban centers, even for the higher income group, was instrumental in the development of housing estates for the speculative market. The luxury housing market flourished. Houses of Rps. 50-Rps. 100 million ($125,000-$250,000) were built, but as the market for these types became increasingly restricted, houses of Rps. 10-30 million ($25,000-$75,000) became the bulk of private speculative housing development. The oil crises, the Pertamina's (State Oil Company) problems in Indonesia and the international economic situation after 1974, have had their repercussions on speculative housing of this nature. Conspicuous housing estates of 300-400 houses in and around Jakarta created a political backlash, which took the form of Government intervention into speculative housing.

The Investment Board issued a regulation whereby real estate developers are now required to build one luxury house: three houses for the middle-income group: 6 "low-cost" houses, whereby a "low-cost" house is defined as a house not costing more than Rps. 3 million ($7,500) per unit including land.

By this regulation commercial real estate developers are now also induced to develop "low-cost" mass-housing. However, as it is now their problems are still not solved as regards acquisition of "cheap" land and "cheap" money to support their programs.

Non-profit housing organizations (NPO) have existed already since before World War II. But their role in housing production is up to now still very limited. The largest NPO such as in Surabaya has an output of only some 200 houses per year.

In view of their potential role in the mobilization of individual families to participate in housing development, there is at present considerable interest with the Government to take the necessary steps to foster the growth of NPOs. Part of housing by NPOs may take the form of organized mass-housing.

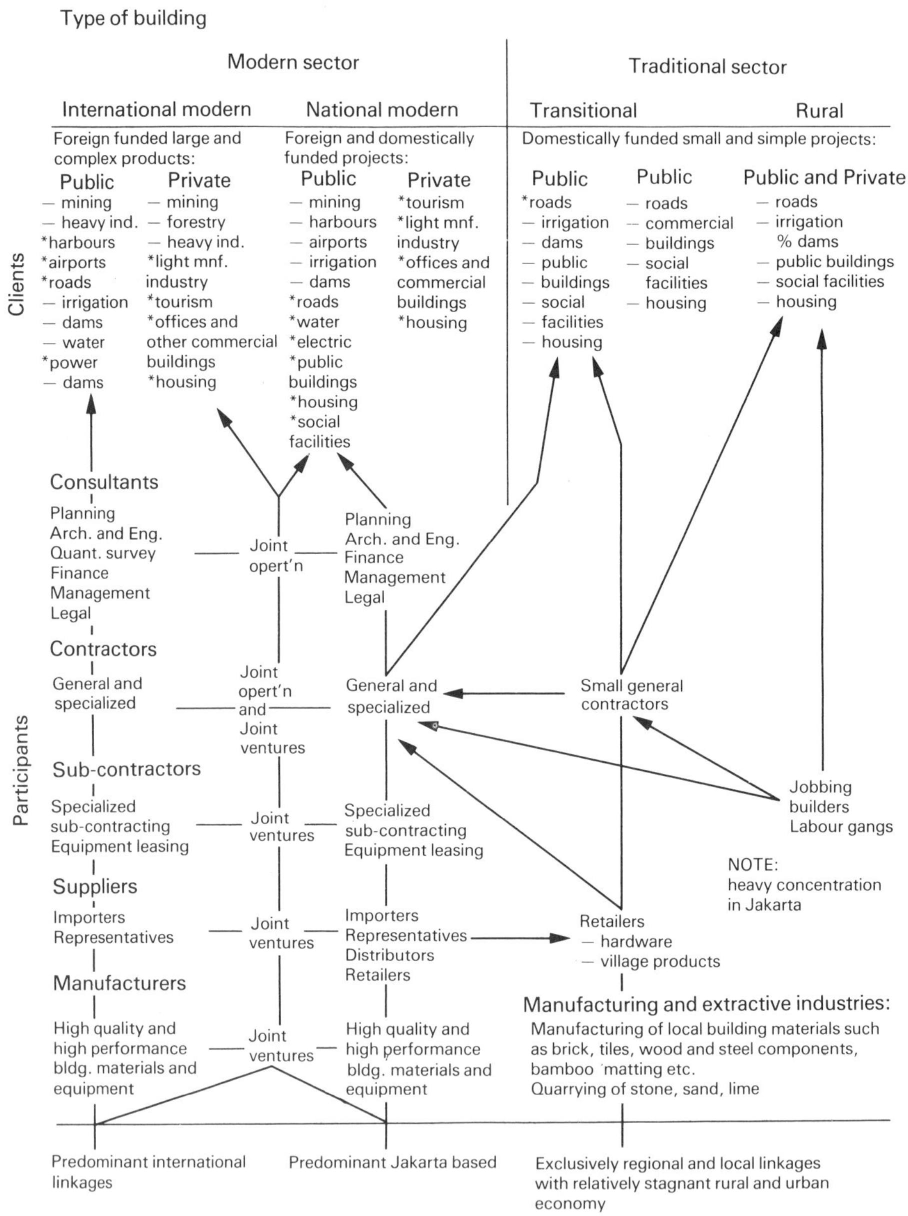

Fig. 3.1. Organization of the Construction Industry in Indonesia

3.2 IN SEARCH OF ALTERNATIVE CONCEPTS OF MASS-HOUSING PRODUCTION

As can be seen above the problem of mass-housing is here to stay for quite a while. It can also be expected that public and private expenditures will increase in this sector. Given this situation, the question is "What concepts are appropriate for Indonesia to get the most out of mass-housing in terms of production?"

Mass-housing vis-à-vis the Building Industry

Mass-housing can be seen as a bulk demand by an organized client system, such as the National Urban Development Corporation, housing cooperatives and other non-profit organizations and commercial real estate developers, for certain building materials, components and skills, within a relatively short period, concentrated at a certain location.

Annual demands have achieved a level of 500-2,000 housing units in different locations. In the next years levels of 1,000-10,000 housing units per year can be expected.

Budget allocations and regional development priorities have caused heavy fluctuations in the demand for goods and services for mass-housing in places where new settlements have been constructed. Project performance has therefore suffered because of delays in logistics. Studies conducted of several projects,[1] surveys[2] and discussions during seminars indicate the existence of fundamental problems in two areas. One is the nature of the building industry, and second the use of inappropriate building processes and technologies.

The building industry in Indonesia has a dualistic character[3] consisting of a modern and a traditional sector. A schematic presentation of the organization of the building industry is given above (Fig. 3.1). Early experiments with "mass-housing" prototypes of 150-200 houses have experienced difficulties in controlling cost, quality and time, using traditional building materials and ordinary building procedures where contractors were selected through competitive bidding without the necessary knowledge and experience in mass-housing production techniques. The unexpected increase in demand for roofing tiles for instance have placed a heavy burden on the capacity for production by local small industries and their delivery mechanisms. It has caused problems in standardization, quality control and delivery schedules. The same problems, and much more severe, were encountered with wooden components. Local timber suppliers could not cope with the demand. Timber has to be brought in from the other islands, causing delays and stagnation. Wood component manufacturing farmed out to various small workshops did not meet specifications. Introduction of new building techniques and materials, such as puzzolan cement hollow blocks and built up roof trusses, designed to reduce costs have only marginal effects.

Inherent inelasticities in the traditional building material industry to adjust itself to these demand fluctuations have been instrumental in the shift to utilize the potentials of the modern sector of the building material industry for mass-housing.

In the early 1970s new and modern manufacturing plants have been established in and around the large cities such as Jakarta and Surabaya, producing new building materials and components which are now competing with traditional materials. The list of selected building materials and component manufacturing plants established since the early 1970s shown below gives a clear indication of emerging potentials in the modern sector of the building industry to support mass-housing (Table 3.2).

Table 3.2. List of Selected Building Materials and Component Manufacturing Plants, established since 1970*

Type of factory	Capacity	Location
Asbestos cement	est. 75,000 tons/year	Jakarta, Surabaya, Medan
Cement blocks	est. 3,000,000 blocks/month	Jakarta
Particle board	5000 prefab houses/year	Sukabumi
Wood wool	7,500 sheets year/month of 1 x 2 m	Bogor
Saw-mills and wooden components	no complete data available	Jakarta, Surabaya, Surroundings
Concrete mix and concrete components (precast)	"	Jakarta
Aluminum sheets and extrusion	"	Jakarta
Steel structures	"	Jakarta, Surabaya
Plastic pipes	"	Jakarta, Surabaya

*Compiled from statistics from the Foreign and Domestic Investment Board and direct interviews with manufacturers.

These are plants which are capable of producing strategic materials and components for mass-housing, such as roofing, walls, floor, doors and windows, replacing traditional materials. Several of these enterprises have singly or combined developed semi- or fully-prefabricated houses to be offered in the mass-housing market. However, not one has been tried out on a scale larger than several hundred. These are typical industry-sponsored building systems with inherent weaknesses common to them, such as monotony, inflexibility and inferior aesthetic qualities. Cost-wise they are nearing the threshold of about Rps. 22,000 ($55/m^2) which can be also reached with ordinary or modern building materials with rationalized management. An example of a prefabricated housing system is the one developed by a consortium of asbestos-cement and aluminum extrusion manufacturers.

Next to these locally developed industry sponsored building systems, other imported ones have also made their entry. A concrete housing prefabrication plant has been tested in Jakarta which produces multi-story flats at a cost per m^2 twice as high as the other systems. Several firms are negotiating to introduce prefabricated wooden houses through joint venture with local counterparts. However, their marketability is doubtful because of their prohibitive cost and social acceptability. People tend to regard wooden houses as inferior and semipermanent.

As yet there is no conscious effort to develop rationalized client sponsored building systems for mass-housing. Those which have been built by the National Urban Development Corporation are perhaps the nearest to this concept. However, rationalization has been until now conceived in structural and technical terms, rather than in the broader sense of management of the whole building process.

Therefore, the performance in terms of cost reduction, quality and time control is still marginal.

On the national scene there is still a lack of evaluative mechanisms to make the best choice or to develop the most appropriate building systems for a certain location and at any given time. The expected growth and distribution of the demand for mass-housing and the dynamic structural changes in the building industry seem to indicate the imperative to develop capacities for continuous evaluation and assessment of the impacts and implications of technological choices in the field of mass-housing. Until now, structural changes in the building industry have been largely the result of market forces. It has introduced a situation whereby the modern sector becomes dominant, replacing the rôle of the traditional sector in the process. In a country where traditional brick and tile making, small scale metal and wood industries, carpentry and other building skills are important contributions to the traditional urban and village economy, these shifts of roles may well have important repercussions, socially and economically. Although it is doubtful whether the contribution of mass-housing will be that important to the whole economy, as a matter of principle any choice of technology for mass-housing should not only benefit the project in a limited sense, but it should also create other benefits external to the project. This is also consistent with current national development policies, which stipulate that construction projects should create new job opportunities and help to redistribute income among local communities. Introduction of industrialized and industry-sponsored building systems should be therefore scrutinized as to their benefits and costs to society and the economy, because of its tendency to be capital intensive and centralized.

The Client-sponsored Building System: A Conceptual Approach

The client-sponsored building system (CSBS) is a tool which serves the best interest of the client. The building system itself can be vizualized as being a building with all its elements and components as a physical entity, as well as a process of production of a building with interdependent activities. The client or client system can be limited to those parties directly involved as patrons of the production services, including the users of the product, or it can also include the supra-client or society at large represented by a public interest oriented client system.

In industrially advanced societies with a well developed industrial organization and abundance of management skills and supporting technologies, the concern in applying the CSBS may be limited to the development of building systems in physical sense to achieve cost reductions, better quality control, better appearance and flexibility to adapt the building to changing user's needs.[5] In Indonesia, structural problems and capacities of the building industry to support mass-housing programs indicate the necessity to conceive CSBS also as a building process and its management. In other words, the development of a supportive building industry which goes hand in hand with creation of new jobs and redistribution of income is part of the objectives of the CSBS.

The need to include the development of the building industry in the building process can be illustrated as follows. Modern industrial plants producing building materials and components are now still operating far below their designed capacities. Idle capacities of 40-60% have been reported and are common phenomena. The problem is inherent with marketing new products in a market which is not yet organized to absorb and sensitized to the potential performance of those products for mass-housing. Demand fluctuations are severe due to institutional instabilities and discontinuities in decision making on the demand side, and the lack of an information system geared to better and long-range forecasting of mass-housing demands. Many of the modern industries suffer from this situation. Inefficiency, ineffectiveness and bankruptcy, even among these modern industries, become ultimately a social cost and a burden to the whole economy. In the traditional sector the situation is worse. Small scale urban and village industries supplying all kinds of building materials and components have not the capacity to aggregate their own market. At all times they rely on middlemen to market their products and to finance their operation. Any sharp fluctuation in demand creates windfall profits for middlemen, who as supply mobilizers have a strategic position to bargain for price increases for their own profit,[6] but feel no responsibility for quality control and strict delivery schedules. Being conservative risk-takers, they also act as a formidable barrier for innovation in building material technology. Proliferation in the building materials industry without regard to product quality has resulted from this situation. Casualty rates among this group can be expected to be high.

The CSBS in Indonesia has been conceived within this setting. As a building process it has the objectives:

(a) To organize demands in bulk for mass-housing so as to make long-range forecasting possible, thereby reducing demand fluctuations to manageable proportions;

(b) To standardize performance specifications for materials, components, skills and services, so as to make rationalization possible which increases productivity;

(c) To develop an information network between the various parties involved or potentially involved in the building process, so as to increase their efficiency and effectiveness in their role;

(d) To develop financial and institutional mechanisms to replace the role of middlemen between client and suppliers of materials, components and services, so as to make technological innovations possible and to ensure that benefits from increasing demands will accrue to those industries and services directly involved in the building process, preferably local industries and services which benefit local communities.

The ultimate objectives are thus briefly to increase productivity, to reduce costs, to increase quality and construction time performance, to create external benefits in mass-housing in the form of new job creation, redistribution of income to local communities, development of viable small enterprises in conjunction with modern large enterprises to support national mass-housing programs.

Productivity, cost reduction, quality and time performance are objectives which can be attained through competitive mechanisms which are well-known problems in ordinary project and construction management. However, employment creation, redistribution of income, transformation and development of small scale urban and village industries into efficient modern enterprises are outside the realm of ordinary responsibilities of contractors and middlemen and may run contrary to their commercial interest.

These latter objectives are developmental in nature which entails development risks, ordinarily born by public bodies. Can contractors and middlemen be entrusted with the role of "development agent" within the above context? Or should other professional bodies be created to act this role? These are pertinent questions which always come up in the conception of management models of building processes in mass-housing of this nature.

A theoretical ideal model of an organization for the management of a building process in mass-housing to attain the above objectives has been evolved from models which are well known in European countries.[7]

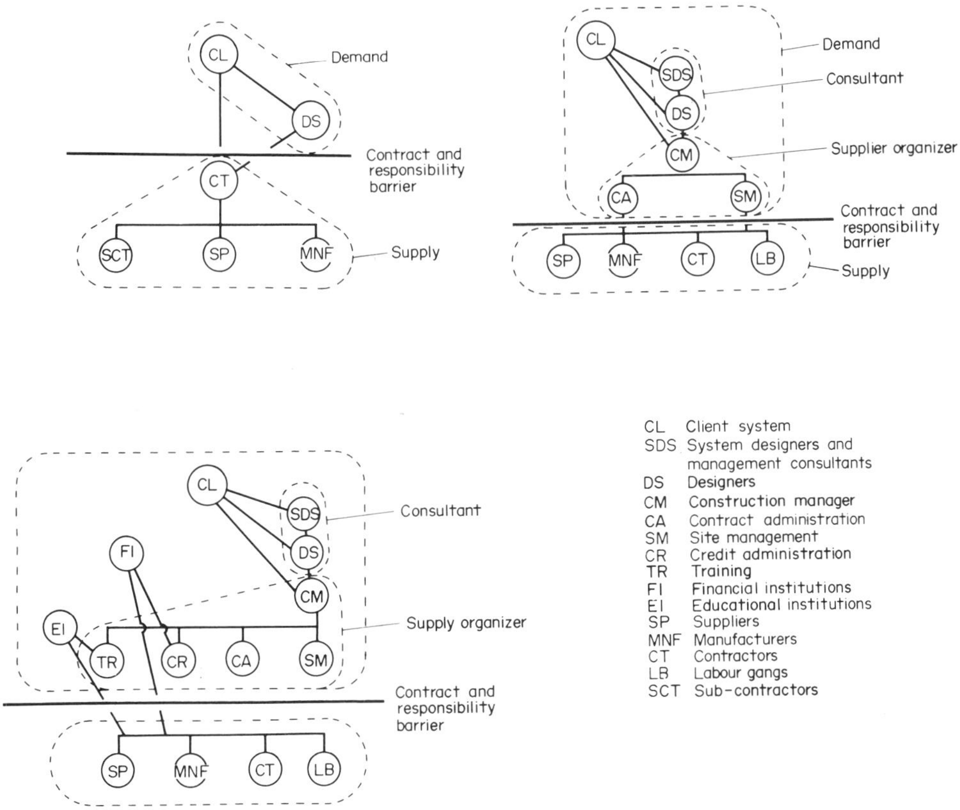

Fig. 3.2a, 3.2b, 3.2c Simplified models of functions and their relationships in mass-housing project management.

Figure 3.2a represents a model of working relationships between client, designer, contractor, supplier and manufacturers. The client, through the services of the designer as consultant, exercises external control on the contractor through contract agreements. The contractor, subcontractor, supplier or middlemen and manufacturers are for practical purposes a closed system, which can only be penetrated by the client and designer with special agreements. In a situation where information flow between all parties involved in the building process is very weak, such as is the case in Indonesia[8] — new building technologies, innovative design, etc. for mass-housing cannot be transmitted between the designer and contractors, supplier and manufacturers during the crucial stages of project planning and design, where basic decisions are made which will affect the outcome of the project. Evaluation of mass-housing projects using this system shows convincingly that the system does not work.[9]

From this experience a theoretical model was drawn up as shown in Fig. 3.2b. The basic aim of this concept is to break down the barriers between all the parties involved in the building process by using the so-called "building team",[10] which consists of a representative of the client, a system designer and management consultant, a designer, contractors and suppliers or manufacturers of strategic materials and components. The role of the contractor can be taken over by a professional construction manager, engaged on a management contract basis. The "building team" should be established as early as possible, preferably at the project planning stage, through competitive selection of candidates. In this way two objectives are aimed at: (a) to generate a free information flow between all parties concerned, and (b) to have direct access to all production factors to attain higher project performance through better utilization and coordination of those factors during the construction stage. In discussions and workshops with members of the building industry the question was persistently asked: Who is taking the risk if the contractor is replaced by a construction manager? Theoretically, the client should be taking the risk in the development of mass-housing techniques, especially in the case of a public body acting as client. The nature of the risk is developmental, rather than financial. A public body such as a National Urban Development Corporation could act the role of such an agent, transferring the results to other mass-housing projects undertaken by itself and other parties which it serves. The experience, as a body of knowledge, is transferable through publication and education.

Figure 3.2c is a further elaboration of the model described in Fig. 3.2b. In demand situation where the volume is large enough and continuous, construction activities can be scheduled in such a way so as to give building materials and component industries to plan ahead. To assist small scale industries, a credit system can be introduced, geared to the need for standardized materials and components for mass-housing. A training program can be attached to the credit system to introduce new products and processes, to upgrade existing products and adjust them to new standards, and to develop management and other skills needed for production and construction. All these responsibilities can be coordinated by a construction management team which cooperates with financial and educational institutions. Fluctuations in demand for goods and services can be neutralized with the creation of regional bufferstocks, separately financed from construction projects.

The Concept of Project Development Cycle as Vehicle for Institution Building and Manpower Development

The concept of "project development cycle" in mass-housing as a continuous activity is also not novel.[11] Within this context the model as described in Fig. 3.2b can be seen as a dynamic model, which in time produces permanent institutions and relationship through continuous exercise.

The construction management team may grow into a Technical Service Organization (TSO) whose function as a professional non-profit organization is to serve and develop non-profit housing organizations, such as housing cooperatives and associations.

It can also absorb the functions of system designer, management consultants and designer. The TSO can be part of the function of the National Urban Development Corporation or be a separate entity.

Continuous exercise can also produce manpower with the necessary skills to fill the needs of the various functions involved in mass-housing, enhancing their combined performance.

3.3 THE SUKALUYU PROJECT: IMPLEMENTATION OF AN IDEA

In the early 1976s the National Urban Development Corporation (NUDC) initiated a housing project in Sukaluyu in the North Eastern part of the city of Bandung. Responsibility for design and supervision was given to the Directorate of Building Research (UN-Regional Housing Center) in Bandung, which cooperated with the Bandung Institute of Technology in planning the project.

A site of 1.5 ha was acquired for this purpose. Land shortage in Bandung and corresponding high land values made it necessary to impose a density of 100 houses/ha or more if possible. Standard two story row-houses of 45 m^2 were required, costing not more than Rps. 1.2 million ($3,000) per unit excluding land and infrastructure or Rps. 2,700/m^2 floor area ($6.75 per m^2). Site works and infrastructure should not exceed Rps. 1,000/m^2 ($2.5 per m^2) of land.

A standard design was used, which was developed by the Directorate of Building Research. Strategic materials and building components which for mass-housing will cut costs and construction time, and still ensure a high degree of quality control were scheduled to be factory made. Thus, asbestos cement roofing materials, pre-cut beams, prefabricated doors, windows, frames and staircase, particle board floor and room dividers were specified. The rest were to be locally made materials and components, which can be produced by small scale industries.

The concept of a "building team" was used. After the preliminary design was finished, a meeting was called which was attended by selected contractors, suppliers and manufacturers who were introduced to the concept of forming a "building team" for the construction of the project. Suppliers and manufacturers of strategic materials and components were asked to submit a proposal which should contain the price of their products, including services rendered during construction. One supplier or manufacturer was selected for every material or component. After they were selected, the contractors were asked to submit their proposal based on the selected materials and component and the corresponding quoted price. Three criteria were used to evaluate the contractor's proposals: (a) their concept of construction management of mass-housing and technical solutions to achieve cost and time savings; (b) cost analysis based on their proposed concept; and (c) estimated building cost based on their analysis. One contractor was selected among five competitors, who happened to be the second lowest bidder, but whose concept of construction management was the most convincing.

After the "building team" was formed, the design was reviewed and the final construction documents prepared. A lump-sum contract was negotiated between client and contractor, based on the final construction document prepared by the "building team". Separate contracts between the contractor and suppliers and manufacturers were drawn up, with the approval of the client and consultant.

The building cost was fixed at Rps. 1.1 million ($2,750) or approximately Rps. 24,000 per m^2 ($60 per m^2) which came down to the same figure as the lowest bid. Construction time was fixed at three months.

Preliminary Results in the Sukaluyu Project

The Sukaluyu Project was an experiment to implement the Client Sponsored Building System and the "building team" concept. It served also as a test for some of the hypotheses formulated in the TDI/EW Center proposal for cooperative research in low-cost housing.[12]

Although the Sukaluyu Project was one of those "crash-programs" which made it next to impossible to have time for good planning, the cooperative spirit and good understanding which was created between the members of the "building team" made experimentation possible.

In terms of mass-housing production, the objectives of the experiment were limited to:

(a) To attain a building cost of not more than Rps. 22,000/m^2 (approximately \$55/m^2) without sacrificing quality;

(b) To improve construction time performance;

(c) To get first-hand experience from contractors, suppliers and manufacturers to get a better insight in future construction planning for mass-housing.

Comparative figures for building cost and construction time in other similar projects are respectively Rps. 22-25,000/m^2 (\$55-\$62.50) and 4-4½ months.

Except for building cost, all the objectives were achieved through direct control of strategic materials and components and rationalization of construction on site. Strategic materials, such as puzzolan cement hollow block for bearing walls, roofing materials, components for flooring and partition walls were factory made.

Since there was no time to develop new products, standard materials and components were used which were already in the market, except for wooden components which were adapted to new specifications. Figure 3.3 shows a breakdown of the cost structure in percentages, and "areas of intervention" through the "building team" system.

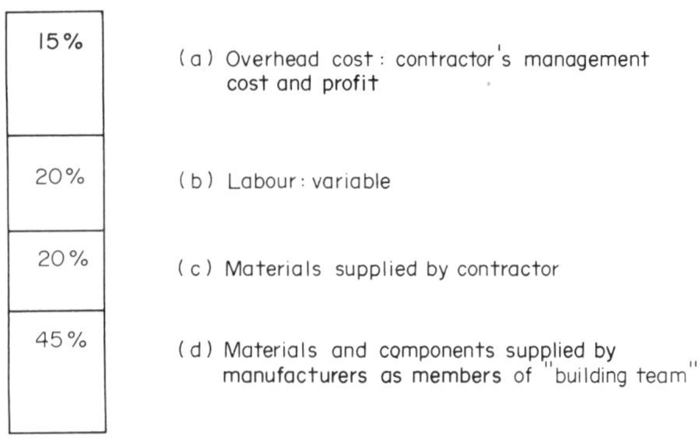

Fig. 3.3. Structure of building cost in %.

Overhead cost was fixed after selection of the contractor. In other words, the selected contractor may retain his overhead costs as proposed in his bid. This is to give him an incentive to participate in the deliberation in the "building team" to reduce construction cost.

Labor is still within the internal control of the contractor and is still a contractor's risk. After the project was finished, the contractor reported that he lost during the first phases of the project, when he had to transform skills from ordinary traditional materials handling into skills for assembling of new materials and components. However, the period was relatively short and he was able to recoup his losses with a considerable margin through an increase in labor productivity.

He also reported that labor productivity could be further increased by differentiation and specialization of his labor gangs, which he could not achieve in this project because of its limited size. A continuous demand with a higher volume could induce differentiation and specialization and higher labor productivity, if it could be coupled with labor mobility between similar projects.

Materials supplied by the contractor consisted of such items as sand, lime, stone, cement and other small items. Control over these materials in the Sukaluyu project was still given over to the contractor, since it was felt that the gain was too marginal to develop a special unit within the client's organization to take care of procurement.

Materials and components supplied by manufacturers as members of the "building team" were fixed through negotiated contracts. These materials and components were already tested and have standard market prices, which in the case of the Sukaluyu project were supplied with discounted prices because of bulk purchase.

The whole process, from project initiation to signing of contracts took only slightly more than one month. Construction itself was a relatively smooth affair. There was a free information flow between all the parties involved in the project from the design stage onwards. This has proved to be an extremely valuable experience in forecasting problems in logistics, so that bottlenecks can be avoided. Building costs could potentially be reduced to Rps. 1 million per unit or Rps. 22,000 per m^2 (\$55 per m^2) by using a different system for the foundation and better detailing. But there was no time to make too many changes in the design.

There are also some disturbing conclusions which can be drawn from the project. The structure and capacity of the building materials and component industry will be indeed a constraint to sustain a mass-housing program as visualized in the national five-year plan. Jakarta and its surroundings already have a building industry which offers a variety of alternative building technologies with a combined capacity which can already sustain a relatively large mass-housing program. But for various building materials which are bulky and relatively cheap, transportation costs, even to Bandung, become already prohibitive. Thus, using puzzolan cement hollow blocks for mass-housing in Bandung, for instance, will put a limit of around 1,000-1,200 houses annually to the program.

Expansion of existing production facilities and improvement of village industries to produce hollow blocks with the required performance specifications will need time, capital and managerial skills. The housing program for Bandung for the next two or three years is already fixed at around 2,000 houses annually, which comes under the responsibility of the NUDC. This does not include demands from other housing sectors. This means a minimum increase in hollow blocks manufacturing of twice the existing capacity within one year. Or to find an alternate component to substitute hollow blocks, or to develop an entirely different building system.

Another potential bottleneck is timber. Bulk demands for wooden components for mass-housing in the larger cities of Java will put a strain on the resources of wood processing industries, if no fundamental structural changes are introduced in timber trade, especially in transportation from the other islands. As it is now, idle capacities exist in the saw-mills and other related wood industries. However, a sudden increase in the demand will force these industries to organize their own procurement in the other islands and the transportation of timber to Java, in order to stabilize prices and to control quality, a task for which they are ill prepared. The Sukaluyu project has had to face these problems already.

All these are factors which are outside the realm of control in project and construction management in the ordinary sense. Given the national housing targets, technological solutions will be dependent upon the capacity of the public and private sectors in industry to develop effective means to control and mobilize their resources in a coordinated effort.

Within this context the relevance of the Sukaluyu experiment can be perceived beyond the immediate objectives such as cost, quality and construction time performance. In a sense it can be seen as a beginning of the development of a mechanism to bring together the demand and supply side to come to a constant dialogue.

An institutionalized monitoring, evaluation and feed back mechanism built into decision-making processes from project up to national policy level, developed through continuous and systematic exercise will make it possible to improve capacities for problem forecasting and policy planning down to project planning and implementation.

The experience in the Sukaluyu project is now being analyzed and "recycled" into another mass-housing project of nearly four times the size of the Sukaluyu project. The same concepts of the CSBS, the "building team" and a two pronged approach to building technology (using off-site pre-fabrication of components and on site labor intensive assembly) are being applied in a labor housing project in Jakarta. Next to this, other activities are now underway, which geared to each other constitute a total systems development in national housing construction which may have far reaching effects, if properly perceived.

3.4 ELEMENTS FOR THE DEVELOPMENT OF A DYNAMIC MANAGEMENT SYSTEM FOR MASS-HOUSING

On the national scene there are now activities, which by design or accidental, can be perceived as future functions which have strategic importance in the development of capacities in the management of national resources for mass-housing.

These activities, perhaps randomly selected, are:

(a) The establishment of a study-committee on non-profit housing organizations;

(b) Establishment of education in the field of housing;

(c) The Small Investment Loan (KIK) and Fixed Working Capital Loan (KMKP) schemes;

(d) The United Nations Industrial Development Organization (UNIDO) project for the development of small scale building materials industry;

(e) Industrialization of Housing Construction;

(f) Establishment of a Registration and Licensing Committee for the Construction Industry.

The Study Committee on Non-profit Housing Organizations

The Study Committee is now being initiated by the Directorate General of Planning, Housing, Building and Construction (PHBC), Department of Public Works and Power. The purpose is to monitor and evaluate selected housing development activities undertaken by non-profit organizations (NPO), in order to systematize their experience for policy and program formulation for the National Housing Authority. The Director General of PHBC is the Secretary General of the Board.

In simplified form the organizational diagram of the Study Committee will be as follows:

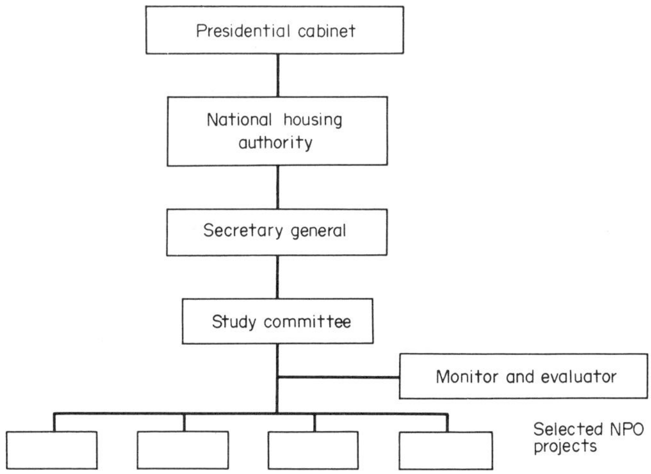

Fig. 3.4

One of the selected NPO projects in the Home Ownership Program (HOP) for ITB staff, administrative and technical personnel, presently estimated to have a potential membership of 300 families and being in the process of implementation. The ultimate institutionalized form of the HOP is still not yet clear, but it can potentially be developed into a Building Association or Housing Cooperative, assisted by a Technical Service Organization (TSO). Other pilot projects will be the HOP for the University of Gajahmada in Jogya, and the Housing Association in Surabaya.

Within this context experience gained in implementing the CSBS, "building team" and two-pronged approach to mass-housing technology can be monitored and evaluated, and fed into national decision mechanisms to institutionalize systems and procedures in mass-housing production through the establishment of NPOs and TSOs.

Establishment of Education in the Field of Housing

An initiative is now being undertaken to establish a Middle Management Course in Housing Management. The courses are expected to start in 1977, and will be of 6 weeks duration, three times a year. Annual output is set at around 200 persons per year. The courses will be organized as a cooperative effort between the Directorate of Building Research in Bandung, the Bouwcentrum in Holland and the Bandung Institute of Technology. Students will be invited from existing practitioners.

The courses are intended to be an extension of research and development activities undertaken by the research and academic staff of the cooperating institutions, using actual projects as field laboratories. With this system it is hoped that students can be directly confronted with actual problems and can develop realistic approaches to solving housing management problems using simulation games based on actual projects.

The relationship of this training program with the study Committee on Non-profit Housing Organization and the development of Technical Service Organizations is obvious. The training and research and development program can use pilot projects under the Study Committee as field-laboratories. New concepts and methods can be tested, monitored, and evaluated, and alternatives tested through simulation games during the training courses. The students, being practitioners in various functions within the housing delivery system, are thus not only trained in the concepts of housing management, but are actually active participants in the development of new concepts.

The "feed-back loop" between policy and implementation can be shortened, both in terms of information as well as in skills. The students can be projected here as change-agents in the development of the housing delivery system.

A Real Estate Management Course will be established within the Technical Faculty, the University of Indonesia. The course is being initiated in cooperation with the Real Estate Association of Indonesia. It is at present still in the preliminary stages of preparation. However, the basic concept is already clear. The objective of the course is to train business managers for real estate enterprises, both public and private. Both courses, the Middle-Management Course in Housing Management and the Real Estate Management Course, will be complementary to each other. The first addresses itself to the problems of project and construction management, while the latter with that of the management of enterprises.

Other graduate courses are being initiated at the Bandung Institute of Technology in the field of Human Settlements and Technology and Industry, in which housing is to be part of them. Although still fluid at present, these courses can be seen as opportunities to prepare manpower for various functions in the housing delivery system, especially research and development, and also manpower with specialized knowledge. The research programs attached to these graduate courses can be integrated with the activities of existing research institutions and other non-degree courses, such as the Middle-management course for Housing Management.

The Small Investment Loan and Fixed Working Capital Loan Scheme

The Small Investment Loan (KIK) and Fixed Working Capital Loan (KMKP) schemes are designed to assist small scale urban and village enterprises in their development. These enterprises are entitled to apply for loans up to Rps. 5 million ($12,500) under favorable conditions.

Within the context of project management such as illustrated in Fig. 3.2, and other conditions permitting, these schemes can be utilized to assist small scale building materials industries with capital to bring their capacity to the necessary levels of production to sustain mass-housing programs. These schemes can also replace the role of the middlemen as financier, so as to make it possible for construction managers or TSOs to have direct access to these enterprises through package loan agreements, consisting of capital loan in conjunction with management or production training, management assistance and other inputs which are deemed necessary.

The UNIDO Project

The United Nations Industrial Development Organization (UNIDO) is now engaged in Indonesia in assisting the Indonesian Government to launch an integrated development program for the manufacture of building materials in support of the national housing program.

The project started from 1975 until 1980, and is carried out through the existing building materials research institutes, such as the Directorate of Building Research (UN-Regional Housing Center), the Ceramics Research Institute and the Materials Testing Laboratory, all of them located in Bandung.

Activities of the project encompasses quite a wide spectrum:

1. Raw materials assessment;
2. Structural clay industry;
3. Application of structural clay for housing;
4. Refractories;
5. Lime industry;
6. Lime-based building materials industry;
7. Wood industries;
8. Production of building elements from agricultural and forest waste products;
9. Concrete technology;
10. Artificial light-weight aggregate industry (ALWA);
11. Mass production and industrialization of low-rise housing components;
12. Fibro-cement industries;
13. Mortar technology;
14. Fine ceramics industries;
15. Building materials for high-rise housing.

From the above list of activities, it is easy to perceive the UNIDO project as an input into the development of a management model as described in Fig. 3.2. Mass-housing programs can be administered as an organized market for new or improved materials, components and building systems. Small scale building materials industries coming within the KIK and KMKP schemes can get technical assistance and training in the use of better production techniques and processes through a loan-package with better marketing prospects.

Establishment of production and marketing cooperatives can be fostered, replacing the middlemen.

Thus, a whole spectrum of new possibilities can be opened to improve the existing building material industries and develop new ones transforming them into modern enterprises.

Industrialization of Housing Construction

The main objectives for the industrialization of housing construction in Indonesia:

1. To increase the labor productivity and to reduce the need for skilled labors;

2. To reduce the construction cost;

3. To expedite the completion of work irrespective of climatic factors;

4. To improve the quality of products and houses;

5. To rationalize and economize in the use of building materials.

The industrialization of housing construction in Indonesia calls for, e.g., the introduction and the development of non-conventional construction systems based on modern techniques; mechanizations of the production process; rational planning and designing; the use of standardization and modular coordination; and training facilities for labor and technical personnel.

It is, therefore, essential to promote and undertake activities to pave the way of the development of industrialization in the field of housing construction.

Although industrialization of housing construction is a way to overcome the lack of progress in the production of houses, a gradual process leading to full industrialization is recommended, taking into account, e.g., the actual conditions of a region or area.

The following steps were considered for implementation:

1. Rationalization;

2. Improved production of building materials and building elements;

3. Production of non-conventional, non-traditional and new building materials element and components;

4. Introduction and utilization of effective hand tools, light machines and site equipment.

Rationalization represents an important step to achieve increased productivity, higher operations, improved quality and better economy.

Rationalization is characterized by improved organization, planning and control and by effective use of materials, tools, machines and equipment.

The application of rationalization measures does not require considerable investment and substantial changes in the existing construction techniques.

Gradual industrialization is closely connected with the development of the production of building materials, elements and components.

Improved production and the production of non-conventional non-traditional and new materials with better or specific mechanical, physical and other properties is indispensable for the development of industrialized construction methods.

Mechanization varies according to the type and standard of machines, equipment and tools used for construction.

The degree of mechanization in operations that can be performed manually should be carefully scrutinized in Indonesia where labor is abundant. It should be noted, however, that in Indonesia there is an acute shortage of construction labor, especially skilled labor, causing, e.g., a low productivity of housing construction and a high cost of construction.

Therefore, mechanization is an important step to increase the productivity of housing construction and thereby absorbing a large proportion of unemployed labor.

The above comments are the key concepts involved in the gradual industrialization of housing construction show that *partial prefabrication* represent an important phase from conventional to industrialized housing construction.

By partial prefabrication, the construction of a house is still by conventional methods but using some prefabricated elements and components such as blocks, beams, columns, lintels, trusses, purlieus, rafters, doors, windows, floor-beams, floor and wall panels, etc.

Research and development on the design of the houses building systems, prefabrication elements and component which were carried out by the Directorate of Building Research (Regional Housing Centre, Bandung) are described in the drawings on the following pages.

The Registration and Licensing Committee

An initiative by the Indonesian Council for Construction to establish a Registration and Licensing Committee for the construction industry is now underway. The function of this committee is to institute Regional Registers and administer them. Licensing will be part of its function, through the powers vested in Local Governments.

Registration can be limited to consultants and contractors only, or extended to include suppliers and manufacturers of building materials and components. In the case of mass-housing where one of the objectives of project management is to develop the building industry through participation in mass-housing projects, registration can be perceived as a tool for selection and monitoring of those who are to be given opportunities to participate in mass-housing projects. Licensing follows after evaluation to meet certain standards of performance required by the project management.

Especially in the traditional sector of the building industry, the Registration and Licensing System can be a great help in keeping track of potentials in terms of skills and production capacities, through which conservation of these potentials can become more effective.

The Model Reconstructed

The activities which have been identified as potential future functions can thus be perceived as being part of the development of the model for mass-housing management as described in Fig. 3.5. A diagrammatic presentation of these functions and their relationship can be as follows:

Thus, the stage is set to set into motion a process of institution building in the field of mass-housing management. As a management system, it is an open system. Its behavior and growth patterns are at present difficult to predict. However, training and education, research and development, and an open information flow between all the functions involved will be undoubtedly a strong leverage to attain stronger cohesion between all these functions, which determines their combined performance.

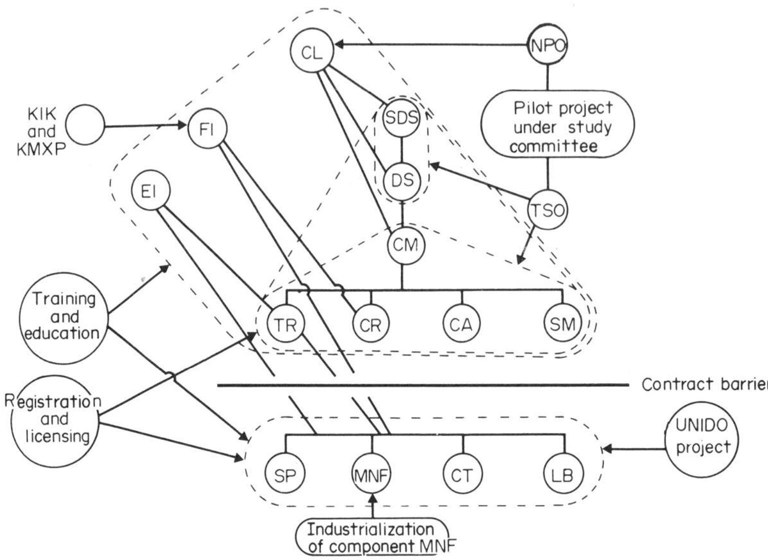

Fig. 3.5. Total Systems Development Model

3.5 CONCLUDING OBSERVATIONS

In retrospect what has been presented was a framework for the management of mass-housing production. Starting from the ordinary system of using Client-Designer-Contractor relationships, it was elaborated into a system which embraces a host of functions with increasing complex relationships, as the spectrum of objectives unfolds to include external ones, ordinarily outside the responsibilities of project management.

This concern about external objectives stems from the observation that mass-housing as a demand is a new phenomenon, which embodied technological imperatives for which, on the one hand, the industrial environment in Indonesia is as yet not ready, on the other hand, there is also a lack of evaluative mechanisms to determine the right choice of housing technologies. The concept as has been presented argues for the adoption of a management system of a building process, which in itself is both a learning and development process, in the sense that through monitoring and evaluation of designed experiences a systematic body of knowledge can be acquired, new skills, institutions and their relationships can be developed, and capacities induced to innovate and develop new technologies.

Although at present our knowledge is still very fragmented as to the nature of the causes which induce changes in total mass-housing production performance, various phenomena suggest that these can be predicted. Various projects using a housing type as has been built in Sukaluyu, have given different results in terms of production performance, indicated in unit cost per m^2. Projects using contractors who are selected through competitive bidding seem to have a tendency to result in high unit cost per m^2. Observation of the low-cost housing projects suggest that the unit cost is around Rps. 25,000 ($62.50) per m^2 or above. The same type of houses constructed by contractors as part of a "building team" have resulted in unit costs of around Rps. 22,000-Rps. 25,000 ($55-$62.50) per m^2. Increasing capacities in internal management of contracting firms through repetition of activities have resulted in a unit cost of Rps. 19,850 ($49) per m^2, as has been shown in the latest tender in Depok. A study which is now underway to build low-cost housing of the same type with improved materials and rationalization of construction, using the CSBS and "building team" concept, is expected to lower the cost further to Rps. 18,000-Rps. 19,000 ($45-$47.50) per m^2. From the above observations a hypothetical performance chart can be reconstructed as follows:

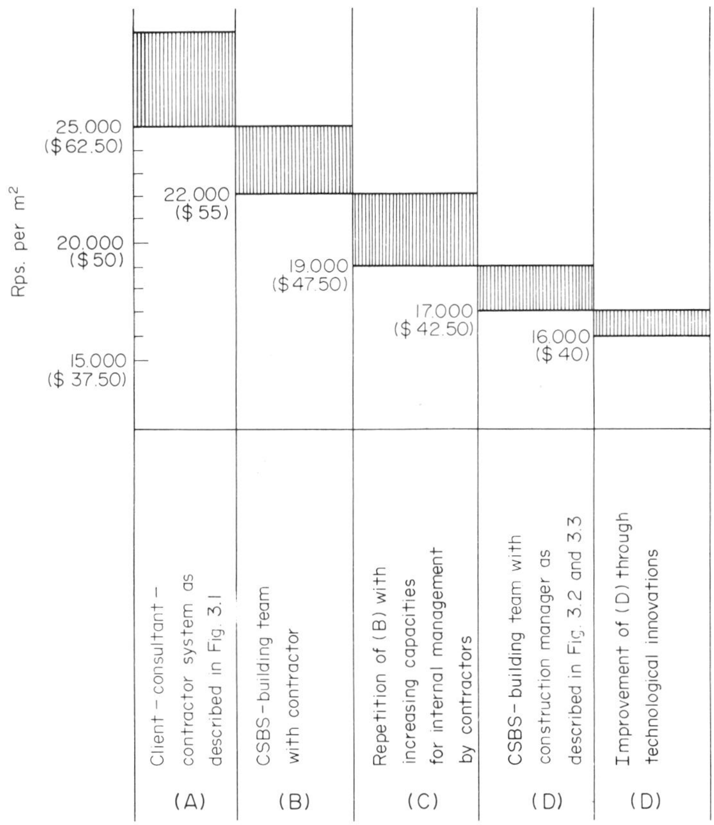

Fig. 3.6. Hypothetical Performance Chart

The chart suggests that in an industrial environment such as can be found in Indonesia, in the early phases when mass-housing is introduced, design and technological innovations have only marginal effects. There are still too many constraints within the project environment itself, which are detrimental for improving project performance. In such a situation innovations in management will yield better results than innovations in design and technology. "Project development cycles" and innovations in design and technology will ultimately result in lowering the cost further coupled with increasing quality.

The Directorate of Building Research (UN-Regional Housing Center) in Bandung has done intensive studies and experimentation in industrialization of housing construction for Indonesia. It can now be anticipated that with the introduction of the CSBS and "building team" concept these experiments can be transferred to the building industry. Hitherto, housing projects lacked the necessary volume and continuity, and an appropriate management system to transfer these experiments and to make them effective. If at this stage of development, mass-housing projects are designed around known technologies and predominantly use manufactured products from modern and relatively large industrial plants, many of the studies and experiments undertaken by the Directorate of Building Research are aimed at the development of small scale enterprises which can be diffused among the traditional sector of the urban/village industries.

Thus, innovations in project and construction management and housing technology combined are expected to lower building costs so that housing can be enjoyed by more people, and construction activities can create more jobs and distribute more income to those sectors.

Fig. 3.7. The Sukaluyu project under construction : pre-manufactured components and labor-intensive assembly on site.

Fig. 3.8. Facade showing asbestos cement roofing, puzzolan cement hollow blocks, and pre-manufactured front walls, windows and doors.

Fig. 3.9. Maisonettes in the Depok project, with landscaped gardens.

Fig. 3.10. Another view of the Depok project.

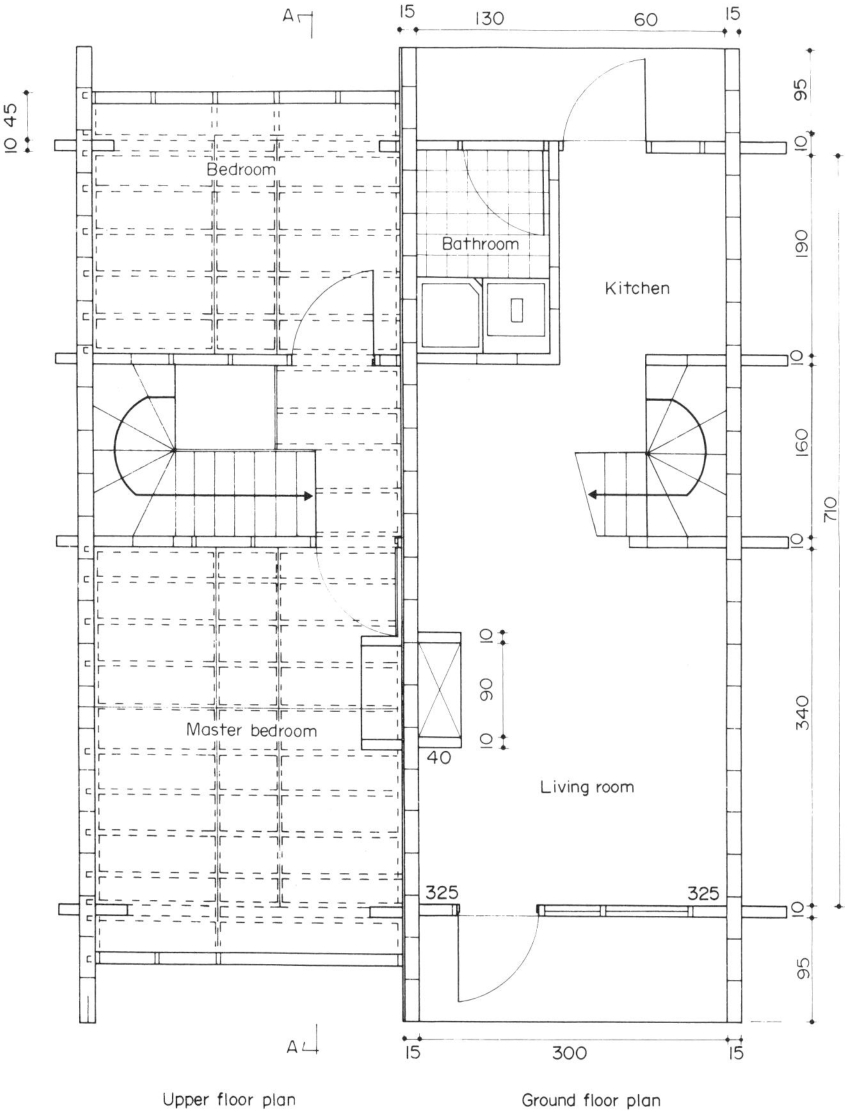

Fig. 3.11. Low-cost housing project Sukaluyu, Bandung.

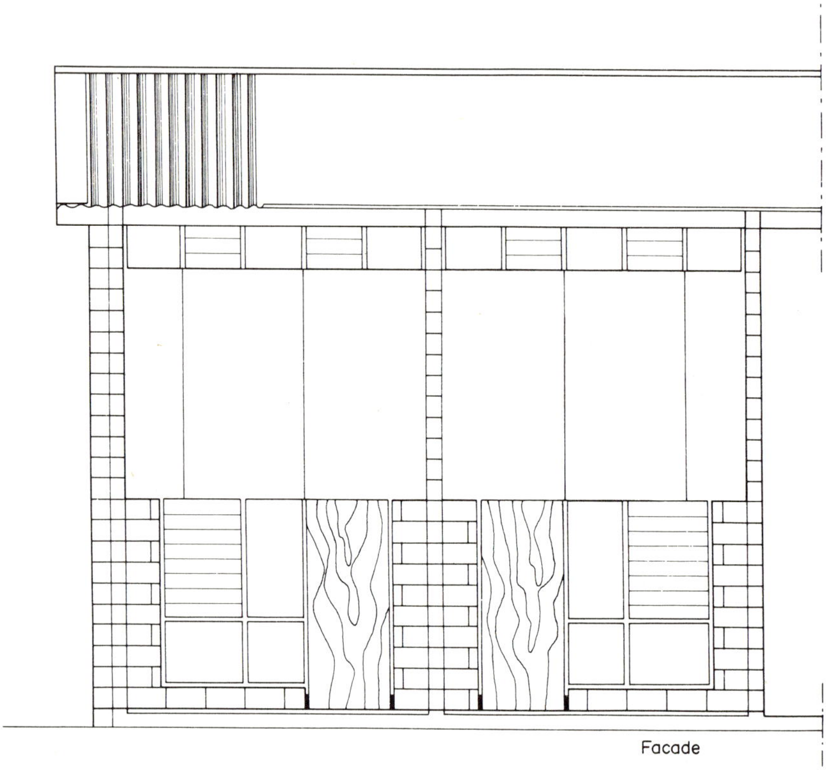

Fig. 3.12. Low-cost housing project Sukaluyu, Bandung.

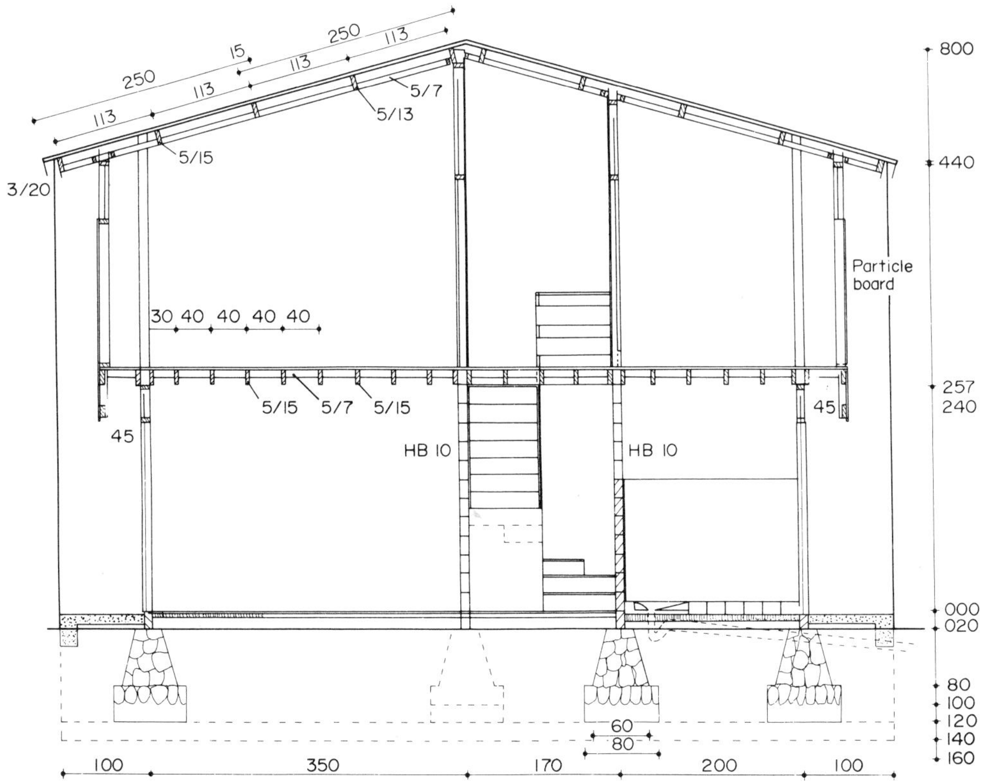

Section

Fig. 3.13. Low-cost housing project Sukaluyu, Bandung.

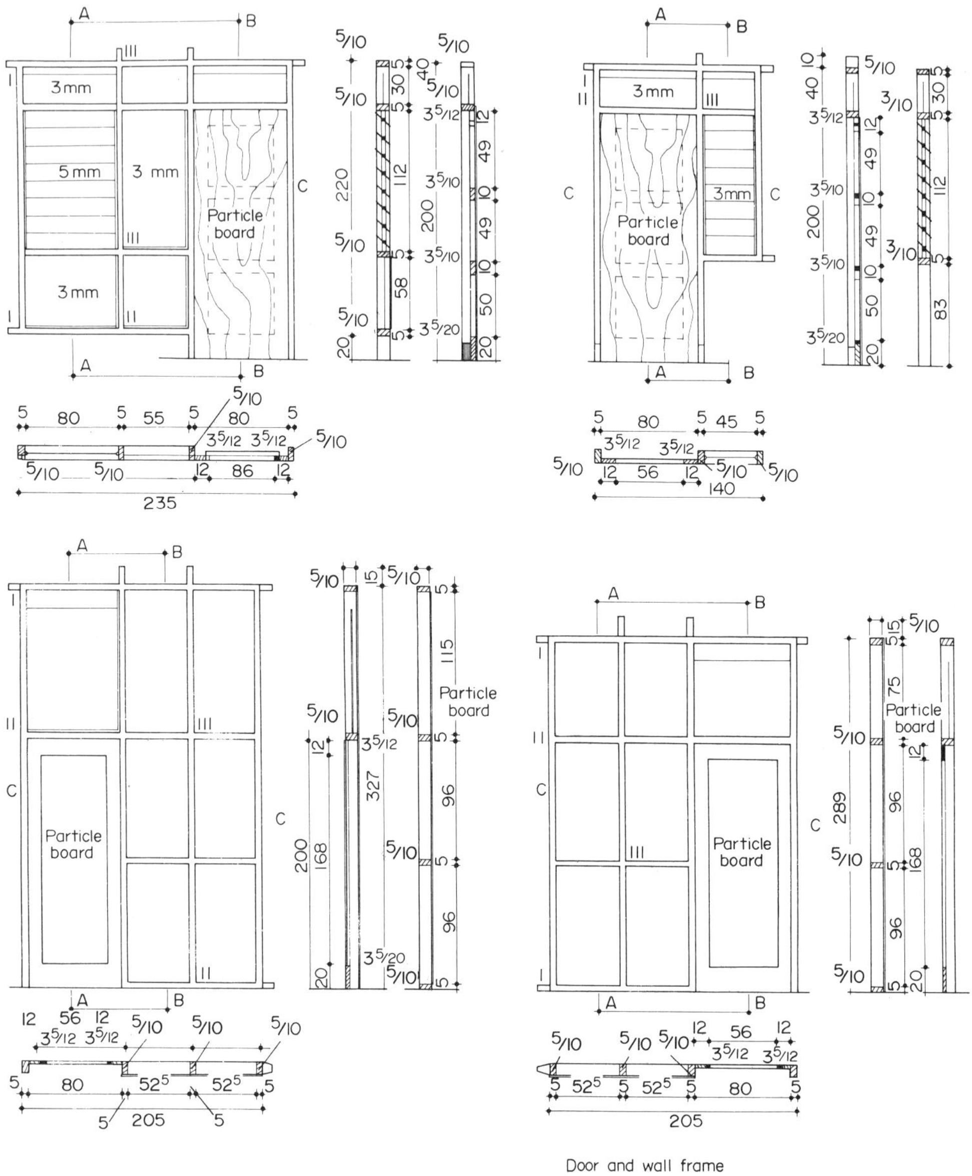

Fig. 3.14. Low-cost housing project
Sukaluyu, Bandung.

Low-cost house construction

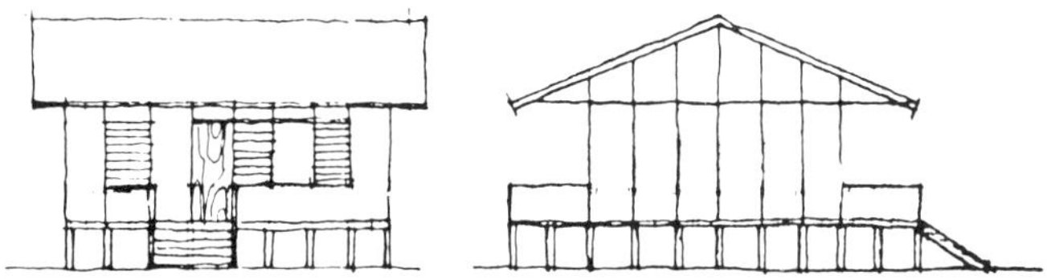

Platform house construction (Malay house)

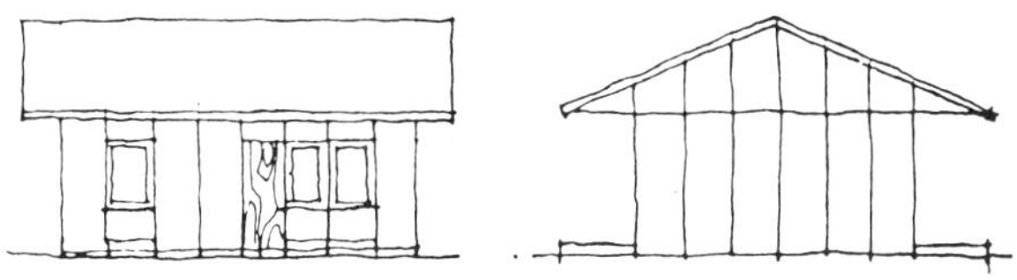

Ground floor house construction

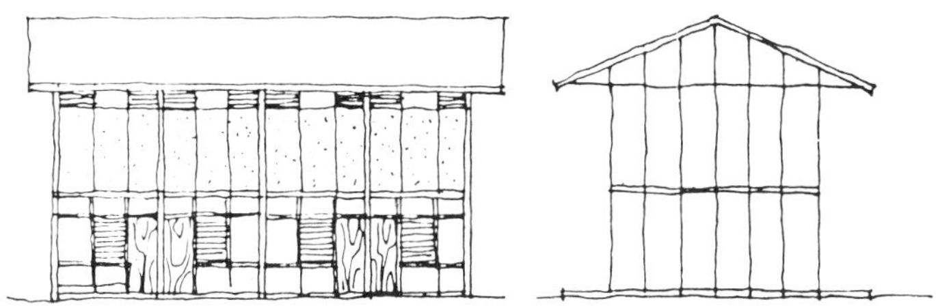

One storey house construction

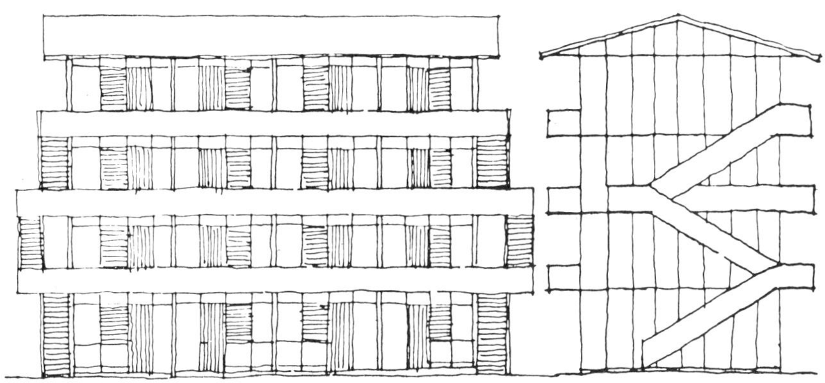

Multi storeys house construction

Fig. 3.15. Low-cost House Construction.

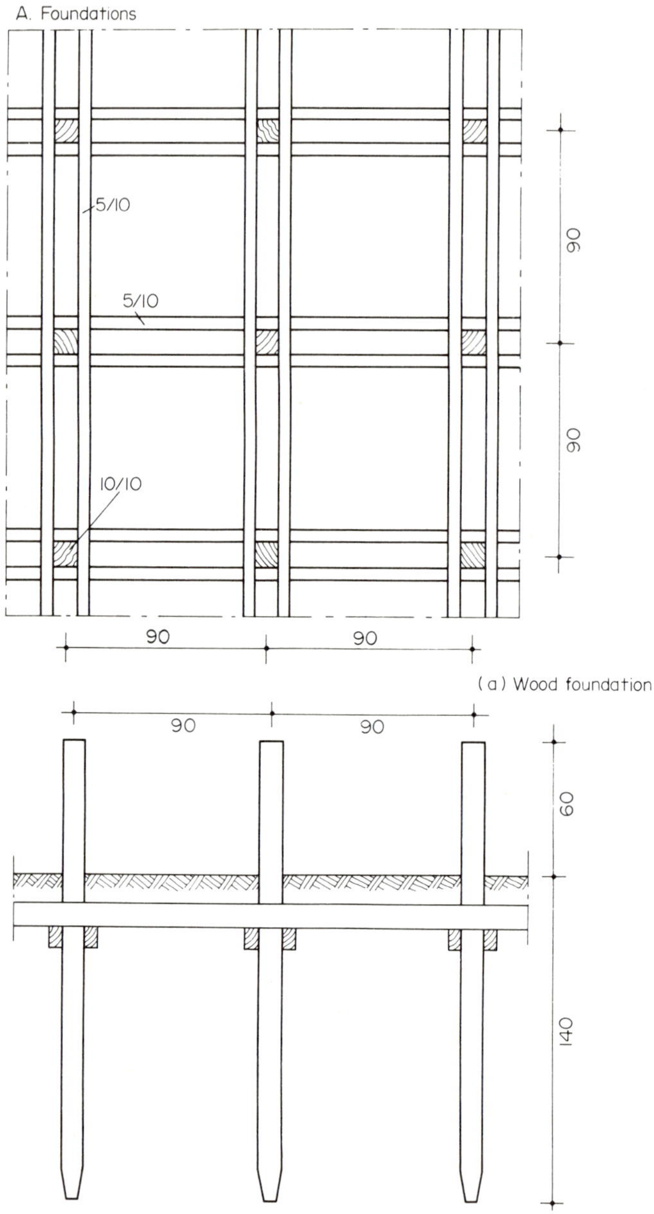
A. Foundations
5/10
5/10
10/10
90
90
90
90
(a) Wood foundation
90
90
60
140

Fig. 3.16

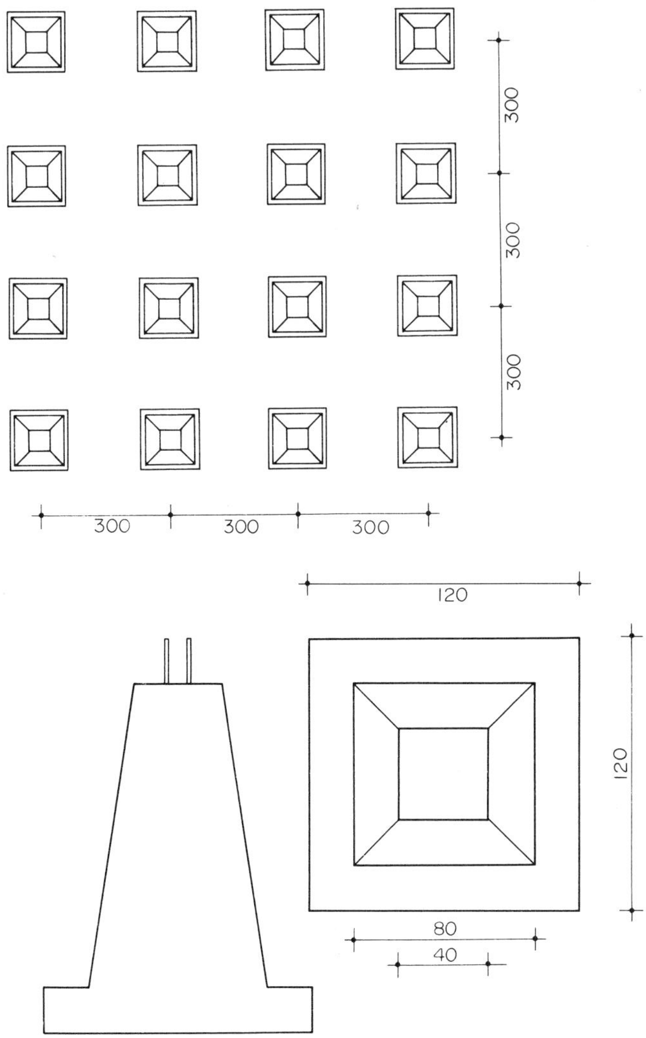

C. Concrete foundation

Fig. 3.17

B. Beam and column

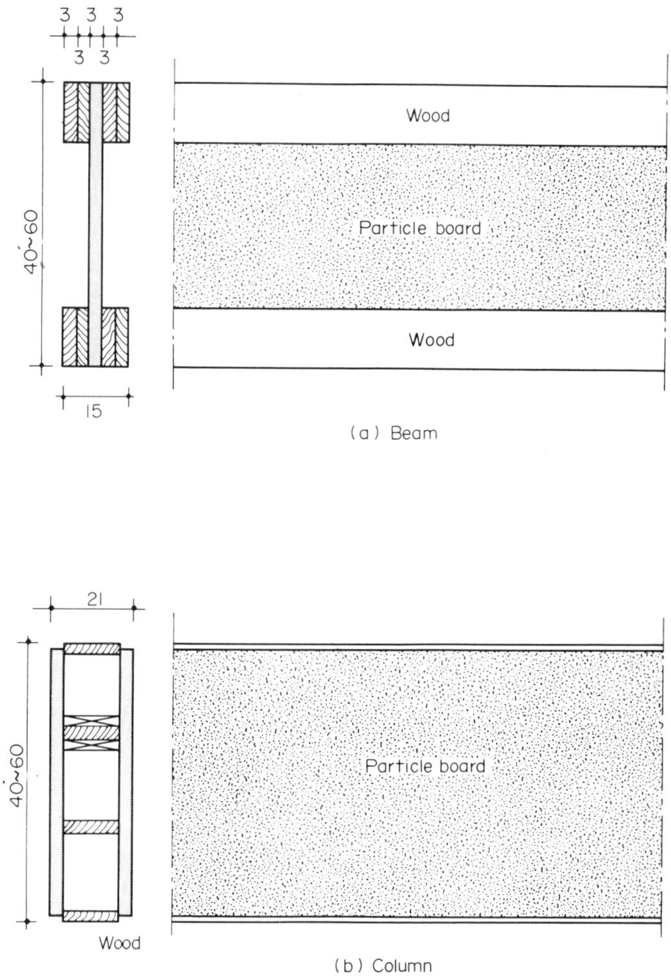

Fig. 3.18

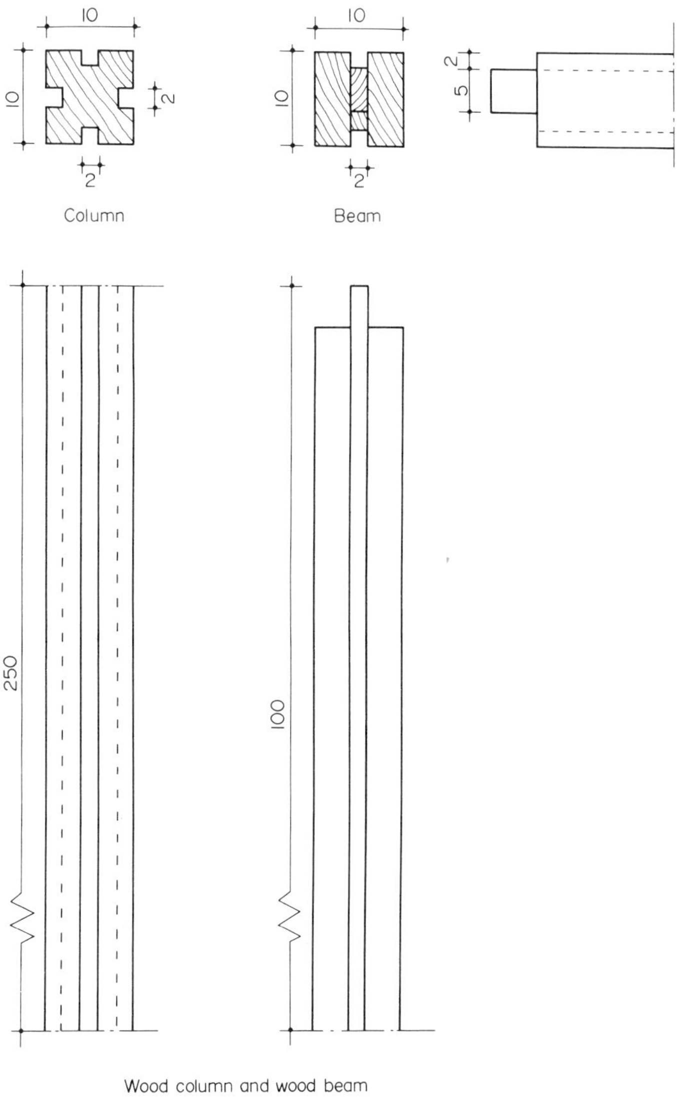

Fig. 3.19. Wood column and wood beam.

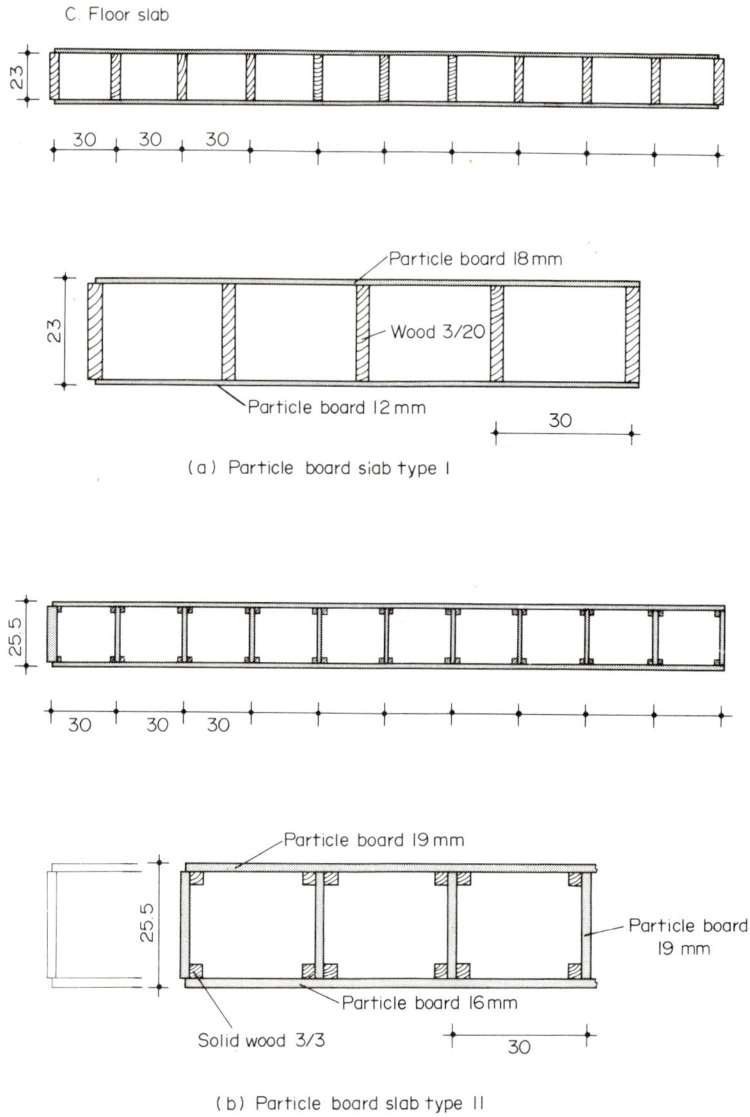
C. Floor slab
23
30
30
30
Particle board 18mm
23
Wood 3/20
Particle board 12mm
30
(a) Particle board slab type I
25.5
30
30
30
Particle board 19mm
25.5
Particle board
19 mm
Particle board 16mm
Solid wood 3/3
30
(b) Particle board slab type II

Fig. 3.20

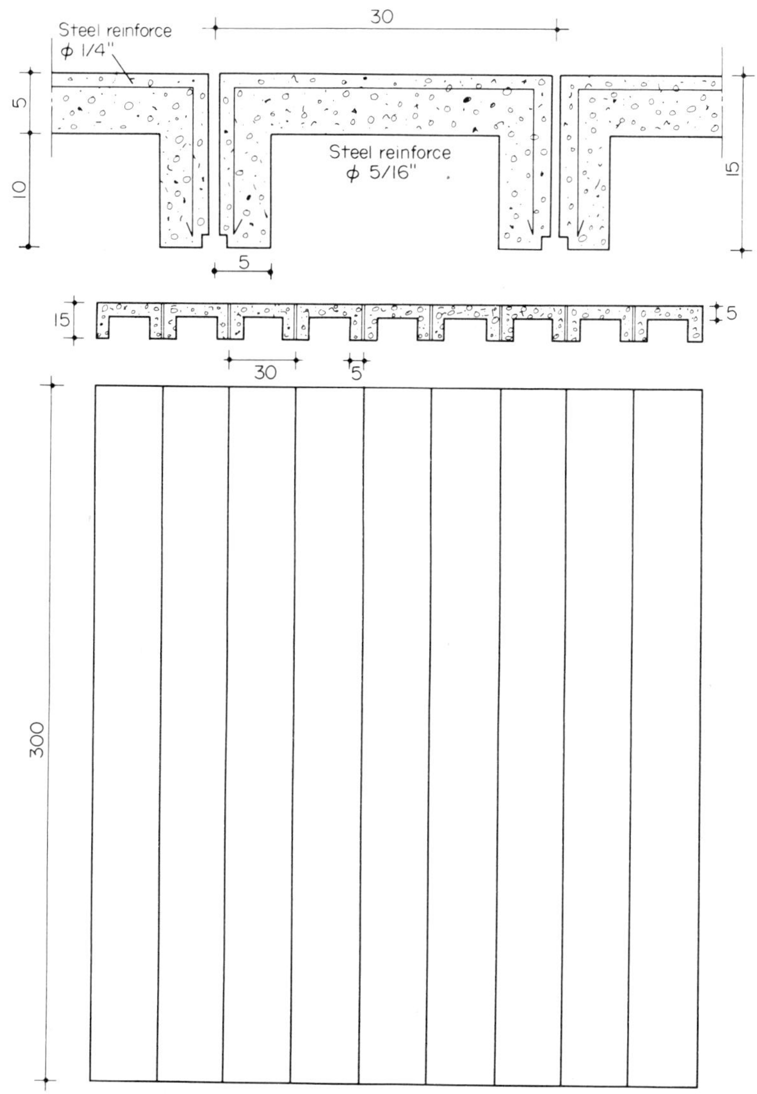

(c) Precast concrete beam

Fig. 3.21

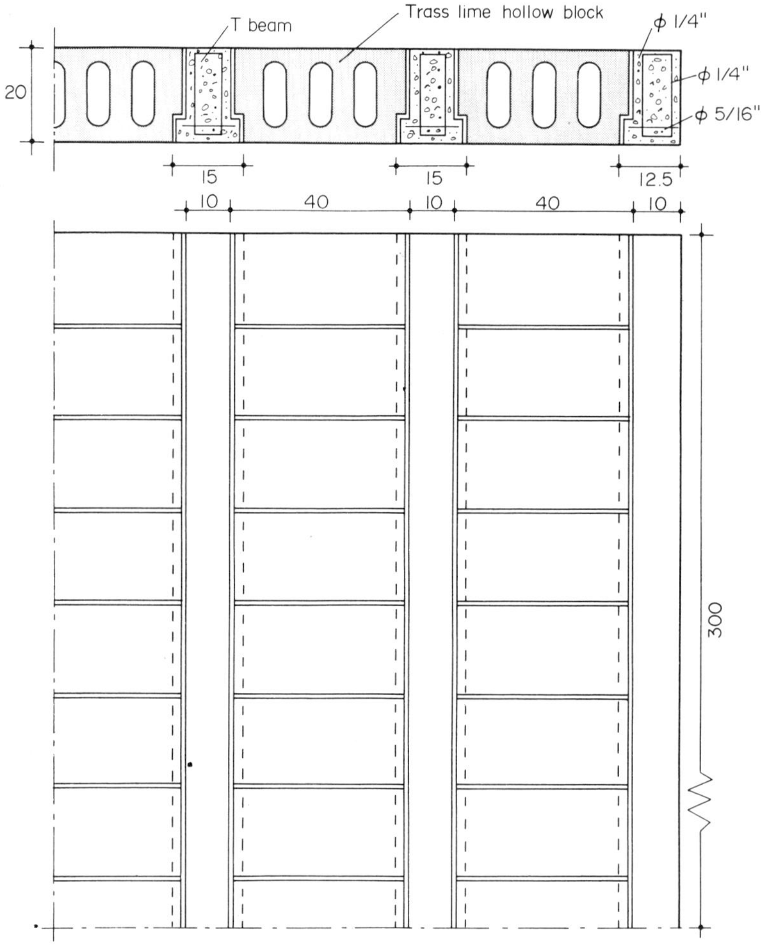

(d) Concrete beam and trass lime hollow block

Fig. 3.22

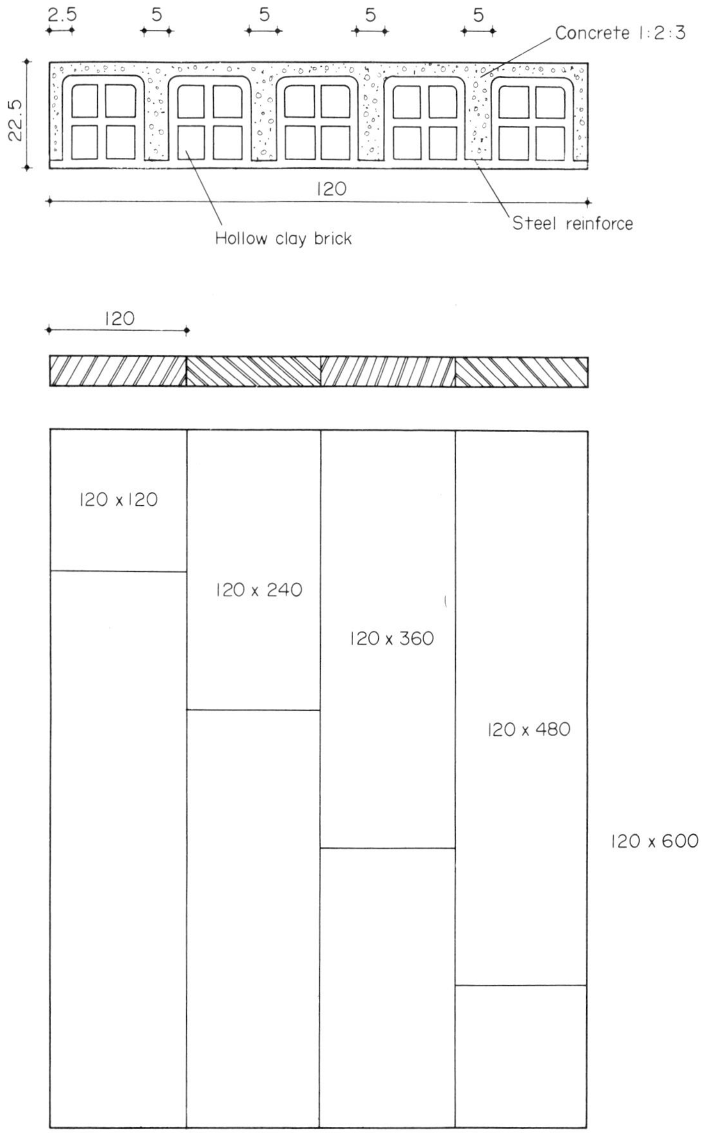

(e) Hollow clay brick slab

Fig. 3.23

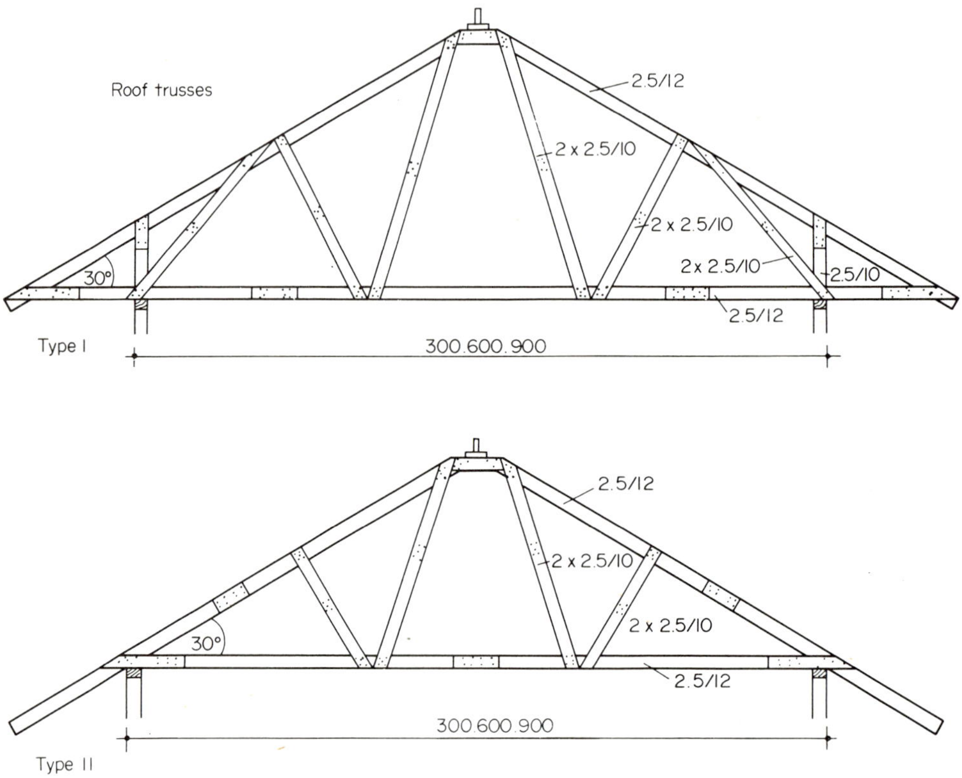
Roof trusses
2.5/12
2 x 2.5/10
2 x 2.5/10
2 x 2.5/10
2.5/10
30°
2.5/12
Type I
300.600.900
2.5/12
2 x 2.5/10
2 x 2.5/10
30°
2.5/12
300.600.900
Type II

Fig. 3.24

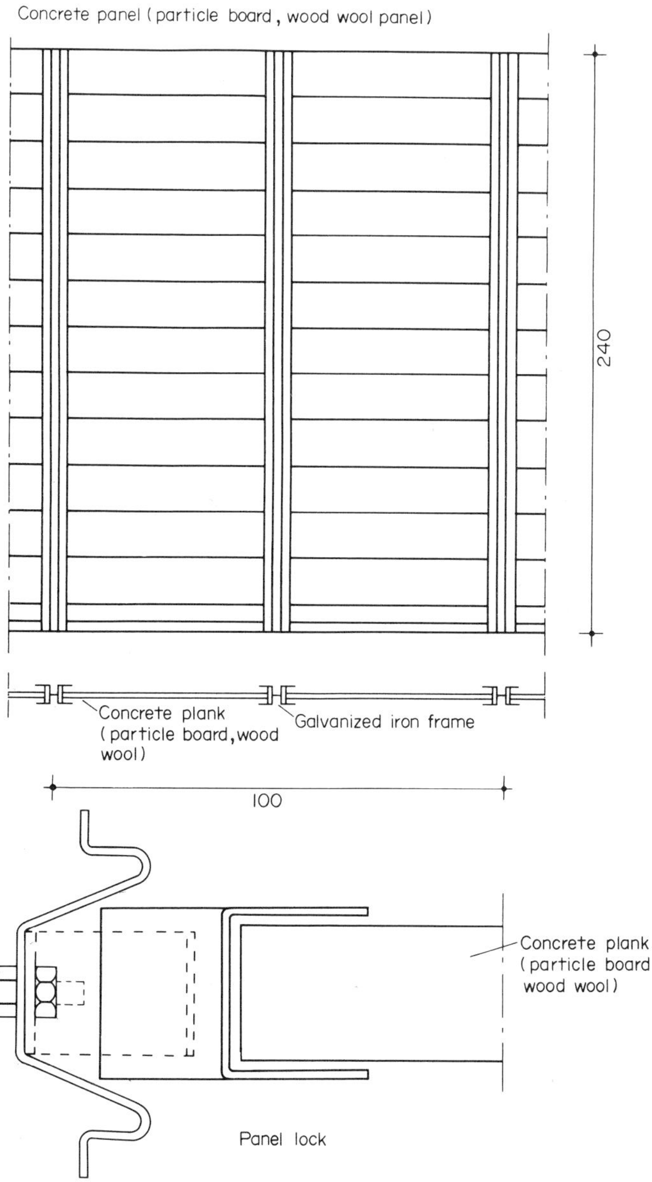
Concrete panel (particle board, wood wool panel)
240
Concrete plank (particle board, wood wool)
Galvanized iron frame
100
Concrete plank (particle board wood wool)
Panel lock

Fig. 3.25

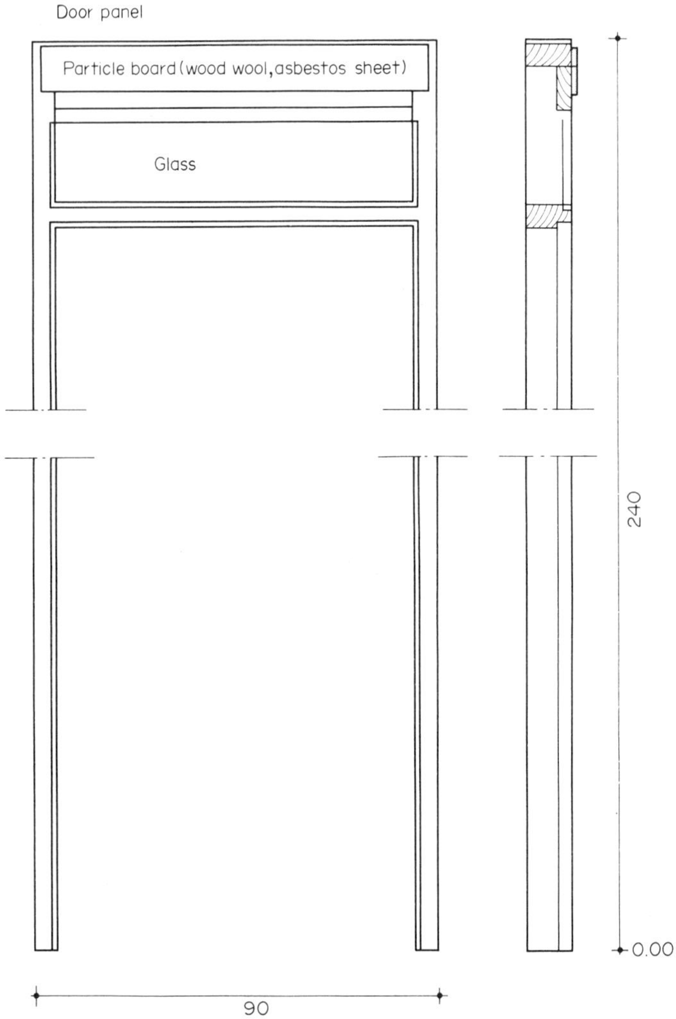
Door panel
Particle board (wood wool, asbestos sheet)
Glass
240
0.00
90

Fig. 3.26

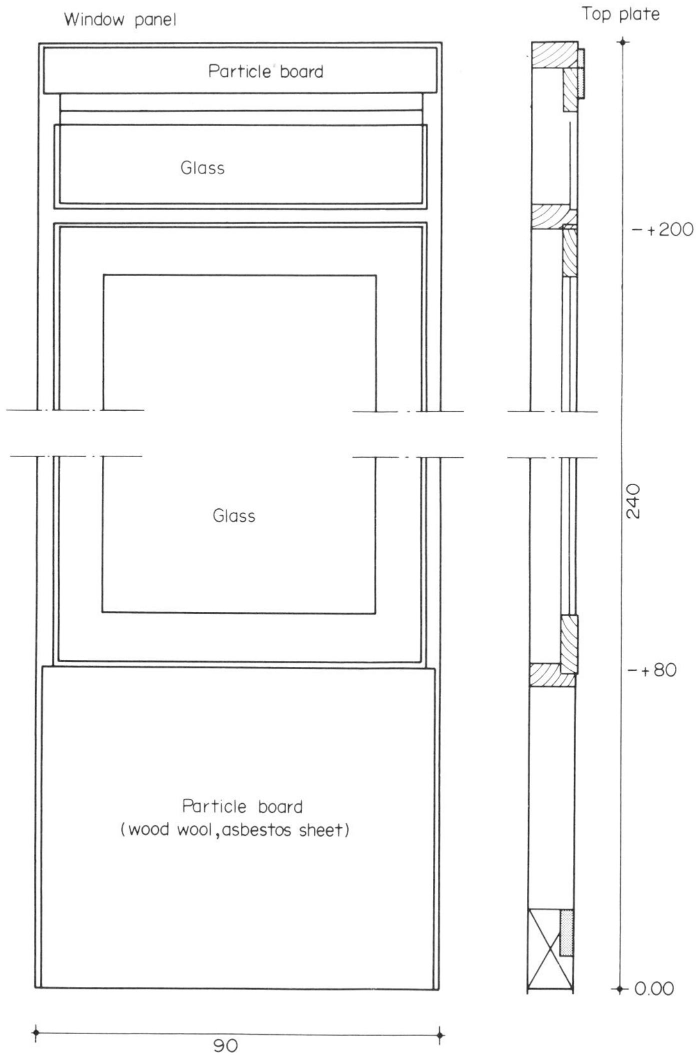
Window panel
Top plate
Particle board
Glass
-+200
Glass
240
-+80
Particle board
(wood wool,asbestos sheet)
0.00
90

Fig. 3.27

REFERENCES

1. Hasan Poerbo and Tjuk Kuswartojo, *et al.*, *Evaluasi Proyek Rumah Murah P. 1000 di Jakarta, Karawang, Bandung, Semarang, Jogyakarta, Klaten, Surabaya, Jember. (Evaluation of Low-Cost Housing Project "P. 1000" in Jakarta, Karawang, Bandung, Semarang, Jogyakarta, Klaten, Surabaya, Jember,* unpublished report, LAPI-ITB, Bandung 1973.)

2. (a) PT. Encona Engineering, Inc., *Laporan Akhir Proyek Site and Services, Klender, Jakarta-Indonesia (Final Report on the Site and Services Project, Klender, Jakarta-Indonesia,* unpublished report, Bandung 1975.).

   (b) Hasan Poerbo and Bambang Panudju, *et al.*, *Survey dan Studi Persiapan Pem bangunan Perumahan Delapan Kota di Indonesia. (Survey and Study for the Preparation of Housing Development of Eight Towns in Indonesia,* unpublished report, LAPI-ITB, Bandung, 1976.)

3. Albert G. H. Dietz and Hasan Poerbo, *Industrialized Housing in Indonesia,* Industrialization Forum, Vol. 6 (1975), number 2.

4. The Sukaluyu Project in Bandung — Rps. 24,000/m^2 (\$60 per m^2).
   The Depok Housing Project — Rps. 20-25,000/m^2 (\$50-\$62.50 per m^2).
   The U.K.A. Labor Housing Project (estimated) Rps. 20,000/m^2 (\$50 per m^2).

5. Dr. Ir. Prof. Jan Delrue, *An Architectural Approach for a Rationalization of the Building Process,* University of Leuven, 1974.

6. No systematic study has yet been made about the role of middlemen in the building trade. However, during the survey for the Sites and Services project for Jakarta, and in discussions with the director of the Ceramics Research Institute, who has a pilot project to develop a production and marketing cooperative for brick and tile manufacturing, it was apparent that the role of middlemen is very dominant. Middlemen act as financiers and marketing agents for brick and tile manufacturing. For instance, the price of bricks is Rps. 5 per piece off factory. The selling price is Rps. 11 (\$0.03), of which transportation and handling cost is not more than Rps. 3 (\$0.01). Thus, the middlemen can have a net profit of around Rps. 3 (\$0.01) per piece. At present there is no other mechanism which can replace the middlemen.

7. J. Van Ettinger, *Towards a Habitable World,* Task-Problems and Methods-Acceleration, Elsevier Publishing Company, Amsterdam, 1960. E.F.L. Brech, *Construction Management in Principle and Practice,* Longman Group Limited, London, 1971.

8. There are at present only two journals with a limited circulation, which can be used as reference in practice. But even these two journals are still not well utilized for product promotion by the industry. Marketing is done by individual firms through their sales-organization, in a market situation which is very fragmented.

9. J. Van Ettinger, Sr. and J. Van Ettinger, Jr., *Problems and Methods of Low-Cost Housing,* Bouwcentrum, Rotterdam, January 1969. Hasan Poerbo and Tjuk Kuswartojo, *et al.*, *Evaluasi Proyek Rumah Murah P. 1000, op. cit.*

10. J. Van Ettinger, Sr. and J. Van Ettinger, Jr., *op. cit.*, p. 15.

11. J. Van Ettinger, Sr. and J. Van Ettinger, Jr., *op. cit.*, p. 16.

12. Gwen Bell, *et al.*, *Strategies for Human Settlements: Habitat and Environment*, The University Press of Hawaii, Honolulu, 1976. Article by: Goodman, Dietz, Poerbo, Burian: "Low-Cost Housing Technology: Problems, Issues, and Proposed Solution".

There are two basic concepts in proposing a solution:

(a) A two-pronged approach in mass-housing production with larger components manufactured in shops and labor-intensive assembly on site.

(b) Aggregating the housing market to achieve continuity in demand.

4

LOW-COST HOUSING IN THAILAND

Seng-Lip Lee, Tongchat Hongladaromp and Ricardo P. Pama

4.1 Overview of Housing Situation in Thailand

Introduction

The Kingdom of Thailand with a population of over 40 million has an annual population growth rate of 3% [1] which is one of the highest in the world. According to the latest census, 21% of the population live in the urban areas and 79% in the rural areas. Bangkok which is the capital has a population of over 4 million and constitutes half of the entire urban dwellers in the country. What is of serious concern is the continuing rapid migration of rural dwellers to Bangkok causing overcrowding in the city. This is evidenced by the presence of slums and blight all over the city, the biggest of which is the one around the port area of Klong Toey where over 5,000 families or about 30,000 people live.[2]

Bangkok with an annual population growth rate of 5% is at present facing a housing shortage of over 100,000 units and it is estimated that in 1986 the housing shortage will exceed 190,000 unless urgent steps are taken now.[1] It is further estimated that more than half of the shortage falls into the low and lower-middle-income group. In Thailand, the low-income group consists of families whose income is less than $75 per month and the lower-middle-income group are those whose income falls within $75-$150 per month as in Table 4.1.

Table 4.1. Classification of Families According to Income.

Monthly income	Group type		Distribution %
Less than $75	A		35
$75 - $150	B		40
$150 - $250	C		10
Greater than $250	D		15
		Total	100

Past Government Action in Housing Development

The Government first decided to participate in the field of housing production in 1940 and the emphasis has been on urban housing. Apart from public housing provided for government employees in the provincial towns and cities, the rest were centered in Bangkok. The rationale being that traditional rural housing is generally self-built and low-resource consuming, thus rural housing shortage is not much of a problem in the same magnitude as that of its urban counterpart especially Bangkok.

Up until late 1972, four government agencies were entrusted with the task of providing housing for the low-income group. These are the Public Housing Bureau of the Ministry of Interior, Public Housing Division of the Department of Public Welfare of the Ministry of Interior, Housing Welfare Bank of the Ministry of Finance and the office of Slum Clearance, Urban Renewal and Housing of the Bangkok Municipality.[3,4]

The Public Housing Division was established in 1940 as a division of the Department of Public Welfare, Ministry of Interior. At that time housing shortage was not a serious problem and the main function of the Public Housing Division was to construct houses for members of self-help settlements in several provincial areas.

In 1942, the Public Housing Bureau was created and was first financed with a meagre $50,000 annual appropriation. World War II was at its peak and the implementation of any housing project was hampered by the war. The Public Housing Bureau was responsible for the following:

(a) Granting mortgage loans to land-owners for construction or renovation houses for their own use;

(b) Constructing houses for hire-purchase;

(c) Building houses for rent.

The size and character of its programs were largely determined by the limited annual budgets. It might be of interest to note that in the same ministry, the Public Housing Division and the Public Housing Bureau co-existed. The former was responsible for technical operations such as architectural and engineering designs, construction, inspection and maintenance while the latter was concerned with business administration, finance, selection of tenants, collection of rents and services.

The third agency involved with public housing is the Government Housing Bank in the Ministry of Finance which was formed in 1953. The responsibilities of the Government Housing Bank are as follows:

(a) To let land and/or place of habitation on a hire-purchase basis to any person not having a place of habitation of his own;

(b) To grant loans for any of the following purposes: (i) to enable the borrower to buy land or place of dwelling for his own habitation, (ii) to enable the borrower to build, extend or repair his own place of habitation, (iii) to enable the borrower to redeem any mortgage encumbering his own land or place of habitation, and (iv) to enable the borrower to redeem his own land or place of habitation sold on redemption;

(c) To purchase immovable property for resale on a hire-purchase basis to persons not having a place of habitation of their own;

(d) To accept transfer of property on pledge or mortgage as security for loans;

(e) To open deposit accounts for a term of two years or more.

The full working capital of the bank was quoted at $25 millions, but the initial capital which was granted by the government was only $1 million. Because of its inadequate working capital, the bank has only financed 453 houses since 1953. One current activity of the Government Housing Bank is to develop large land subdivisions for resale as lots on a hire-purchase basis. This activity unfortunately did not directly benefit the low-income families.

The office of Slum Clearance, Urban Renewal and Housing was created in 1960 to act as a temporary clearance agency to assist persons who were displaced at short notice to provide a site for the headquarters of the Southeast Asia Treaty Organization. About 1,000 families or 10,000 people were involved. Since then this agency has cleared away two more slum areas involving about 10,000 people. Other significant activities are the development of two land subdivisions with about 3,000 lots for hire-purchase.

While each of these agencies actually carried out different aspects of housing activities, it was obvious that such fragmentation was not conducive to effective housing development. There was no realistic housing policy formulated, no long-term planning, no research or innovative experiments conducted, and most serious of all was the fact that these agencies were not able to command the interest of the decision makers who determine the resource allocation patterns and legislations which could bring about the removal of serious obstacles to the housing production process. Consequently, the record shows that from 1949 to 1973, the resultant public housing produced averaged less than 500 units/year as can be seen in Table 4.2.

Table 4.2. Public Housing Constructed Until 1972.

Year	No. of units
1949 - 1952	680
1953 - 1956	1,592
1957 - 1962	2,612
1963 - 1967	1,394
1968	480
1969	630
1970	1,088
1971	412
1972	1,472
1973	336

The Formation of the National Housing Authority

On December 13, 1972, the National Executive Council established the National Housing Authority (NHA) as the sole agency on housing. It was set up to unify all housing activities and to make a concerted effort at solving the urban housing problem. Except for the Government Housing Bank, all functions including assets and personnel of the previous agencies were inherited by the NHA.

A Board of Housing Committee was appointed, with the Deputy Minister of Interior as Chairman and the members from the Budget Bureau, Ministry of Finance, Governor of Bangkok, National Economic and Social Development Board, Town and Country Planning Department and a number of prominent bankers.

The NHA, besides being a consolidated body of the former four agencies, has assumed a greater stature and an enlarged authority. Financially, it has a much broader base and is more flexible in execution. As a state enterprise, NHA can raise loans from local and international sources as well as utilize government budget allotments in its project.

Initially, the NHA operated on guidelines provided by the National Plan[5] which estimates the housing needs of the greater Bangkok area to be 170,000 units in 1982. A more up-to-date projection by the NHA of the housing needs is shown in Table 4.3. The NHA is presently carrying out a program of surveys in provincial urban areas to determine the extent of those cities' housing situation for policy formulation and programming purposes.

Table 4.3. Distribution of Housing Needs by Income Groups.

Year	No. of units needed			Total
	Group Type A	Group Type B	Group Type C	
1974	36,500	53,200	14,600	104,300
1980	47,800	69,100	19,200	136,700
1986	65,900	97,500	26,800	191,200

Despite the preplanning of the NHA involving organization arrangements, legislative powers of operation, policy and program articulation, the NHA has been encountering numerous difficulties since its inception. This is all too clearly reflected by the NHA being able to produce only 3,000 housing units during its first two years of existence. Some of the problems of the NHA are attributed to the following:

(a) Lack of significant amount of revolving funds at its inception;

(b) As a public enterprise, the NHA is self-sufficient, thus compounding its financial problems as most of the housing units produced require some form of subsidy;

(c) Lack of resources for land acquisition and powers of eminent domain;

(d) Lack of support by the authorized agencies for infrastructural development;

(e) Delays caused by interference and unnecessary obstruction by project scrutinization agencies.

Current and Future NHA Development Program

In March 1975, the government emphasized in no uncertain terms that solving the housing shortage is one of its primary goals. In response to this, changes were made within the NHA in an effort to make it more fitted to handle the mamoth tasks ahead. The Ministry of Finance has promised the NHA full support in mobilizing an investment capital amounting to $80 million for the construction program in 1976, of developing 24,000 units while an additional 24,000 units are presently being programmed to start in 1977. The overall target is to solve the housing backlog by the year 1980 by building 24,000 units per year for five years commencing in 1976.

The NHA plan essentially identifies three major target groups with income up to $250 per month as shown in Table 4.1. This pattern of income group distribution is based on a socioeconomic survey conducted in the Bangkok Metropolis in 1972.[6]

Capital amortization is fully subsidized by the Government for Group A (the low-income group), one-half subsidized for Group B (the lower-middle-income group), while Group C is not subsidized. Monthly payments made by tenants will cover interest charges of 5-10% including estate management and maintenance charges.

Building types will consist of walk-up flats, duplex and row houses including incremental housing or site and services dwellings, determined according to socio-economic considerations of the target group. For the NHA to achieve their goal in 1980, full commitment and support from the government is essential in many aspects. Some of these are outlined as follows:[7]

(a) Capital requirements for the Five-Year Plan involve not only government guarantee on bonds to be issued as well as other loans, but also direct government subsidy amounting to some $300 million;

(b) Land required for the 1977 program has yet to be acquired. Due to the pressing time factor, the government must take a firm stand in persuading other government agencies who own idle lands to make them available for immediate use by NHA;

(c) On the longer aspect of land acquisition, the NHA must be provided with powers of compulsory purchase;

(d) Property tax legislation should be promulgated to induce the release of idle lands into the market. This could be done by levying heavy taxes on idle lands;

(e) External loans from organizations such as the World Bank must be fully endorsed by the government;

(f) Budget programming for infrastructural projects such as roads, water supply, electricity, sewerage system, telephones, etc., must be coordinated to ensure that the NHA receives the full cooperation of the agencies concerned. This will contribute to the substantial reduction in costs of the project of the NHA.

The Role of the Private Sector in Housing Development

So far only the role of the government has been discussed in housing development. The private sectors have recognized the increasing housing shortage and the housing industry began to grow rapidly in the 1960s. By 1969 the business was booming and in 1972 there were over 45 major housing projects in existence in Bangkok and the surrounding suburbs. The Thai Investment and Securities Co. (TISCO) conducted a study of 37 of these projects[8] which involves a total of 9,982 units covering an area of 5,368,000 m^2. The houses provided are generally two to three bedroom units and range in cost from $3,500 to $50,000 with a mean cost of around $10,000 excluding the land. The land cost ranged from as low as $8.75/$m^2$ to $75/$m^2$ with an average price of about $16/$m^2$. Considering an average size of land to be 460 m^2/house, the total cost is $7,360.

With the exception of a few prefabricated systems, these houses are generally built on wooden piles with an *in situ* concrete structure for the ground floor. Cement blocks are generally used for the ground floor exterior walls as well as the fence surrounding the yard. The second floor is generally of wood construction and has asbestos sheet roofing. Plywood or chipboard is used for interior walls.

The financing for these projects can be classified into two types, i.e., private financing and loan financing. The capital source for the so-called private financing is generally the developers' own capital and borrowed funds, on short- to medium-term loans, supplemented by deposits from prospective home buyers and credit from contractors and material suppliers. In the latter case, the developer is refinanced by a commercial bank or other financial institutions. Loan financing which allows prospective home owners without cash to purchase their homes is shifting from hire-purchase arrangements to mortgage financing. The buyer must generally pay 25% deposit and pay monthly installments of approximately $125 for 10 years or more. Interest rate for mortgage financing varies from 11-12% per annum.

The TISCO study[8] estimated that the number of housing units built by private developers is 2,000 units per annum. It forecasts an increase of 3,000-6,000 units per annum in this decade.

Clearly, with the kind of houses developed by the private sector, only families in Groups C and D categories can afford houses in the open market. In order to remedy the situation, the Government began a program to include the private investors in their plans to provide low-cost public housing. Under this program, private investors and developers are to build low-cost houses according to the Government's specifications. The Government will in turn buy them on long-term payment plans and rent them to qualified applicants. The investors are, however, experiencing difficulty in implementing this scheme because of the low budget of the Government.

Very recently, some attention has been directed towards the World Bank's Site and Services scheme whereby only land, infrastructure and loans are made available. The construction of the houses are left to the prospective home-owners. Some criticisms about the site-and-services type of housing are centered on the possibility of low standard of construction if it is left entirely to the home-owners themselves. This, however, need not be the case if standardized, but very flexible housing systems could be adapted in conjunction with the site-and-services scheme. The NHAs plans for research into better housing prototypes and for promoting technical training of manpower could be coordinated to promote standardized building systems and private investors must be encouraged to use them.

In the TISCO study mentioned earlier, a brief review of the figures revealed the extent of what private investors can and are willing to spend on housing development. Tax incentives may be used to encourage housing systems development but financial institutions must still be willing to back the investors.

The location of housing subdivisions as well as industries by private investors must be properly planned and controlled. Over the past ten years uncontrolled mushrooming of housing subdivisions and other light industries in the suburbs of Bangkok has had an alarming effect on the general urban character of the metropolis. It has caused tremendous congestion of traffic and a general inadequacy of service facilities. This is partly due to the lack of coordination between the Town and Country Planning Department working out master plans for city and regional development, the Board of Investment promoting industries and the Municipal Government which grants the construction permits. It is now hoped that the NHA will provide these effective tools in locating housing projects in Bangkok as well as carrying out its aim to promote the development of other cities and new satellite towns. It is worth noting that the first of these satellite towns to be built is Nava Nakorn situated 43 km north of Bangkok and this project is undertaken by a private group.

The Problem of Finance

A. Issues concerning private financial institutions

Housing as an industry in Thailand is a relatively new enterprise and the increasing housing demand over the past decade has made the industry more competitive than ever before. A recent housing survey shows 78 housing projects underway which will provide 10,336 new housing units over the next two years. This does not include the units planned by the NHA and private developers operating outside the Bangkok metropolis. A rough estimate of the money involved in these housing developments is estimated at $70,000,000 per annum. By Thailand standards this is a sizable sum of money.

It is further estimated that about 27% of all project costs come from the developer's own financial fund which includes money borrowed from private financial sources, personal savings and less loans from friends and relatives. Another 18% of the project costs are derived from down payments or deposits by prospective home buyers and another 13% of the costs are financed by material suppliers and contractors through various kinds of credit facilities. This leaves a balance of 42% which is financed directly or indirectly by commercial banks and non-banking financial institutions.[9] These figures show the important role played by financial institutions in the housing industry in Thailand. To ensure a continuous flow of loanable funds to the housing industry, there are both institutional and legal constraints which must be dealt with by the government:

1. The first of these constraints is the acceptance of long-term deposits by banks. The restrictions imposed on banks to accept public deposits on an annual basis have actually prevented them from offering substantial long-term credit for development projects — particularly in the case of funds to home buyers. Since the public has already shown its confidence in the banking system as evidenced by the growth of fixed deposits over the years, therefore, it is deemed essential that some provision allowing commercial banks to accept long-term deposits would generate a very positive response from depositors.

2. There is a need to promote the establishment of specialized financial institutions geared to accomodating the overall needs of the home building industry. This is because of the presence of a large number of financial institutions which are unable to provide sufficient funds and technical assistance to developers. It is well known that in the building industry, the greatest financial risk lies in the pre-construction stage. In order to minimize this risk, specialized financial entities could be established which would take more than just a passing interest on its clients. Such institution could take the form of "Building Societies" or "Savings Loan Association" whose sole purpose is to mobilize public funds and channel them towards the specific purpose of promoting home ownership.

3. There is the question of effective law enforcement with regard to existing civil and commercial laws especially those concerning claims procedure. Many financial institutions view home building financing as a low priority as far as investment ventures are concerned. In Thailand, the process of making a claim against a borrower who defaults on his payments is particularly complicated, and potential lenders are reluctant to provide funds for home building finance. The time lag involved between making a claim and obtaining satisfaction is unduly long. To allow the system to continue at its current sluggish rate benefits a very small group of individuals who are adept at employing stalling tactics to avoid penalties. This single factor alone is enough to put-off interested investors.

4. The government must consider tax concessions to borrowers. Since home ownership is viewed as a civil right of individual in most countries, therefore, the government must assist individuals in attaining this goal. One example could be the granting of income tax concessions to individuals who pay interest on housing loans. Another example is the reduction of the 1% direct tax paid on mortgages collected by the Land Department and the 2% fee involved when transferring land ownership. To benefit the low-income earners, these existing taxes could be halved when small plots of land are involved say less than 400 m^2.

5. There is a great need to reduce interest rates on loan borrowings. The prevailing rate makes up a very high portion of development costs and, in many instances, directly prevents low-income earners from making a decision to purchase a house even if long term credit is available. The government could correct the situation by coming to the aid of both the supplier of credit and the customer as well. With regard to assistance to supplier of credit, the central Bank could consider extending its facilities to the suppliers of housing loans either directly or by re-discounting notes issued by developers and even home buyers. As a condition of lending, the rate of interest and the type of customers could then be effectively regulated.

Ideally, the Government Housing Bank should undertake the responsibility of guaranteeing finances for small home buyers. Low interest on this type of secured loan would, perhaps, be the right catalyst for involving more people in the business of housing and securing greater interest from the public at large. The Government Housing Bank could charge a nominal fee for the issuance of a letter of guarantee and the amount collected could be used to cover expenses that might arise in the case of a claim. On a large scale, it is expected that the venture will be a profitable one and that funds accumulated in this way over a period of years could be made available for other welfare schemes in the future.

B. Issues concerning the government housing bank

The aims of the Government Housing Bank have been enumerated earlier. Basically, the government's participation *vis-à-vis* housing development finance followed two basic patterns. Firstly, by giving direct grants to the public housing implementation agencies, namely the Public Housing Division, Public Housing Bureau and the Office of Slum Clearance, Urban Renewal and Housing before the NHA was established. These grants are now given to the NHA. However, this practice will be terminated in the future following a recent government directive that no public enterprise shall in the future be allocated funds from the annual budget. The other form of government financial resource allocation for housing development was through the Government Housing Bank.

The Bank's activities are mainly aimed at the middle income home seekers; whether in the form of short term finance for housing developer or long term loans for individual home seekers. Prior to the unification of the various housing agencies by the establishment of the NHA, the Government Housing Bank was entitled to purchase property and resell to home seekers such as land sub-division operations. However, this activity and remaining assets have now been transferred to the NHA.

Due to its low working capital allocated by the government, the record shows that from 1953 to 1975, the Government Housing Bank has made 2,546 loans amounting to $10,000,000. Although the total number and value of loans have increased significantly during the last few years, the fact remains that the Government Housing Bank has been operating in a somewhat constrained situation. This is mainly due to the following factors:[10]

1. The lack of policy guidance, real objectives, target and forward planning in general. As a result, the Bank has in the past 23 years been operating on a day-to-day basis with no particular ends to meet;

2. The volume of funds that the Bank can mobilize through existing means and from existing sources has been inadequate;

3. The range of activities that the Bank can undertake is very limited compared with what other financial institutions can do;

4. Lack of public awareness of the Bank's role and the lack of more up-country branches.

With the formation of the NHA which sets itself the task of building 24,000 housing units/year over a five-year period until 1980, the Government Savings Bank is expected to play a very significant role. This housing program will involve a capital mobilization of $570 million. Within the realm of administrative feasibility, the following measures that could be instituted effectively over the next few years after a thorough investigation is made are as follows:[10]

1. To develop and to increase existing sources of capital. These might include the expansion of Government Housing Bank branches and a campaign launched in order to attract savers irrespective of whether they are home seekers or not. It is worth noting that at present, in spite of the restricted savings facilities of Commercial Banks, almost half the total asset of the Banks come from savings deposits. This may be considered a direct competition to the Government Savings Bank, but since most of the latter's accumulated funds are claimed by the central Government for relocation and with housing receiving little or no benefit from this process, it may be deemed proper that the Government Savings Bank should undertake the vigorous tapping of this source by itself.

2. Other potential sources of fund for housing finance include the life insurance and social security funds. With the former, certain government regulations requiring these insurance companies to keep a fairly large amount of liquid fund in hand may need adjustment, in order to allow their release for housing investment. The volume of fund that might be mobilized from this source is as yet uncertain, but could well reach the $50 million level.

3. Housing bond issues could raise a certain amount of funds for the NHA and Bank operations but the limitations of this fund mobilization method are not yet known, particularly after the initial $80 million has been accumulated from commercial bank reserves as proposed for the 1976 housing development of the NHA.

4. During the first few years while the NHA will not have used up the entire $80 million, the Government Housing Bank may generate incomes from the interest of short term loans with this fund.

To implement these measures, the Government Savings Bank will need the full backing of the central government.

The Need for Overall Policy Consideration

High hopes are placed on the NHA to solve the housing shortage in Thailand. It is widely believed that, unless an overall national policy is promulgated, the NHA will not be very effective in overcoming the housing shortage which it has set for itself in its Five-Year Plan. Definite government policy which will serve as the basis of the NHA should center on the following:[11]

1. Need for an overall urbanization strategy in Thailand. The government should formulate a national strategy in which Bangkok is looked at in relation to the country as a whole, and other urban centers are allowed to expand and develop. Such strategy would take the pressure off Bangkok and help alleviate its problems. As of now, Bangkok serves as the only reception center for low-income migrants from the rural areas.

2. Need for a metropolitan plan to provide overall guidelines for growth. These guidelines must establish basic principles and methods of control over the use and

cost of land in and around Bangkok. To achieve this, laws and regulations concerning land use, land price and rental control, urban renewal and real estate business control must be promulgated. Stringent laws and regulations concerning building construction and control must be passed.

3. Need to give high priority to low-cost housing development. As the private sector is already participating in the housing development program for the middle-income group, it is left to the government to make an effort to meet the housing needs of the lower and lowest income group. Top priority must be given to housing development for the low-income housing. Estimate of housing needs and nature of low-income housing must be made. These require the formulation of an overall policy on public subsidy, land acquisition, taxes and squatter settlements. It is crucial that the Government must decide which existing slums or squatter settlements must be cleared, rebuilt, saved or rehabilitated. The Government must keep in mind that the process of public housing, slum clearance, saving, up-grading, resettlement and urban renewal are interrelated.

Slum or squatter clearance entails massive resettlement of affected families. This in turn creates a tremendous strain on the limited public housing. Therefore, for a successful slum clearance program, there must be a parallel successful public housing program to relocate the families affected.

4. Need of a policy concerning public participation. Beside public subsidy, the program of urban renewal should be seen not only as a social improvement, but also as a means to generate economic development and provide for environmental improvement. The elements of urban renewal that has to be taken into consideration are:

(a) Conservation — the authorities must identify the areas worth preserving;

(b) Rehabilitation — the authorities must identify the areas needing improvement or up-grading;

(c) Rebuilding — the authorities must identify the areas needing demolishing and rebuilding.

4.2 RESEARCH AND DEVELOPMENT ON VARIOUS ASPECTS OF LOW-COST HOUSING AT THE ASIAN INSTITUTE OF TECHNOLOGY

As pointed out in Chapter 1, the Asian Institute of Technology is one of the founding sponsors of the Low-Cost Housing Technology Network and has several academic divisions actively involved in research projects related to low-cost housing technology and policy. Most of these studies have been undertaken by the faculty and students in the Divisions of Environmental Engineering, Community and Regional Development and Structural Engineering. These studies are briefly summarized under four main sections: Slums and Squatter Settlements, Housing Industry, Design and Evaluation Criteria and Utilization of Indigenous Materials.

Studies on Slums and Squatter Settlements in Thailand

A. A study of the structure, function and magnitude of mini-squatters in Bangkok

Squatters have become an increasing feature of cities in developing countries. They have penetrated in large numbers to the hearts of the cities occupying any vacant spot they can find; deep in residential areas, along the roads and canals, or interspersed between developed sections. While these small patches of "slum like" shanty structures have always been considered undesirable, no attempt has ever been made to study their structure and function, and their magnitude, especially when they are scattered in small groups or in the form of isolated structures all over the city.

A study was conducted in AIT by Agrawal[12] to determine the structure and function of these small groups of "slum like" shanty structures or mini-squatters in Bangkok and to appraise the magnitude of the problem. This study provided the following contributions:

(a) An identification of mini-squatter problem in developing countries;

(b) A rationale for the distinction between mini-squatters and slums;

(c) An insight into the extent of mini-squatter problem in Bangkok and its major characteristics.

B. *Workable principles for the control of squatter settlements in Bangkok*

The squatter problem is one of complexity and diversity, and lasting solutions to the problems have been elusive, despite the numerous techniques and approaches that have been tried out. This is primarily because the squatter problem has not been comprehensively understood or defined.

The study conducted by Mutunayagam[13] is aimed at finding a comprehensive definition of the squatter problem with specific reference to Bangkok and to recommend workable principles as a basis for their control and progressive elimination.

The study shows how proper reorganization of squatter control mechanism can contribute to the orderly spatial growth in the city, and help promote the general well-being of low-income people, while preserving the legal rights of private property ownership.

The study provides the following:

(a) A rationale for the proper identification of, and distinction between slums and squatter settlements;

(b) An identification of the squatter problem in Bangkok with its entire ramifications;

(c) Guidelines for the comprehensive policy for the control of squatter settlements in Bangkok;

(d) Recommendation of outline programs for the successful implementation of such policy.

C. *Low-cost environmental systems for squatter settlements in Bangkok*

The aims of this research work which was conducted by Philip[14] are fourfold: firstly, to review and evaluate low-cost environmental systems (water supply, wastewater disposal and solid waste disposal); secondly, to test such low-cost environmental systems by application to a particular Bangkok problem; thirdly, to draw conclusions as to the most suitable environmental system for Bangkok and fourthly, to draw out of this exercise, an approach to policy for such environmental systems.

In the first part of the study, after defining the target population, a general discussion of the present environmental systems are undertaken, as well as the socio-economic conditions of the affected target population. The resulting conclusions are tested in a case study to confirm, reject or expand upon the original findings. A general agreement of the findings were found to reinforce the original conclusions.

In the second part of the study, various alternatives for various low-cost environmental systems are evaluated. The major criteria used for evaluation are sanitation, ecology, health, nuisances, capital costs and operational expenses.

The implementation mechanics of the system are assessed in the third part of the study. In analyzing the present situation as related to decision making, an overview of the collectivities, parties involved, and thereby, a review of the present decision making mechanisms, is outlined.

Finally, the above three phases are further analyzed to single out the most suitable action to bring about the implementation of the appropriate low-cost environmental system. In addition, policy recommendations are drawn for such low-cost environmental systems. The findings from such studies are then tested within the confines of a test area; in this instance, the Din Daeng squatter settlement in the municipality of Bangkok.

The impact of these low-cost environmental systems on the daily life styles of the target population is also studied.

D. *A comparative cost-benefit study of the high-rise building for the low-income residents of Huay-Kuang, Bangkok.*

Riensuwarn[15] made an evaluation of the Huay-Kuang highrise building project by means of a benefit-cost-analysis. Attention was directed to the gross tenant benefit according to income distribution of the tenants regardless of all other benefits generated by the project. The variation of the gross tenant benefit — cost ratio with the rent income ratio before occupying the flat and the income of the occupants was studied. The results showed that the middle income people would benefit more than the low income-people at every rent-income ratio.

The second part of this study has compared the initial project cost of the Huay-Kuang highrise buildings at various land costs and densities with single-story row houses for equal numbers of dwelling units. It was found that the single-story row houses were 32% cheaper if the same density of dwelling units was maintained, but this requires a density of row houses higher than that permitted by the National Building Research and Development Center of Thailand. If the single-story row houses were built on a plot of land large enough to meet the standard, Reinsuwarn[15] concluded that at a land cost of $\$21.25/m^2$, the project would still be cheaper than the highrise building built in Huay-Kuang by the National Government.

E. *Peoples' housing resources: preconditions for the stimulation of domestic resources in popular housing in Bangkok.*

Few would disagree that public housing programs offer conditions that never suit the people for whom they are meant. Past experience is evidence of the futility of such housing programs. The primary reasons for past failures are the following: misconception that a house is a product rather than an environment, misunderstanding about people's priorities, how they change with the stages of urban settlement process and what kind of housing they can afford; and under-estimation of people's potential for self-housing.

The first part of the study conducted by Igbal[16] deals with housing and establishing the urban settlement process in Bangkok. A model of urban settlement process, contrary to established models, is presented.

The second part of the study deal particularly with popular housing areas in Bangkok. The study areas were selected on tenure security criteria, prevalent in various settlements. A comparative study of the components of environmental security and housing input in each study area was made to understand what conditions and amount of effort people have put in their housing. People's investment and under-investment at settlement and city level were estimated. It was found that under prevalent environmental security was about $4.5 million are being under-invested by the people in their housing.

To gain an insight on the preconditions for the stimulation of people's domestic resources into their housing, various components of environmental security, i.e., tenure security, employment security, physical security and their variables, were analyzed by multiple regression analysis to determine which variables affect the variance in housing input. It is believed that the results are of great value to officials and agencies dealing with the low-income housing problem to implement a policy which determines how and under what conditions people would be encouraged to invest their domestic resources in their housing.

The last part of the study deals with the people's potential saving systems; how people save, how much they can save, and identifying their saving systems.

F. A cost indicator for low-income shelter

As a result of the none too impressive performance of the conventional housing policies in ameliorating the housing problem in most of the cities of the developing countries, the authorities concerned have exhibited gradual concessions and adaptations in reframing their policies more realistically. A significant concession is the acceptance of the fact that the majority of the housing stock for the low-income people in urban areas is provided by these people themselves, who build their shelters without any help or guidance from the established institutions. It is in this context that prompted this study by Selvanathan[17] to gain greater knowledge of the people's ability to build their own shelter.

This study also aims at gaining greater understanding of the rather obscure informal housing sector involving the urban low income people. The effect of the escalating cost of construction materials and wages for labor on the capacity of these people in building their shelter is demonstrated and cost indicator for the same is suggested. It is hoped that the methodologies suggested in this study will serve as a useful tool for the Public Authorities in gauging the low income people's ability to build their shelter and in measuring the magnitude of the situation, based on which, suitable housing policies can be framed.

Studies on the Housing Industry in Thailand

A. Prefabrication in housing construction in Thailand

A study was conducted by Rerkshanandana[18] to evaluate the capabilities of local builders in prefabrication and to determine the most suitable system of construction for the local conditions. Capabilities were considered in terms of manpower, equipment and experience.

The analysis was carried out in three stages. The first stage was cost analysis yielding the minimum unit cost for each system of prefabrication. In the second stage of analysis, the results obtained in the first part were used in the comparison of the performance of the systems in terms of cost, construction time, structural safety, and flexibility in design and maintenance. The economics of prefabricated buildings were then studied in the last stage. The economic study emphasized the determination of shape and sizes of building having the minimum cost.

The study concluded that the shortage of heavy lifting equipment and lack of experience are the main obstacles to prefabricated construction in Thailand, and that the beam-column prefabricated system is the most suitable one at present.

B. Survey of available construction materials, equipment and labor in Thailand

A survey was conducted by Huntrakul[19] with the aim of building an adequate data and information on the construction industry in Thailand, with special emphasis on how and where to get major construction materials, equipment and labor forces. The past and present prices of materials and daily wages of labors were compiled and studied and an attempt is made to forecast their future values. Equipment available on rental basis in Bangkok and its rental rate are compiled in a suitable form for general use. The available sizes and types of construction materials, the material costs and labor wage rates in each region of Thailand are also studied.

C. The opportunity for systems building in developing countries

The work of Wadhwa[20] was aimed at the formulation of a housing program for the population in new cities, with special emphasis on housing environment, users or human requirements, and provision of economical and well planned community facilities. The present and future state of the art in prefabrication industry, as well as anticipated technological advances, have been projected and discussed.

Several European and American building systems have been studied and the opportunity of systems building in developing countries has been investigated. It is suggested that a proper and efficient system for various countries, with a view of the labor, materials and technology situation in that country, be designed and developed.

Finally, the feasibility of setting up systems building industries in the archipelago of new city modules is discussed and necessary suggestions are made.

Studies on Design and Evaluation Criteria for Low-Cost Housing

A. Design and evaluation criteria for low-cost housing

This study conducted by Kukreti[21] was aimed at the formulation of design and evaluation criteria for low-cost housing. User needs and functional requirements for low-cost housing are specified and performance concept used to formulate the criteria. The design and evaluation criteria were established for different physical subsystems of the housing system. Several minimum standards suggested in various countries were studied and used as the basis to specify the different criteria.

An evaluation methodology is suggested consisting of the formulation of an evaluation matrix which represents the relation of the functional requirements of the user to each of the physical subsystems of the dwelling unit. To quantify the evaluation process the methodology includes the determination of indices which indicate the performance of the individual subsystems, and the extents to which individual functional requirements of the user and the housing system as a whole are satisfied.

B. Performance evaluation of a low-income multi-family housing system in Bangkok

An evaluation of a low-income multi-family housing system in Din Daeng, Bangkok was made by Pothiapinyavisuth[22] with the use of an evaluation criteria and methodology established on the basis of user's needs as described by Kukreti[21] An interpretation and discussion of the results of the evaluation is made, from which an improved design is suggested. A review of the established evaluation criteria and evaluation methodology is made in view of their relevance and practicality and an improved evaluation criteria and methodology for low-income multi-family housing system is recommended.

C. Design and construction of low-cost houses for low-income families in Thailand

In an attempt to help solve Thailand's low-income housing problems, two experimental low-cost asbestos cement houses have been designed and constructed at AIT by Lee, Pathomkulmai and Hongladaromp[23] for use in suburban and rural areas. Attractive, sturdy and comfortable, these one-storey asbestos cement houses are simple to build with the employment of low-skilled labor and prefabricated construction materials. The unit cost of the duplex at the time of construction is about \$25 per m^2 of floor area. With floor areas of 50 m^2 per unit, the total cost is \$1,250 per unit. The other house is a single detached unit with floor areas of about 60 m^2, the total cost being \$1,600. These experimental houses were designed to meet the requirements of the Thai National Standard Specifications. Figs 4.1 and 4.2 show these experimental asbestos cement houses.

Fig. 4.1. Front View of the Single Detached Unit Prototype Asbestos-Cement Low Cost House.

D. Performance evaluation of experimental asbestos cement low-cost houses

A performance evaluation of the two asbestos cement experimental houses built in AIT was made by Uymatiao.[24] The evaluation criteria and methodology suggested by Kukreti[21] was used. Recommendations for the improvment of these units were made based on the results of the evaluation and from interviews made with the occupants. New designs are suggested which ensure better housing for the satisfaction of the dwellers. A review of the evaluation criteria and methodology is made and modifications are suggested to make them relevant, practical and suitable to the needs of the users.

Fig. 4.2. Views of the Duplex Unit and Single Detached Unit Prototype Asbestos-Cement Low Cost Houses.

E. Design of low-cost houses for lower-middle income families in Bangkok

In order to help solve the problems of housing for the lower-middle income families, two housing systems were designed based on a design criteria for low-cost houses which was modified to suit the lower-middle income group. These housing systems as suggested by Tungpanitansook[25] utilize locally mass-produced materials and the construction does not require sophisticated construction equipment. The housing systems were also designed to meet the requirements of the Thai National Standard Specifications, Minimum Standard for Housing and Environment. The two types of houses are simple to build and employ semi-skilled labor and use locally prefabricated materials. Load bearing concrete block wall with asbestos cement panel roof is adopted for the first type while a hybrid system of concrete block wall and corrugated asbestos cement panel is selected for the other system. Each house covers a plinth area of 73.5 m^2 and the total cost is about \$3,000 per unit.

F. Design of a low-cost rural school building

This study which was conducted by Lowsunthorn[26] is an attempt to present a solution to the problem of the inadequacy of rural school buildings, a problem confronting the development of rural education. School building design, being one of the interdependent elements in the planning of an educational system, is affected by social, political, cultural, demographic and climatic factors as well as educational and economic policies. These factors are briefly reviewed, and the concept and criteria for the design of a low-cost rural school building are presented. Anthropometric data, different arrangement of classrooms, sun shading diagrams and natural methods of environmental control are collected and provided in the form of figures and tables, which may be used as a design guide.

The proposed design, which contains two classrooms, each 6 m x 8.6 m in plan, is basically a structure with two longitudinal load bearing walls resting on reinforced concrete slab on grade. To attain the 6 m clear span required without intermediate columns, the asbestos car port tile roof is supported by the two load bearing walls and a central timber truss which is supported in turn at the central partition and the two end walls. The lateral stiffness of the system is strengthened by a horizontal bracing system at the ceiling level. The cost of the proposed school building is approximately $26,50 per m^2, an equivalent of $24.50 per pupil place with a ratio of 0.92 m^2 per pupil place.

Utilization of Indigenous Materials

A. Development of bamboo pulp and fiber boards

In Thailand, asbestos fibers are widely used as reinforcement in the production of building boards. Asbestos fibers, however, are scarce and costly and hence, there is a need to develop indigenous substitute. It is hoped that natural organic fiber such as bamboo which is abundant and cheap may prove to be a good substitute for asbestos fibers. The country imports $8 million worth of asbestos fibers every year from Canada and South Africa and this constitutes a tremendous drain of much needed foreign exchange which otherwise could be used for other urgent development projects.

An investigation was conducted to examine the feasibility of using bamboo pulp as a substitute for asbestos fibers in order to produce building boards which are relatively cheap and light, durable, workable, has good thermal and acoustic properties, fire resistance and freedom from emission of toxic gases. The shortcomings which some of these bamboo pulp boards have are their relatively large moisture movement and lower durability as compared with asbestos fiber boards. However this is completely acceptable in the context of semi-permanent single story buildings for which there is a vast need in Thailand.

In developing this new material, the use of available technology and facilities for mass production was an important factor considered since investment for design and construction of a new production line is costly and impractical.

Portland cement was chosen as the binder material or matrix as it is readily available and relatively cheap in Thailand. It is durable and possesses a certain degree of fire resistance. Moreover, in Asia and in particular Thailand, cement as a material has a high degree of social acceptability. This factor is very important in selecting materials for use in low-cost housing. Type III rapid hardening portland cement was selected in order to reduce curing time and thereby speed up the production process.

Various types of natural organic fibrous materials were considered such as water hyacinth, wood and bamboo. The preliminary investigation revealed that bamboo fibers appeared more promising than the other two on account of its superior strength and surface texture. Moreover, it possesses good buoyancy characteristics which help prevent bundling of the fibers during the mixing operation.

Two types of bamboo reinforcements were used; bamboo fibers and bamboo pulp. The term bamboo fibers refer to those fibers which were obtained by mechanically hammering dried short bamboo sticks of specified length. The fiber length varies from 1.5 to 2.5 inches with a mean diameter of 0.014 in. Pulp in this context refers to those fibers which were obtained by cooking bamboo chips in a 20% NaOH solution at approximately 170°C under a steam pressure of 120 psi for 6 hours. The average length and diameter of the pulp used are 0.106 in and 1.08 milli-inch respectively.

The pulp is used to arrest the cracks thereby raising the tensile resistance of the composite. The bamboo fibers serve as crack retarders to enable the composite to attain a certain degree of post-cracking ductility.

Various mechanical and physical properties of the boards were studied and these are discussed in detail by Pakotipraphа, Pama and Lee.[27] The strength and rigidities of the composite in axial compression, direct tension, flexure and torsion were studied. The physical properties of the composite which were studied include water absorption, drying shrinkage, fire resistance, impact resistance, permeability and durability.

The investigation showed that bamboo pulp and fiber boards are suitable substitute for asbestos fibers. With cement paste binder, the resulting composite yields building boards which can be used for walls, ceilings and roofs. The flexure strength satisfy ASTM requirements for building boards. Bamboo pulp and fiber boards are impervious, non-combustible and possess good impact properties and exhibit post-cracking ductility. The cost analysis also showed that bamboo pulp and fiber boards are competitive in cost compared with asbestos cement boards. Moreover, the production process is such that existing plants can be used with minor modification. It is hoped that the use of natural organic fibers such as bamboo will save much needed foreign exchange and at the same time reduce the basic cost of the material.

B. Development of coir fiber boards

Coconut palm is a plant that grows in most of the islands and coasts of the tropics. Fresh coconut is an important item in the world diet and copra is the most important product in world trade. One of the most important by-products of the copra industry is the husk, the outer covering of the nut, from which coir fiber is extracted. Due to its abundance and relatively low cost, the use of coir fiber was considered as reinforcement for low-cost building boards.

The boards are formed by mixing coir fiber with cement and water and then molded under a lateral pressure of 280 psi. The various mechanical properties of the boards in direct tension, axial compression, flexure and torsion were studied. The physical properties of the board such as water absorption, permeability, expansion, durability impact resistance, creep, shrinkage and fire resistance were also studied in detail and these are reported by Pama, Cook and Oranratnachai.[28]

The test showed that coir fiber boards satisfy all of the requirements for building boards except water absorption. This suggested that unless protective covering is provided on the surface exposed to the weather, coir fiber boards could only be used as interior boards. They possess very good impact resistance and also showed post-cracking ductility. They are, however, combustible and disintegrate after burning. More investigations on way of improving their water absorption properties and fire resistance are needed.

C. A study of wood-wool boards

There are two board types of products which can be made with cellulosic material such as wood in combination with cement. The first type is a product where the cellulosic material is used as aggregate in the concrete. The second type is where the cellulosic material is in the form of long strands such as in wood-wool boards. The wood fiber serves as reinforcement and arrest the cracks thus increasing the resistance of the composite to tensile stresses.

Factories for producing wood-wool boards have been built in most countries in Asia. These are widely used in building construction as partition walls, ceilings, concrete formworks, etc. The manufacturing process normally starts by stripping the wood into small strips approximately 1 mm x 4 mm in cross section and 40 mm long by means of a planing machine. The wood fibers are passed through a water spray, and cement is squirted on the wet wood fibers. The mixture is molded to the required dimension, trimmed to size, pressed and then left to dry for at least seven days. Due to the simple production process and the availability of the constituent materials, wood-wool boards can be produced cheaply. An investigation was conducted to understand the various mechanical and physical properties of wood-wool boards in order to encourage the efficient use of the material under appropriate condition thereby reducing the cost of the structure. More details are presented and discussed by Pama, Bovornsombat and Nimityoungskul[29]

D. A study of bamboo as reinforcement for concrete

Worldwide inflation and the scarcity of steel in Thailand and other developing countries in Asia have necessitated the search for suitable indigenous materials as expedient reinforcement for concrete. Bamboo which is cheap and available in abundance in most parts of Asia has attracted the attention of research workers for the last three decades or more.

Previous studies by Glenn[30] have indicated that the use of bamboo as reinforcement is feasible but its use is inhibited by its poor characteristics such as bond, volume changes and degradation. Wide cracks and large deflections have been reported in roof beams, girders and slabs of bamboo reinforced experimental structures. As such its wide scale use has been temporarily discontinued until more is known about its true behavior. However, its use can be safely attempted in slabs on grades. During World War II, some roads in Southeast Asia are known to have been built with bamboo as reinforcement but no definite references are available about their life and serviceability. This points to the need for investigating the possible use of bamboo as reinforcements for structures such as slabs on grade where bond is not critical.

An investigation was conducted at AIT by Pama, Durrani and Lee[31] on the behavior of bamboo as related to its use as reinforcement for slabs on grades and to compile such basic information so as to produce a design and construction guide for bamboo reinforced concrete structures.

Pai ruak (*Thyrsostachys oliveri* Gamble), a common and cheap variety of bamboo found in Thailand was used in the investigation. Various mechanical and physical properties of the bamboo were studied such as tensile strength, bond strength, compressive strength, modulus of elasticity, moisture absorption and dimensional changes. A design criteria for bamboo reinforced concrete slab on grade is developed and various construction procedures are suggested. An experimental slab on grade was also built to observe the long term behavior of the structure.

E. A study of rice husk ash as a pozzolanic material

Rice husks comprise nearly 20% of the weight of harvested and dried crop and constitute the largest by-product of the rice milling industry. They have low value in terms of food nutrients for agricultural purposes, high bulk-abrasive property, resistance to weathering and high ash content for which large economic uses have long been sought in all rice growing countries. The ash constitutes about 20% of the total by-product of rice husk. Its chemical composition compared with fly ash and pulverized fuel ash is somewhat identical except in the proportion of its elements.

With the increasing cost of portland cement largely due to the ever-increasing cost of oil, there is a considerable need in developing countries such as Thailand to produce a cementing material chapter than portland cement. A research program[32,33] was started at AIT to investigate the pozzolanic activity of rice husk. Two types of ash were used in the investigation. The first was obtained from the rice threshing site where the husks are burnt for disposal purposes. The second was produced by burning the husk at controlled temperatures in a furnace and grinding the resulting ash to the desired fineness in a ball mill.

Strength and volume change tests were carried out using up to 50% cement replacement with the "village burnt" ash. The results indicated that the ash, in general, had poor pozzolanic activity. The ash produced under controlled conditions, however, conformed to ASTM requirements for pozzolanas. The results of setting time, strength and volume change tests indicated that up to 20% cement replacement could be achieved without any significant adverse effect on the properties of the concrete.

The study is being extended to determine the pozzolanic activity of rice husk ash mixed with lime. It is hoped that this material will be useful especially in rural areas where rice husk ash is abundant.

F. *Use of lateritic soil for low-cost housing*

Lateritic soil from Chantaburi and Hua Hin provinces in Thailand were studied by Nasir[34] to investigate its use for low-cost housing. Soil-cement blocks were cast in a CINVA-Ram machine. The effect of soil properties, moisture content, cement concentration, compacting pressure, curing time and method of curing on the compressive strength, durability, and moisture absorption of the soil-cement blocks were studied.

Results showed that the response to cement stabilization of soil blocks made from the two soils was excellent. Criteria were established to evaluate the compressive strength, durability, and moisture absorption of the soil-cement blocks. It was observed that, to satisfy these criteria, Hua Hin soil, which had a higher activity required a greater amount of cement.

4.3 STUDIES ON CONDITIONS OF LOW-INCOME PEOPLE CONDUCTED BY OTHER INSTITUTIONS IN THAILAND

A. *Housing conditions and problems of the people in Manangasila slum areas*

In 1970, the Faculty of Social Administration of Thammasat University conducted a survey to determine the housing conditions and problems of the people living in Manangasila slum in Bangkok. A report[35] in Thai language was published which presents the general problems faced by the inhabitants of this particular slum. The conditions and characteristics of this slum area are described and short-term and long-term solutions to the problems of the slum dwellers in Manangasila are suggested.

B. *Social welfare research survey in the slum areas of Klong Toey*

A comprehensive report[36] was prepared by the Faculty of Social Administration of Thammasat University on the conditions of the slum in Klong Toey. The report describes the actual living conditions of the Klong Toey inhabitants. It presents statistical data regarding the structural and social background of the families and examines the factors that made these families migrate to Bangkok. The report also presents data on the educational attainment of the inhabitants, vocation and income, property ownership and debt obligations. It also describes the family planning practices of the people and their attitudes on moving out of the area.

The report includes recommendations on how to reduce migration of rural inhabitants to Bangkok and on ways of alleviating the housing conditions of the present inhabitants of Kong Toey.

C. *Demography of Bangkok: A case study of differentials between big city and rural populations*

The Institute of Population Studies at Chulalongkorn University has conducted a study of the demography of Bangkok.[37] The report reviews urbanization in Asia in general and presents a population projection up to the year 2000. The report also deals with the historical perspectives of Thailand's population. It analyzes urbanization patterns using the municipal area as the equivalent of urban population. The report contains data on population composition, age — sex composition, labor force, occupation, literacy, education, ethnicity and religion. It also presents housing and urban conditions and estimates future population growth.

D. *Internal migration in Thailand*

A report published by the Institute of Population Studies at Chulalongkorn University[38] examines the internal migration in Thailand covering the period 1947-1972. The report presents background information on internal migration noting problems such as unemployment, housing shortage and other environmental factors. It also examines the volume and patterns of migration, reasons for migration, characteristics and problems of adjustment and determines the relationship between migration and other factors.

E. *Survey of the slum of Soi Charurat, Makkasan*

Prasarn Mitr College of Education has conducted a survey of the conditions of the slum in Soi Charurat, Makkasan. This report[39] describes the geographical condition and background of the area. It analyzes the causes of migration to this area and describes the social conditions of the area. It presents data on income, vocation, and educational attainment of the inhabitants. It describes the conditions of existing services as well as the housing conditions, attitudes of the people towards the area and examines the needs and problems of the inhabitants.

F. *Social development in peri-urban areas: a study of the needs and problems of children and youth in four slums in Bangkok*

This study[40] was conducted by the Applied Scientific Research Corporation of Thailand to determine the needs and problems of the children and youth in four slum areas in Bangkok. The report describes the rise of slums in Bangkok and focuses on the physical and social characteristics of the four slums being studied. It contains data concerning the place of origin of slum dwellers, family structure, types and size of households and age structure of the inhabitants. It describes the means of livelihood and enumerates the major occupation of the people. It contains data on income, unemployment, underemployment and describes the results of the survey on the satisfaction of the people with their jobs. It gives details on the pattern of expenditures and examines their debts and savings. It also considers in detail the education, employment, juvenile delinquency and welfare of the children and youth in these areas.

G. *Problems of life and options for action of the six slums in Bangkok*

This study[41] which was conducted for UNICEF describes the physical characteristics of the houses and examines the rent and tenancy, utilities and sanitary facilities and life styles of six slums. It studies the demography, ethnicity, migration and reasons for migration. It discusses the family structure, child-rearing, delinquency and material and child health in these six areas. The report also deals with the health services, family planning and community welfare in these areas. Finally, it presents a proposal for an urban community development program as a means of alleviation of the conditions of the people in these slums.

4.4 PROGNOSIS FOR THE FUTURE

The establishment of the National Housing Authority should bring about significant and efficient efforts in tackling the housing problem which has risen to the top of the nation's priority list. For the NHA to succeed, full cooperation from the National Government is needed as discussed in Section 3.1 — Private "low-cost housing". Various legal and institutional constraints discussed previously must be dealt with by the Government in order to encourage a greater degree of participation by Commercial Banks in the housing industry. The Government Housing Bank must also be allowed by the Government to have greater flexibility in its operation in order to meet the financial requirements of the NHA.

The NHA as the sole agency in Housing must utilize the resources of the various academic and research institutions in the country in conducting various studies concerning low-cost housing. To avoid duplication, a network of institutions working on various aspects of low-cost housing must be formed and these institutions should be supported by research and development funds which the Government could raise from industrial agencies with a stake in housing development.

REFERENCES

1. *The Third Five-Year National Economic and Social Development Plan, 1972-1976,* National Economic Development Board, Office of the Prime Minister, Bangkok, Thailand, 1972.

2. Hongladaromp, T.. *Klong Toey — 1973,* Report submitted to National Housing Authority of Thailand, Asian Institute of Technology, Bangkok, Thailand, 1973.

3. *The Development and Sequence of Housing Project of the Ministry of Interior,* Social Service in Thailand (Department of Public Welfare), Mahaadthai Press, Bangkok, Thailand, 1960.

4. *Review of Current Housing Situation in Thailand,* Social Service in Thailand, (Development of Public Welfare), Mahaadthai Press, Bangkok, Thailand, 1960.

5. *Five-Year Development Plan, 1976-1980,* National Housing Authority, Bangkok, Thailand, 1976.

6. *Draft Policy Plan,* National Housing Authority, Bangkok, Thailand, 1974.

7. Narongdej, P., *The Housing Situation in Thailand,* Roving Workshop on Low-Cost Housing, Asian Institute of Technology, Bangkok, Thailand, October, 1975.

8. *A Study of Housing Sub-divisions in the Bangkok-Thonburi Metropolis,* Thai Investment Securities Co., Bangkok, Thailand, 1972.

9. Simapichaicheth, P., *Low-Cost Housing Finance in Thailand: A View from the Private Sector,* Roving Workshop on Low-Cost Housing, Asian Institute of Technology, Bangkok, Thailand, October 1975.

10. Intarakomalyasut, M., *Low-Cost Housing Finance in Thailand: A View from the Government Sector*, Roving Workshop on Low-Cost Housing, Asian Institute of Technology, Bangkok, Thailand, October 1975.

11. Sakornpan, C., *Socio-Economic Consideration for Low-Cost Housing Development*, Roving Workshop on Low-Cost Housing Development, Asian Institute of Technology, Bangkok, October 1975.

12. Agrawal, P. K., *Mini-Squatters: Their Structure, Function and Magnitude Case Study of Bangkok*, Thailand, Thesis, Asian Institute of Technology, Bangkok, Thailand, 1975.

13. Mutunayagam, N. B., *Towards Principles for the Control of Squatter Settlements in Bangkok*, Thailand Thesis No. 626, Asian Institute of Technology, Bangkok, Thailand, 1974.

14. Philip, M., *Low Cost Environmental Systems for Squatter Settlements in Bangkok*, Thesis No. 625, Asian Institute of Technology, Bangkok, Thailand, 1974.

15. Riensuwarn, D., *Highrise Building for Low-Income People: A comparative Cost-Benefit Study of the Huay-Kuang Project*, Special Study Report No. 40, Asian Institute of Technology, Bangkok, Thailand, 1975.

16. Iqbal, P., *People's Housing Resources: Preconditions for the Stimulation of Domestic Resources in Popular Housing in Bangkok, Thailand*, Thesis Asian Institute of Technology, Bangkok, Thailand, 1975.

17. Selvanathan, A. R., *A Cost Indicator for Low Income Shelter*, Thesis, Asian Institute of Technology, Bangkok, Thailand, 1975.

18. Rerkshanandana, N., *Prefabrication in Housing Construction*, Thesis No. 688, Asian Institute of Technology, Bangkok, Thailand, 1974.

19. Huntrakul, K., *Study and Survey of Available Construction Materials, Equipment and Labor in Thailand*, Thesis No. 804, Asian Institute of Technology, Bangkok, Thailand, 1975.

20. Wadwa, L. C., *The Opportunity for Systems Building in Developing Countries*, Thesis No. 470, Asian Institute of Technology, Bangkok, Thailand, 1972.

21. Kukreti, A. R., *Design and Evaluation Criteria for Low-Cost Housing*, Thesis No. 683, Asian Institute of Technology, Bangkok, Thailand, 1974.

22. Pothiapinyavisuth, A., *A Performance Evaluation of a Low-Income Multi-Family Housing System*, Thesis No. 811, Asian Institute of Technology, Bangkok, Thailand, 1975.

23. Lee, S. L., Pathomkulmai, S. and Hongladaromp, T., *Design of Low-Cost Houses for Low-Income Families*, Research Report No. 14, Asian Institute of Technology, Bangkok, Thailand, 1974.

24. Uymatiao, G. L., Jr., *Performance Evaluation and Design of Asbestos Cement Low-Cost Houses*, Thesis No. 814, Asian Institute of Technology, Bangkok, Thailand, 1975.

25. Tungpanitansook, V., *Design of Low-Cost Houses for Lower-Middle Income Families*, Thesis No. 813, Asian Institute of Technology, Bangkok, Thailand, 1975.

26. Lowsunthorn, P., *Design of a Low-Cost Rural School Building*, Thesis No. 807, Asian Institute of Technology, Bangkok, Thailand, 1975.

27. Pakotiprapha, P., Pama, R. P. and Lee, S. L., *Development of Bamboo Pulp Boards for Low-Cost Housing*, Proceedings, IAHS Symposium on Housing Problems — 1976, Clemson University, South Carolina, U.S.A., May 1976.

28. Pama, R. P., Cook, D. J. and Oranratnachai, A., *Mechanical and Physical Properties of Coir-Fibre Boards*, Proceedings, Symposium on New Horizons in Construction Material, Lehigh University, Bethlehem, Pa., U.S.A., November 1976.

29. Pama, R. P., Bovornsombat, S. and Nimityongskul, P., *Mechanical and Physical Properties of Wood-Wool Slabs*, RILEM (Accepted for Publication), Paris, 1977.

30. Glenn, H. E., *Bamboo Reinforcement in Portland Cement Concrete*, Bulletin No. 14, Clemson Agricultural College, North Carolina, U.S.A., 1950.

31. Pama, R. P., Durrani, A. J. and Lee, S. L., *A Study of Bamboo as Reinforcement for Concrete Pavements*, Proceedings, The 1st Conference of the Road Engineering Association of Asia and Australia, Bangkok, Thailand, February 1976.

32. Columna, B., *The Effect of Rice Hull Ash in Cement and Concrete Mixes*, Thesis No. 678, Asian Institute of Technology, Bangkok, Thailand, 1974.

33. Cook, D. J., Pama, R. P. and Damer, S. A., *Rice Husk Ash as a Pozzolanic Material*, Proceedings, Symposium on New Horizons on Construction Material, Lehigh University, Bethlehem, Pa., U.S.A., November 1976.

34. Nasir, S. H., *Use of Lateretic Soil for Low Cost Housing*, Thesis No. 217, Asian Institute of Technology, 1969.

35. *Survey Project on Housing Conditions and Problems of the People in Manangasila Slum Areas, Bangkok, 1970*, Faculty of Social Administration, Thammasat University, 1970.

36. *Report on the Social Welfare Research Survey in the Slum Areas of the Port Authority, Klong Toey, Bangkok*, Faculty of Social Administration, Thammasat University, 1971.

37. Goldstein, S., *The Demography of Bangkok: A Case Study of Differentials Between Big City and Rural Populations*, Institute of Population Studies, Chulalongkorn University, 1972.

38. Prachuabmoh, V. and Tirasawat, P., *Internal Migration In Thailand, 1947-1972*, Institute of Population Studies, Chulalongkorn University, 1974.

39. *A Survey of the Slum of Soi Charurat, Tambon Makkasan, Bangkok*, Prasarn Mitr College of Education, 1966.

40. *Social Development in Peri-Urban Areas: A Study of the Needs and Problems of Children and Youth in 4 Slums in Bangkok*, Applied Scientific Research Corporation of Thailand, 1972.

41. Morell, S. and Morell, D., *Problems of Life and Options for Action*, Report Submitted to the UNICEF, 1972.

5

LOW-COST HOUSING IN KOREA

Sung Do Jang and Hang Koo Cho

ACKNOWLEDGEMENTS

The authors wish to express their sincere acknowledgements to those listed in the reference appended to the end of this chapter for the citation of parts of their papers. Thanks are also extended to Mr. K. S. Suh, head of Survey and Statistics Section, Housing Research Institute, Korea National Housing Cooperation for his assistance to collect the relevant materials for editing and for his preparation of the section on In-country Information Exchange System.

5.1 COUNTRY OVERVIEW

The territory of the Republic of Korea, a southern half of the Korean Peninsula stretching from north to south in Northeast Asia, comprises a land area slightly less than 100,000 km^2, and approximately 30% of which is either arable or habitable. The rest is mountainous. There is sharp seasonal fluctuation in temperature. For instance, in Seoul, the capital city of Korea, the highest temperature is over 32.8°C in summer while it goes down as low as -14°C at mid-winter. Annual precipitation ranges from 500 to 1,400 mm depending on localities.

The country is administratively divided into nine provinces called "do", and urban areas are classified into two categories, cities *(shi)* and towns *(eub)*. Cities are the ones with more than 50,000 population and towns are of between 20,000 and 50,000 population.

The population of the Republic of Korea in 1975 stood at about 34.7 million with an annual average growth rate of 1.8% during the period from 1970 to 1975, which represents a 0.47% decrease in the growth rate over the preceding period from 1966 to 1970 of 2.27%.

The ratio of population between urban and rural areas was 28-72% in 1966. However, the ratio remarkably changed to a proportion of 48.4-51.6% in 1975 due to rural out-migration. The population increase in cities was accelerated mainly by increased employment opportunities and better social welfare facilities in urban areas since successful implementation of the First Five-Year Economic Development Plant during 1962-66.

The country launched in 1962 the initial series of Five-Year Economic Development Plans, which has transformed the country from an agricultural society into a major industrial one. By 1975, the real GNP, which grew at a rate of 9.8% per annum since 1962, had risen from $3.9 billion to $13.3 billion at 1970 steady price; while *per capita* real GNP* had increased from $150 to $377, or from $87 to $532 at 1975 prices, which represents one of the highest rates of sustained growth in the modern economic history.

The steady improvement in the living standard of the average Korean family over the past decade is manifested by the substantial expansion and increased utilization of goods and services. A change in the pattern of household comsumption is reflected in the increasing share of expenditures and other amenities for a better living.

5.2 CURRENT HOUSING SITUATION AND HOUSING POLICY

Housing Situation[1]

The housing situation in Korea poses a growing problem because of the relatively slow increase of supply against the rapidly rising demand for new houses. Ever since 1960, the housing demand has kept rising sharply mainly due to the natural growth of population, the increasing number of households owing to a trend toward the nuclear family system and the rural out-migration in addition to the natural loss by decay.

In 1960, the number of existing houses was 3,464,000 for the total population of 24,980,000 and the total of 4,198,000 households. This figure represented a housing capacity of 82.5%. The housing supply rate in urban areas was 66.6%, while that in rural areas was 88.7%. In other words, 3.4 households out of 10 were without their own houses in urban areas, whereas the homeless households were 1.3 in rural areas.

The housing supply rate, which was 82.5% in 1960, decreased to 78.6% in 1966. This again dropped gradually to 77.8% in 1970 and 75.2% in 1975.

Table 5.1. Shift in Housing Situation between 1960-1975.

Classification	1960	1966	1970	1975		Average annual growth
				Whole Country	Urban	
Population (in 1,000 persons)	24,980	29,193	31,435	34,681	16,699	2.2%
Family Size (persons)	5.95	5.94	5.54	5.13	4.89	-
Number of households (in 1,000 units)	4,198	4,917	5,576	6,757	3,415	2.9%
Number of housing units (in 1,000 units)	3,464	3,867	4,334	4,869	1,834	2.3%
Housing supply rate (%)	82.5	78.6	77.8	75.2	53.7	-

Source: Population and Housing Census Reports.

*The Korean Economy-Growth Equity and Structural Change, Economic Planning Board, Republic of Korea (1976).

In case of urban areas, the number of households increased from 2,525,000 in 1970 to 3,415,000 in 1975. This represents a 35.2% increase, while in rural areas it increased only by 0.3% during the same period. This was the result of the continuous rural out-migration.

As for population increase in cities[2] during the period of 1970-1975, all cities showed 27.8% increase on the average.

The capital city of Seoul contained 13% of the total population of the country in 1966. This increased from 17.6% in 1970 to 19.8% in 1975. Seoul showed also the highest rate of population increase throughout the country.

The number of housing units in urban areas increased by 31.2% from 1,398,000 units in 1970 to 1,834,000 units in 1975, while the number in rural areas dropped by 0.23% to 2,955,000 units in 1975. The decline in rural areas appeared to be due to failure of the new production to recover the loss by obsolescence. The housing demand in rural areas shrunk due to the continuous rural out-migration.

On the other hand, the rate of housing shortage increased from 22.2% in 1970 to 24.8% in 1975, leaving total of 1,880,000 homeless households. Such an increasing trend of housing shortage resulted from a growing popularity of the nuclear family system and insufficient production of new houses. The rate of housing shortage in urban areas showed 46.3%, while less than 10% for rural areas. This indicates a serious housing shortage developing in urban areas. In case of larger cities of over one million population like Seoul, Busan, and Daegu, the housing shortage exceeded an average rate of 46.3%; 47.0% in Seoul, 48.9% in Busan and 52.1% in Daegu.

Table 5.2. Housing Supply Rate by area in 1975.

Classification	No. of household (in 1,000)	No. of houses (in 1,000)	Supply rate (%)
Seoul	1,408	747	53.0
Busan	503	257	51.1
Other urban areas	1,503	830	55.2
Rural areas	3,342	3,035	90.8

According to the criteria* of substandard houses given by the Ministry of Construction, houses in urban areas are more obsolescent than those in rural areas. The number of substandard houses in urban areas reached the height of 18.8% in contrast to 3.2% in rural areas.

In case of houses of low-income people in Korea, the dining room, being the space for collective life, is concurrently being used as the space for private life, i.e., as bedroom and also as studyroom. The kitchen and toilet are in some cases not large enough.

According to the 1970 housing census, the housing standards of the country are low as a whole. The number of houses with floor space of less than 50 m^2 accounted for 64.9% (houses of less than 30 m^2, accounted for 32.7%) of the total number of houses.

The average floor space per house was 45.4 m^2 and the average number of persons

*Substandard housing: with floor area less than 21 m^2, built without building permission on public land, with improper public utilities, or physically in poor shape.

sharing one room was 2.4. These indicate a very low standard of housing in comparison with those in advanced countries.

Housing Construction Trend

According to the housing construction plan of 1975, 70,000 housing units were expected to be erected by public sector; i.e., 44,000 units by the central government, 1,000 units by local governments, 11,000 units by the Korean National Housing Corporation, and 15,000 units by the Korea Housing Bank.

The priorities of financing for implementation of the housing construction plan were set by the following principles:

(1) Construction of houses for rent rather than for sale;

(2) Preferential construction of houses for workers employed in manufacturing industries;

(3) Construction of apartments or row houses rather than detached houses.

The breakdown of housing units by type is shown in Table 5.3. It is noteworthy that most houses constructed by public sector have floor areas of 40 to 80 square which can be identified as low-cost housing units.

Table 5.3. Housing Construction Plan by Type for 1975.

	Type of house	Housing units	Fund (in million won*)
Public Sector		70,000	88,823
Central government		42,950	50,743
	Apartment	19,778	22,200
	Rented	7,300	17,400
	Sale	12,478	14,800
	Row house	7,334	57,779
	Detached house	8,168	6,590
	Rural house	3,670	1,434
	Employee's house	2,000	2,100
	House for flood disaster	2,000	740
	Renewal houses		1,900
Local government		1,250	680
	Apartment for sale	900	450
	Row house	30	17
	Detached house	240	185
	Rural house	80	28
KNHC		10,800	19,400
	Apartment for sale	10,800	19,400
KHB		15,000	18,000
	Apartment for sale	1,500	2,250
	Detached for sale	13,500	15,750
Private sector		130,000	

*Exchange rate at $1 = 400 won.

As for the actual trend of construction for the year 1975, the total constructions registered were 189,000 units, or 94.4% of the planned 200,000 units. Public sector established 89.7% of the planning figure while private sector registered 96.9% of the goal.

Table 5.4 Comparison of Plan and Accomplishing in 1975.

Projector	Planned	Implemented	
		Housing units	Percentage
Total	200,000	188,851	94.4
Public Sector	70,000	62,851	89.7
Central Government	9,116	9,046	99.2
Local Government	22,584	22,056	97.7
KNHC	19,700	18,216	92.5
Company Houses	5,000	1,190	28.8
KHB	10,400	9,066	87.2
Others	3,200	3,277	102.4
Private Sector	130,000	126,000	96.9

Housing Policy and Long-Term Construction Plan

Since the First Five-Year Economic Development Plan launched in 1962, the government placed a greater emphasis in use of limited national resources for the development of industry, leaving the housing project on a low-priority level primarily concerned with the private sector. However, the success of the plan for industrialization triggered a rapid growth of urbanization. In order to cope with such concentration of inhabitants and housing deficit in large cities, the central government exerted various efforts such as diffusion of public institutes and government enterprises over provincial areas, scattering of urban industry into rural areas and improvement of rural income level. Furthermore, to promote housing development, the government formulated "Long-Term Housing Construction Plan (1971-1981)" along with enactment of "Housing Construction Promotion Law" to ensure effective implementation of the plan. The plan, however, failed due to economic recession by various elements during the first phase. The government began to modify the original plan in 1974 and worked out a new plan entitled "Housing Policy and Long-Term Construction Plan".

By the end of 1976, the final year of the Third Five-Year Economic Development Plan period (1972-1976), the long-term construction plan had to be further revised to achieve effective implementation of the scheduled goal during the Fourth Economic Development Plan period (1977-1981). Under the current Fourth Five-Year Plan emphasizing improvement of social welfare as one of the key themes, the government investment for housing sector is to be increased by a substantial degree.

During the Fourth Five-Year Plan period a total of 2,640 billion won* will be invested for construction and supply of 1,330,000 dwelling units. This means that approximately $5.4 billion will have to be invested in housing each year during 1977-81. During the period, the public sector will invest a total of 600 billion won, being equivalent to 23% of the total investment and with which approximately 38.5% of the total projected dwelling units is to be constructed.

*Exchange rate at $1 = 485 won in 1976.

The objectives of housing policy for implementation of the housing plan during the coming five years are as follows:

1. Priority should be given to urban housing construction because the absolute population in rural areas will continue to decrease: rural areas are not in desperate need of new housing. Therefore, housing construction should be concentrated in the urban areas where the housing shortage is most severe;

2. The government and other public entities shall support housing construction for low-income people who cannot finance their own housing by private means;

3. Different lending systems at different interest rates and for different repayment periods shall be implemented for each beneficiary group;

4. Sample household surveys shall be made on a yearly basis to determine the nature of these groups who do not own homes and the optimum housing sizes for them;

5. Priority shall be given to financial assistance for retired military and police officials, war veterans and flood victims;

6. Various institutional measures shall be implemented for the provision of a stable supply of low-priced high quality materials and housing sites in order to stimulate the construction during the period;

7. Private self-help organizations for mutual cooperation shall be encouraged to facilitate home purchases from commercial house builders;

8. The conservation of housing stock should be maintained to the fullest possible extent to insure a sustained increase in the housing stock and to mitigate the housing shortage;

9. Imposing restrictions on the spontaneous removal of existing houses meeting prescribed standards;

10. Legalization of unlicensed housing to the possible extent in consistence with legal criteria;

11. Promoting home improvement through private self-help efforts in line with the Saemaul Movement;*

12. Preventing the imprudent removal of squatter housings for large-scale urban redevelopment,

13. Local authorities shall assume a leading role to carry out the promotion of home improvement programs.

However, some difficulties are expected in the process of implementing the whole plan in raising the necessary housing fund, particularly through private participation,and in view of heavy reliance of the plan on private industry.

In order to achieve the target, the plan envisages the following measures:

1. Increased supply of public funds, encouragement of private fund mobilization and increased inducement of foreign loans for maximization of housing fund;

---

*Refer to section 5.4.

Table 5.5. Ten-Year Housing Construction Plan.*

Housing Units in 1,000, Fund (in billion won)

	72-81[1]		72-74[2]		75-76[2]		77-81[3]	
Sector	Housing Units	Fund	Housing Units	Fund	Housing Units	Fund	Housing Units	Fund
Total	2,163	3,939	413	596	420	703	1,330	2,640
Public Sector	762	854	106	100	144	154	512	600
Private Sector	1,401	3,085	307	496	276	549	818	2,040

**Source:* The Ministry of Construction (1976).

Note: [1] Sum of Supplement plan for 72-76 and Revised plan for 77-81.

[2] Supplement plan at 1974 market price.

[3] Revised plan for the Five-Year Economic Development Plan period at 1976 price.

Table 5.6. Amount and Sources of Housing Funds (1975).

Sector	Amount of funds invested (won)	Sources of funds	Conditions	
			Terms (yrs)	Interest rate (%)
Public	74 billion (22.7%)	National Budget Appropriation		
		National Housing Bond	5	6
		Housing Lottery	20	4
		Bank Funds		
		Debentures	15	8
		Installment of Deposits	20	14
		Housing Special Accounts in Local Governments		
		Foreign Loans		
		AID	20-25	8-10
		IBRD	20	8
Private	252 billion (77.3%)	"Non-systemized" Funds		
		Key Fund, Saving		
		Real Estate Disposal,		
		Retirement Fund,		
		"Kye" Fund,*		
		"Systemized" Funds (Borrowings)		
		Private Loan		
		Company Loan		
		Bank Loan		
		Other Financial Agencies		

Note: *"Kye" is a sort of prevailing private installment deposit system consisting of 10-20 members of relatives or friends.

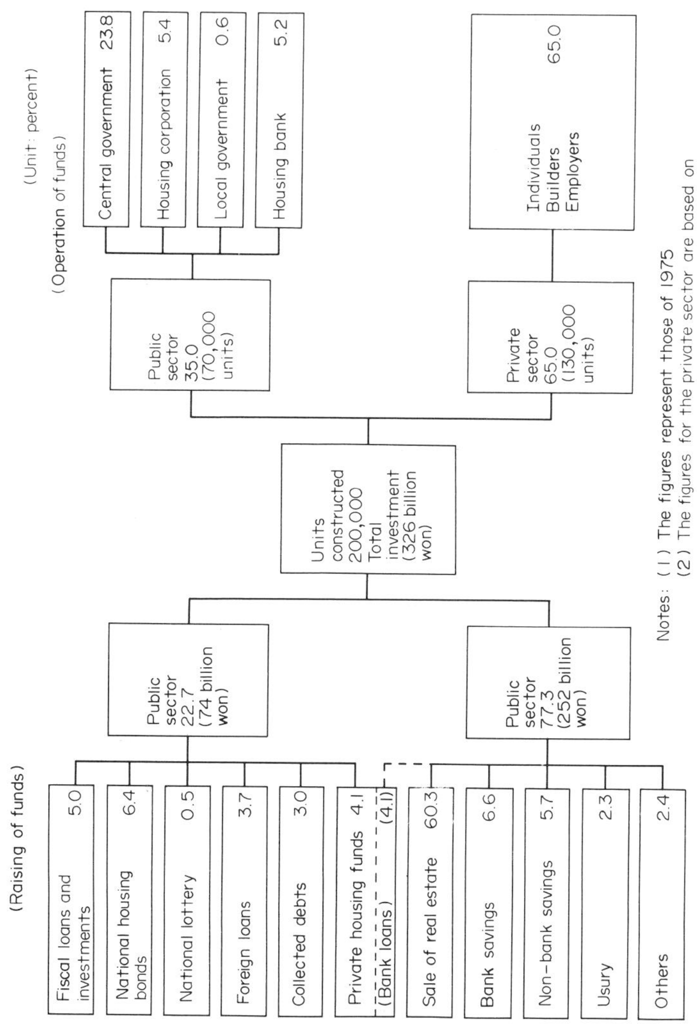

Notes: (1) The figures represent those of 1975
(2) The figures for the private sector are based on KID survey
(3) Source: Study on raising of housing funds and housing financing system (summary) by KID, 1976.

Fig. 5.1. Flow chart of housing funds.

2. Fostering of the housing-related industries and improvement of marketing channel for construction materials in the interest of better housing at lower cost;

3. Collective supply of residential area, designation of apartment area and development of city outskirts;

4. Improvement of housing administration including unification of central and local government housing administrations for promotion of private housing construction.

Financing Policy for Housing Construction

A. Status of housing investment and supply of funds

The government of Korea has been so far concentrating all its efforts on developing the key industries, and consequently, low priority was laid on housing sector. This was evidenced by the fact that the average ratio[3] of investment in housing construction to the GNP during the last decade was 3.7%, as compared with 6-8% recommended by the United Nations for the Escap region.

In general, housing investment in Korea is largely classified into those of public and private sectors. The amount of housing investment in the public sector has been too small as compared with private investments. This indicates that the private sector has been playing a leading role in the field of housing construction.

Table 5.6 shows the amount and sources of housing funds in the public and private sectors in 1975 and Fig. 5.1 illustrates the schematic diagram of housing fund flow.

B. Improvement of housing financing system

In order to meet the effective housing demand, a vast amount of capital is required for investment in housing construction. Nevertheless, there are numerous problems encountered in the raising of housing funds such as the low priority of public investment in housing construction and the low-bank interest rate in the face of progressing inflation. The measures for effective promotion of raising the housing funds with these problems suggested by the government are:

1. To give a high priority to public investment in housing construction;

2. To expand the National Housing Bond sales;

3. To issue the housing debentures from the Korea National Housing Corporation;

4. To seek a substantial amount of loans from abroad;

5. To utilize the social welfare funds, such as the reserve funds for government officials pension;

6. To absorb the private idle funds for housing construction by means of raising the bank interest rate to a realistic level, while giving subsidies to the low-income households;*

7. To develop the secondary market system which will allow transfer of the mortgage loans.*

---

*Suggested to the government by Dr. R. T. Pratt, Professor of University of Utah, U.S.A. in December, 1976.

If the housing funds are operated or supplied ineffectively, it would be difficult to achieve the goal of meeting of the effective housing demand for most part of homeless low- and middle-income groups. However, there are also a number of problems encountered in operating and supplying the housing funds to low-income groups in view of the practice of transferring the mortgage credit house to other people at premium, and of the irrational loan system of public funds between different types of loans offered. The measures for improvement of the operating and supplying the housing funds suggested by Dr. Pratt are:

1. To give large loans with long-term, low-interest rates to the low-income households;

2. To improve the method of selection of occupants;

3. To apply the indexation system to the loan repayment formula to insure an automatic adjustment of interest rates in accordance with the rate of inflation.

5.3 PUBLIC HOUSING ADMINISTRATION SYSTEM AND HOUSING INDUSTRY IN PRIVATE SECTOR

Operation of Public Housing Agencies

A. Central housing administration organization

As the central housing administration organizations, the Ministry of Construction (MOC) carries out the direct housing administrative functions with reference to the national housing policy formulation, and the Economic Planning Board (EPB), the Ministry of Finance (MOF) and the Ministry of Home Affairs (MOH) execute the indirect housing administrative functions in cooperation with MOC. And also, the Annuity Bureau of the Ministry of General Affairs carries out special administrative functions on housing.

MOC carries out the administrative functions pertinent to housing survey, long-term housing demand estimation, establishment of housing construction plan, and supervision and control over various housing projects. EPB has the administrative function of establishing a housing investment plan, in addition to the establishment and execution of an economic development plan, budgeting, resources mobilization and investment fund allocation, technical development, international economic cooperation, etc.

MOF controls and supervises the Korea Housing Bank (KHB) in addition to its main role of management of currency and financial affairs, taxation system, foreign exchange and public properties, and MOH has the function of supervising and controlling over the housing projects being carried out by the local autonomies.

In addition, the Annuity Bureau of the Ministry of General Affairs deals with the housing projects for homeless civil servants, and the Office of Veterans deals with provision of housing for soldiers, both active and retired.

In this framework of horizontally decentralized administrative system, MOC establishes and executes housing construction plans in cooperation with other central administrative agencies. For instance, the establishment of comprehensive housing plan, preservation of housing survey and statistic data, and the functions of supervision and control, design and technical development on housing are the principal functions vested in the MOC. The function of design and technical development stands for (1) approval of housing project design; (2) preparation and diffusion of standard designs; (3) supervision and control over technicalities on housing construction operation. Miscellaneous functions are (1) urban development and redevelopment plan, construction and remodel; (2) inspection and technical examination of housing complex plans; (3) approval and permission of housing complex development projects; (4) designation of housing complexes.

On the other hand, the local housing administrative functions differ by area and level of local government because of the different housing situations and sizes of budget.

As for the housing administrative functions on local government level, they are carried out by the Housing Bureau of the Seoul City government, by Construction Bureaus of each provincial governments and Busan City governments, and by Construction Section of each city and county offices.

B. *Korea national housing corporation (KNHC)*[4]

The Korea National Housing Corporation was established in July 1962 with the purpose of ensuring housing stability of the people and stepping up public welfare in conformity with the government's welfare policy. Since then the Korea National Housing Corporation has been playing a leading role in alleviating the housing shortage.

The KNHC has constructed 86,000 units of apartment-houses with individual floor area of 40-85 m^2, 19,695 units of which were supplied on rent basis throughout the country during the period from 1962 through 1976, with a total investment of 60.5 billion won financed from public housing funds, national housing funds, and AID loan funds. These apartment-houses have been distributed to 300,000 homeless people.

The Korea National Housing Corporation constructed only 1,000-2,000 houses on a small scale until 1972. However, after the turning point when it constructed 10,000 houses in 1974 with the active assistance of the government, it has established a massive construction system and has constructed and supplied a total of 26,076 houses in 25 major cities throughout the country by 1976. These housing units built by KNHC represent 35% of the target of 74,000 houses which have been assigned to public sector per annum during the Third Five-Year Economic Development Plan period. The KNHC plans to have a capacity of 45,000 units per annum for construction and supply of housing by 1980. That is, before the end of the Fourth Five-Year Economic Development Plan.

KNHC subsidiary companies. The Korea National Housing Corporation has established two subsidiary companies to supply a large number of housing construction components, precast panels, and building bricks.

The Hansung Prefab. Company was established in 1971 to supply such prefabricated components as PC panels and other components exclusively to the KNHC. The annual production capacity of the PC panels is equivalent to 5,200 housing units of 13 pyong (42.9 m^2) type apartment.

The Korea Silicate Bricks Co. was established in 1975 with capital of 2,500 million won invested wholly by the KNHC and a loan of 13 million D.M. in kind from the Dorstener Company, West Germany. The Company has annual capacity of approximately 154 million bricks composed of N.F. (Normal Format) and DF (Dunn Format) sizes in order to construct 10,000 housing units of 15 pyong (49.5 m^2) type.

C. *Korea Housing Bank*

The Korea Housing Bank was established as a special corporate in 1967 in accordance with the enactment of the KHB Law aiming at supporting and promoting housing construction for low- and lower-middle-income people. Its main functions are to loan funds and deal with the related matters of management of funds for housing construction, housing transactions and site development.

The major sources of funds are its capital assets, contract savings, issuance of housing bond, borrowings from government, foreign loan and issuance of housing lottery tickets. The funds secured from these sources are directly loaned to the low- and middle-income groups or indirectly to the KNHC, local autonomies, public agencies and private commercial builders. National housing funds which are usually secured from the issuance of housing bond and introduction of foreign loan are sub-loaned at an annual interest rate of 8% with a repayment period of 15 years. Private welfare housing funds which are secured from the operation of KHB fund is loaned to those who have the means of loan repayment and capable of bearing partial financial burdens in construction and purchasing of houses. However, the conditions of loan have often the effect of isolating the potential customers in view of both the size of amount involved and in the repayment term, thus resulting in heavy financial burden to the low-income group. In addition, undue limitations in the operation of KHB fund would hamper positive and broader performance of its function aimed at furthering housing construction.

The status of loans made by the Korea Housing Bank during 1967-1975 is shown in Table 5.7.

Housing Industry in Private Sector

A. Housing industry outlook[5]

Housing industry is to emerge as a large-scale integrated system industry not merely through natural expansion or diversification of traditional house building activities but through gradual organization and integration of many small business activities on a national basis chiefly because of its profit sharing component enterprises such as housing developers, house builders, material suppliers and financial agencies who are all to maintain close interactions under a leading role played by a pivotal enterprise with good operational standing. Housing industry, in view of its foregoing nature, comprises a wide variety of related industries which are to be developed into "group industry" or "system industry". However, in case of Korea, the component industries are not only overly sub-divided with unbalanced sizes but generally lack development of the integration system between them, thus, providing a weak basis for growth as a sector of the economy.

In other words, small private house builders carry out a considerable amount of work without being an organized enterprise even though they represent the most critical sectors of housing industry. As for housing material supply, material production is characterized by general lack of rational system for mass production standardization and modular coordination. And the site development and supply are in most cases carried out by public sector, while private enterprises exploit them as a means of speculation. Housing distribution is also still in the primary stage of development with no systematic market channel, and the important function of after-service is carried out separately by various specialists and technicians who are not directly related to building industry. And moreover, there is no enterprise carrying out all the integrated functions as a consistent business activity. The reasons why housing industry in Korea is in a structurally weak condition lie partly in the lack of government's positive measures for fostering housing industry and partly in the distraughted functions of KNHC (Korea National Housing Corporation) which is still short of assuming the role of laying the foundation for housing industry.

Table 5.7. KHB Loans by Type.

As of August 1975 — Price unit: in million won; ( ): housing units

Classification	1967-1973	1974	1975	Total
National Housing Fund Loan	8,278 (19,167)	20,279 (30,349)	13,180 (21,033)	42,196 (70,459)
Public Loan Fund	4,686 (18,850)	35 (118)	13 (50)	4,736 (19,024)
Private Housing Construction Loan	57,040 (82,333)	8,412 (7,950)	5,862 (4,874)	71,314 (95,157)
Private Industrial Loan	- -	- -	19 (24)	19 (24)
Housing Lottery Fund Loan*	1,430 (1,330)	- -	- -	1,430 (1,330)
Site Development Loan	2,442 (882)**	- -	- -	2,442 (882)
Housing Material Production Loan	273	100		373
Total	74,160 (121,686)	29,276 (38,417)	19,079 (25,918)	122,510 (168,084)

*Since 1973, Housing Lottery Fund Loans are included in National Housing Fund Loan.

**Figure in parenthesis means size of sites developed; unit in thousand "Pyong" or 3,300 m^2.

The recent participation of several large-scale enterprises in the housing industry since 1970 is considered a most favorable tendency for the systematization of housing industry which has so far been characterized by small-scale splinter operation and lack of integration system.

B. Problems in housing industries

Analysis and examinations have been made herein on the basis of subdivision of the industry into four sectors; housing construction, material supply, housing distribution and site supply, based on the result of housing industry sample survey* on 21 housing related enterprises in Seoul.

1. House-building industry. House-building activities are broadly classified into two fields. One is the self-help house building activities performed by end-users, and the other, the housing supply by commercial house builders and various public entities including the central and local government agencies. The former is the case of the individual customers who acquire their own housing sites and have the houses built by contract builders, and the latter is the case that public agencies and commercial house builders construct houses through direct and contract operations for sale to the real customers.

House-building activities in Korea have traditionally been conducted by the end-users themselves except the cases of supply by public entities. However, the increasing participation of large-scale enterprises and small to medium contractors in the house-building activities and development of housing market since the beginning of 1970 has made a considerable contribution to the increase in housing supply. But the private house-building activities in general are still in the primary stage of development from the viewpoint of the number of enterprises and their managerial abilities.

After all, the private house-building enterprises in Korea are for the moment characterized by inefficient industrial structure in the light of the organizational form, size of the capital assets and production structure, and thus, are still incapable of providing the basis for industrial development.

2. Building materials industry. The proportion of the material costs in house building in Korea accounts for average 70% of the total costs as a whole with some variations depending on the types and quality of housing.

Due to the insufficient development of integration system in material industry, most of the house-building materials are produced and supplied independently, and incidentally, with other construction materials without inter-dependent production and supply system.

The imbalance between supply and demand and the price instability of major housing materials involving the lower prices compared with general commodity price indices and the rise of general housing costs have often impeded the house-building activities of the private sector to a considerable degree. For major items of building materials which are now being mass produced, the demand-supply position is comparatively secure with price competitiveness, while it is unstable for other materials produced by small-scale industries chiefly because of the uneven development of production capacity.

On the other hand, the distribution process of building materials is very complicated through multi-stage channel from manufacturers to users, e.g., 3 stages for cement-asbestos board, plywood, plate glass, and 4 stages for cement and tile, thus compelling the consumers to bear heavy financial burdens in purchase of materials paying over 200% margins in some cases.

---

*Survey made by Korea Industrial Development Research Institute task group in October 1974.

3. Housing distribution. Most of the housing transactions are conducted through the intermediary role of brokers who are called Bok-duk-bang in Korea. Most of the brokers are aged people, ill-qualified and inexperienced, mostly not fit for any other productive job.

The related legal codes do not regulate their qualifications and credit standing thus making it possible for them to open up shops with simple reports to the authorities concerned.

They cannot properly assume the role of introducers in view of the fact that they are not able to ensure the shift of ownership rights in the housing transactions. Quite often they pose serious social problems causing heavy damages to the buyers.

From a sample survey data* in 1974 on the attitudes of customers as to the Bok-duk-bang, it was disclosed that 78% of the sampled households were very dissatisfied with the Bok-duk-bang.

As mentioned above, fostering qualified and reliable housing brokers is presented here as an imminent issue for the development of systematic housing distribution industry in Korea, and appropriate measures should be taken to make them function as responsible brokers capable of controlling transfer of ownership rights in housing transactions and further contribution to the development of sound housing distribution system.

4. Housing site supply.[3] Major sources of housing sites are the government land readjustment plan and also the site development project undertaken by KNHC.

This system has played the important role of basic site supply channel as a means of a new town development project aimed at preventing over-concentration of population and influx of industrial facilities into big cities.

Another channel of housing site supply, small as it is in size as compared with the public sources as stated above, is the housing site development activities carried out by private real estate dealers.

The area of housing sites supplied by the private dealers in Seoul City is disproportionally small compared with the areas supplied by the public institutes, and moreover, most of the private site supplying enterprises deal in the activities not as a continuous business but as short term speculations.

A look at the trend of housing site price based on 1967 constant price indicates the land-price index of 170.1% which is two-fold higher than the general commodity wholesale price index of 66.8% as of 1973, and which has played a leading role of cost push in house production. This fact is reflected by the fact that the price† of housing site accounts for 52.5% of the total housing cost excluding design, which is far higher than 20% in Europe and America, and 50% in Japan.

The elements of land-price increase, in general, will be the natural scarcity of land and the economic safety in possessing land entailing no costs or risk of losses, however, the most important factor in the case of Korea may be the practice of taking land for speculations, and also to some extent the demand supply instability. As the land supply especially in big cities tends to be restricted by the limited availability of absolute land surface, the price increase seems to be

---

*Survey made by Lee, T.K., Real Estate Marketing (1974), and sample survey as to 217 sampled households in Seoul.

†Japanese Long-Term Credit Banks, Urban Development and its Precondition, p. 47.

inevitably accelerated in the future along with influx of populations and growing demand for housing sites.

The following are considered the major factors affecting the land supply problems:

(1) Difficulty of land acquisition in large lots due to minor subdivisions of land ownership;

(2) Limitation of potential residential areas because of the enactment of Green-Belt decree;

(3) Inefficient utilization of reserved land due to the prevailing land speculation;

(4) Chronic land price-hike impeding land acquisition for low-income people.

5.4 CASE STUDY ON HOUSING DEVELOPMENT

Improvement of Rural Housing and Living Environment Through "Saemaul Undong"[7]

A. Saemaul undong and its concept

A remarkable improvement in the rural environments and housing situations in recent years has been made through the nation-wide movement, the so-called "Saemaul Undong" (New Community Movement).

The New Community Movement is designed to help villagers create better places to live by themselves, improving their living environments and boosting their incomes with the spirit of diligence, self-help and cooperation. The primary emphasis is placed on the individual person's sense of active affirmation that he is a member not only of the country but also of his local community. There are three distinct notions embodied in the movement: (1) spiritual reform movement to practice the spirit of diligence, self-help and cooperation in order to nurture the progressive and productive spirit of the nation; (2) social development movement to bring beneficial cultural change into family life as a basic unit of society; (3) economic development movement to increase employment opportunities and income.

In early 1970, the President of Korea proposed the Saemaul Movement to prompt modernization and betterment of rural communities by increasing the income of the farmers and fishermen. However, since the government was under the heavy burden of the pressing problems of expediting industrialization of the country when the Saemaul Movement was introduced in the final year of the Second Five-Year Economic Development Plan (1966-1971), no substantial support could be provided for the movement.

B. Government support for self-helping village

The government has provided essential construction materials, such as portland cement and reinforcing bars, to some of development-minded villages for their effective implementation. This kind of minimum financial support by the government stimulated the competitive spirit among the villages. The administrative assistance was given by providing workers for on-site cooperations with the villagers. They give technical assistance and guidance in the different fields of agriculture, engineering and construction.

During the first short-term planning period which ends in 1981, the Saemaul programmes are to be implemented for the community on various levels: local autonomous bodies and villages. The programmes are formed by villagers voluntarily and the plans are communicated to the central government through official channels in a democratic fashion.

During an experimental stage of the Saemaul Movement in 1970, the government invested $10 million to provide each of a total 30,000 farming and fishing villages with 335 bags of cement, while the villages contributed with local labor, land lots and other expenses. The second stage was implemented in late 1971 with an input of $7.5 million assistance from the government providing 500 bags of cement and a ton of steel bars to each of the villages selected to participate in the movement. These materials were used mainly to improve village feeder roads, lined irrigation systems, communal wells and laundries, etc. The resulting economic return in the early stages is estimated at twice the government investment.

The success of these experimental projects during 1970-72 has greatly encouraged not only the government leaders but also the villagers themselves, enabling them to undertake the ambitious major goals of the movement. The most significant change in the improvement of rural housing achieved by the movement was replacement of the traditional straw thatched roofs with cement-asbestos or cement roofing tiles. In consequence, the time, labor and material straw normally required for annual replacement thatching are greatly saved. Thus, the labor and conserved straw can be utilized elsewhere.

C. *Development of the movement*

In 1973 all the villages in the rural area of Korea, totaling 34,665, were categorized into three groups based on their development stages: underdeveloped, developing and developed. The underdeveloped villages are rudimentary ones with no positive endeavors of inhabitants for the cooperative development of their community. The developing villages represent those which are developing with growing efforts of the villagers under the leadership of Saemaul leaders. The developed villages are in a satisfactory stage at which the people can develop their village solely by their own endeavors and enjoy high income in adequate environment.

Development projects are undertaken in accordance with each village's need and capabilities. The projects place emphasis on improving the environment, increasing employment and income level in rural areas; thrift, helping the neighbors, simplification of traditional family rites, and modernizing markets in urban areas.

Table 5.8. Development Stage of All Villages.

Stage	1973	1974	1975	1976
Under-developed	18,415 (53%)	10,656 (30%)	4,046 (11%)	302 (1%)
Developing	13,943 (40%)	13,763 (60%)	20,936 (60%)	19,049 (54%)
Developed	2,307 (7%)	4,246 (10%)	10,049 (29%)	15,680 (45%)
Total	34,665(100%)	34,665(100%)	35,031(100%)	45,031(100%)

D. *Effects of Saemaul Undong*

During the past four years since the movement has started in 1971, the aggregate total number of participants in the movement in 34,665 villages and 35 cities totalled to approximately 215 million. The total number of projects completed during this period reached 2,897,000 cases. The total projects under this movement produced an output of 4.6 times as great as the total amount of the government input. Table 5.9 shows the annual achievements materialized through the Saemaul Undong.

The major items related to betterment of housing and living environment projects during 1971-76 are summarized in Table 5.10.

Thus, the betterment of rural environment and housing as a foundation for the Saemaul Movement has been well proved. The roofs and walls of the rural houses, toilets and barns have been reconditioned. The improvement of farming and village roads provided rural transportation network which has been a top priority throughout the whole projects. The construction of rural transportation network spanned for a linear distance of 42.686 km by 1976, which accounts for 86.8% achievement of the goal.

The welfare facilities include village assembly halls, public bathrooms, public washing places, children's parks and play grounds. The establishments recorded a total of 29,752 village halls out of a targeted 35,608 halls, which is 83% of the target.

Replacement of traditional straw-thatched roofs of 2 million houses by cement tiles or cement-asbestos corrugated roofings was voluntarily done by villagers, which accounts for 82% of targeted 43 million roofs.

E. Standard housing construction project for rural villages[7]

In 1976, the governement, encouraged by the rapidly increasing pace of rural economic growth, launched a new project to construct standard village housing for farming families in an annual income bracket of over one million won. This trial project plan was to construct 680 standard housing units in eight provinces. Of the projected housing units, 453 were for single separate units and the rest for complex units in the form of a rearrangement of the existing poor village housing. A financial support has been given to each unit builders at an amount ranging from 750,000 to 850,000 won depending on the type of housing. The cost of each housing unit was estimated at 1.1 million won for 15 pyong* (48 m^2) floor area. Therefore, each builder could contribute with comparatively small investment of 250-350,000 won. The size of the standard housing units are limited to 15, 18 and 20 pyong types of floor area in accordance with owners' need and their income levels.

All the necessary building components such as doors, window-frames, roofing materials, flooring panels were standardized with reference to floor sizes and types of the units, and manufactured by designated makers and supplied by single channel through the local government.

The standard-housing construction project initiated in 1976 has been so far successful in terms of construction cost and time taken due to the self-help and cooperative nature of the scheme. The actual cost of a trial housing unit having a floor-area of 15 pyong exceeded the original estimate of 1.1 million won by 150 thousand won, however, it still remained competitive against the prevailing level of 1.45 million won. The prototype housing construction needed only 26 days to complete, which is 11 days less than the average practice in the country. A more striking effect was that construction of the housinq units required little skill and few professional builders. This is mainly attributable to the concise guide booklet "How to Construct Housing Units" issued by the government for the rural housing builders and also the provision of standardized building components.

---

*1 pyong is equivalent to approximately 3.3 m^2.

With the success of the 1976 trial project and with the rural households' positive response to the scheme, the government has set a new target of 100,000 housing units to be constructed by 1981 with all possible support to rural households.

F. Long-term prospects

The ultimate goal of the Saemaul Movement is to make the village richer and better. In the early stages of the movement it was concerned merely with improvement of living environment as already started. However, the long-term plan looks ahead toward increase of productivity and income with the Saemaul spirit. The long-term Saemaul development plan is scheduled as follows:

1971-73	Basic foundation	Development of Saemaul spirit and improvement of the environment
1974-76	Self-support	Foundation of the rural standard productive base
1977-81	Self-sufficient	Expansion of production and income capacity

By 1981 it is planned to raise the average farm household income up to 1.4 million won per annum with additional sources of income derived from off-farm operations.

The Jamsil Low-Cost Housing Project: View from Financing, Structural, Design and Material Concept

A. Objective

The increasing population in Seoul, as a result of rapid industrialization, is facing a number of serious problems such as urban sprawl, pollution, and traffic congestion. From this standpoint, the Jamsil Housing estate is now developing as one of the suburban centers designed to release the population pressure in Seoul.

This housing estate already accommodated a population of over 50,000 by 1975, and 100,000 by 1976, and expects 300,000 by 1980 through continuous development. And this estate is divided into 15-story-high-rise apartment district of 320 acres, medium-rise apartment (4-5 story) district of 416 acres, and single family-housing district of 1,037 acres as well as business district of 517 acres and open space of 185 acres.

One of the serious problems in Seoul is about the settlements developed in a disorderly manner in the outlying areas of the city by homeless low-income families and those removed by the redevelopment of urban centers and the slum clearance projects.

One of the major development objectives of the Jamsil area is to accommodate these homeless families into this improved living environment. As a matter of fact, of the total of 19,180 housing units constructed under the 1975 annual housing construction program, 8,710 units were allocated for these low-income families. Because the current prices of those houses commonly built are beyond the reach of the low-income families, the housing development in the Jamsil area focuses an attention on the housing size suitable to the home-buying capacity of the low-income families.

Attention is also paid to the use of new materials and construction methods to reduce

Table 5.9. Number of Participating Villages and Projects Accomplished by Years.

Year	Number of participating villages	No. of projects (thousand)	No. of project per village	Government input (billion won)	Village output (billion won)	Ratio output/input
1971	33,267	385	12	4.1	12.2	3.0
1972	22,708	320	14	3.3	31.3	8.8
1973	34,665 (31,641)	1,093	32	21.5	98.4	5.6
1974	34,665 (33,513)	1,099	32	30.8	132.8	4.3
1975	36,547 (36,341)	1,598	44	165.3	295.9	1.8
1976	36,557 (36,227)	887	24	165.1	322.6	2.0

Table 5.10. Major Items of Saemaul Project for Betterment of Housing and Living Environment

Project	Goal	Accomplishment		
		71-75	76	Total
Farming and village roads	75,499 Km	81,120	2,336	83,455
Housing and living environment				
Replacement of straw-thatch roofing	2,428,000 Units	1,628,000	360	1,988,000
Standard housing for farming family	100,000		680	680
Remodification of sewage system	8,654 Km	19,538	1,178	11,716
Cultural welfare accommodation				
Public laundry	71,020	56,162	2,140	58,302
Sanitary water supply system	27,599	11,235	3,860	15,095
Community wells	1,092,000	112,168	955	113,123
Village hall	35,608 Units	27,051	2,701	29,752
Public bathroom	36,143 Units	6,372	251	6,623

construction costs through the results of research activities.

B. Low-cost housing program

Figure 5.2 shows the cumulative relative household distribution by income groups in Seoul estimated by the Bureau of Statistics of the Economic Planning Board as of the end of 1975.

Rank	Monthly income (won)	C.H.D (%)
D-1	29,000	3.8
D-2	29,000 – 43,500	11.8
D-3	43,500 – 58,000	26.7
D-4	58,000 – 72,500	43.5
D-5	72,500 – 87,000	58.1
Median	87,000 – 101,500	69.4
D-6	101,500 – 116,000	77.0
D-7	116,000 – 130,500	88.0
D-8	130,500 – 145,000	86.9
D-9	145,000 – 159,000	90.0
D-10	159,000 –	100

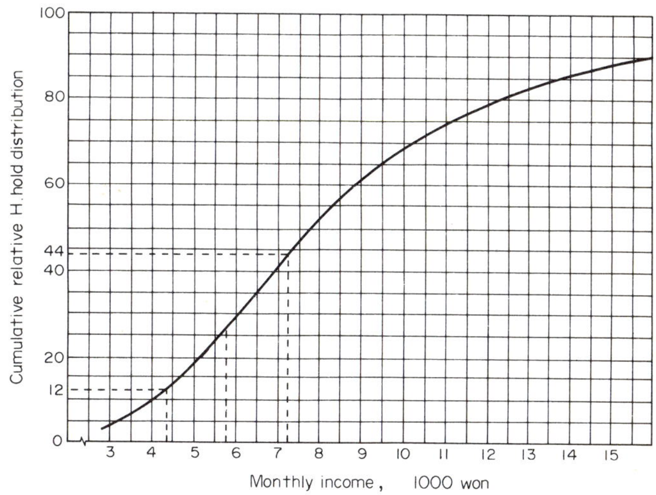

Fig. 5.2. Cumulative Relative Household, Distribution by Income Group in Seoul (Estimated by EPB/BOS).

The low-cost housing units in the Jamsil area have been decided to house the income groups falling between D-3 and D-4 in the household distribution shown in this figure. The household income of these groups mostly range from $90.70 to $150.50 on a monthly average. With this in mind, an attempt was made to estimate and review the respective optimum housing sizes for the groups of monthly average incomes of $90.70-$150.50 and the medium value of $119.60 for determining the housing sizes for the selected groups.

The housing price which a household can afford to pay can be judged through a study of its repayment capacity of the loan funds now being released to low-income families and its saving funds can be counted upon for the initial payment. Table 5.11 shows the figures obtained as a result of a study of the housing expenditure of each household disbursed from 1968 to 1975.

As shown in this table, the housing expenditure accounts for about 17% of the total income. Table 5.12 shows the figures obtained by breaking down the housing expenditure.

From this analytical table, an amount equivalent to 20% of their monthly income was identified as repayment capacity of each household for the loan fund.

Table 5.11. Monthly Income and Housing Expenditure of Salary and Wage Earners' Households in Seoul.*

	Year	1968	1969	1970	1971	1972	1973	1974	1975
(A)	Average monthly income (won)	27,110	32,450	38,630	44,400	53,140	56,520	57,520	74,260
(B)	Housing expenditure (won)	4,260	5,540	6,610	7,370	8,860	9,460	9,730	11,210
	(B)/(A) (%)	15.70	17.10	17.10	16.60	16.70	16.70	16.90	15.10

Sources: *Monthly Statistics of Korea, EPB.

Table 5.12. Breakdown of the Housing Expenditures.

Price unit: won

Year		1973	1974	1975
Monthly income		56,520	57,520	74,260
Monthly expenditure		50,710	51,920	67,740
Balance		5,810 (10.3)	5,600 (10.8)	6,520 (8.7)
Items of housing expenditure	Room rent	330 (0.6)	660 (1.1)	630 (0.8)
	Self-evaluated amount	8,020 (14.2)	7,130 (12.4)	6,970 (9.4)
	Water bill	230 (0.4)	300 (0.5)	290 (0.4)
	House repair	280 (0.4)	560 (1.0)	620 (0.8)
	Furniture & fixtures	600 (1.1)	1,080 (1.9)	2,630 (3.5)
	Others	-	-	70 (0.1)
	Total	9,460 (16.7)	9,730 (16.9)	11,210 (15.1)

Note: Figures in ( ) represent the rate of each item to monthly income by percent.

The conditions of housing loan funds are an essential factor for deciding the ceiling of the amount loanable to a household in relation to its monthly repayment capacity. The housing funds presently available for low-income households are the national housing funds (repayment in 20 years at 8% per annum) and AID Guaranteed Loan funds (repayment in 25 years at 9.4% per annum in the case of the Jamsil project in 1975).

Table 5.13 shows the respective optimum housing sizes estimated for the households with the monthly average income of 90.70, 119.60 and 150.50 dollars, taking into account all the said factors.

Figure 5.3 shows the inter-relations between the income, loan conditions, housing prices, and housing size.

Table 5.13. Estimate of Housing Sizes.

Price unit: dollars

Monthly income	Annual repayment capacity* of loans	Loanable amount† AID guaranteed funds	Housing‡ price	Housing§ price per m^2	Housing size m^2
90.70	217.70	2,057.60	4,115.10	174.90[1]	25
119.60	287	2,712.20	5,424.50	174.90[1]	33
150.50	361.20	3,413.70	6,827.40	159.30[2]	43

Notes: *20% of annual income.
†50% downpayment, repayment in 25 years at 21% per annum.
‡Twice loan fund.
§The estimated price per m^2 is based on the apartment. Housing price survey conducted by the Korea National Housing Corporation.

[1]With central heating system.
[2]With traditional floor heating system.

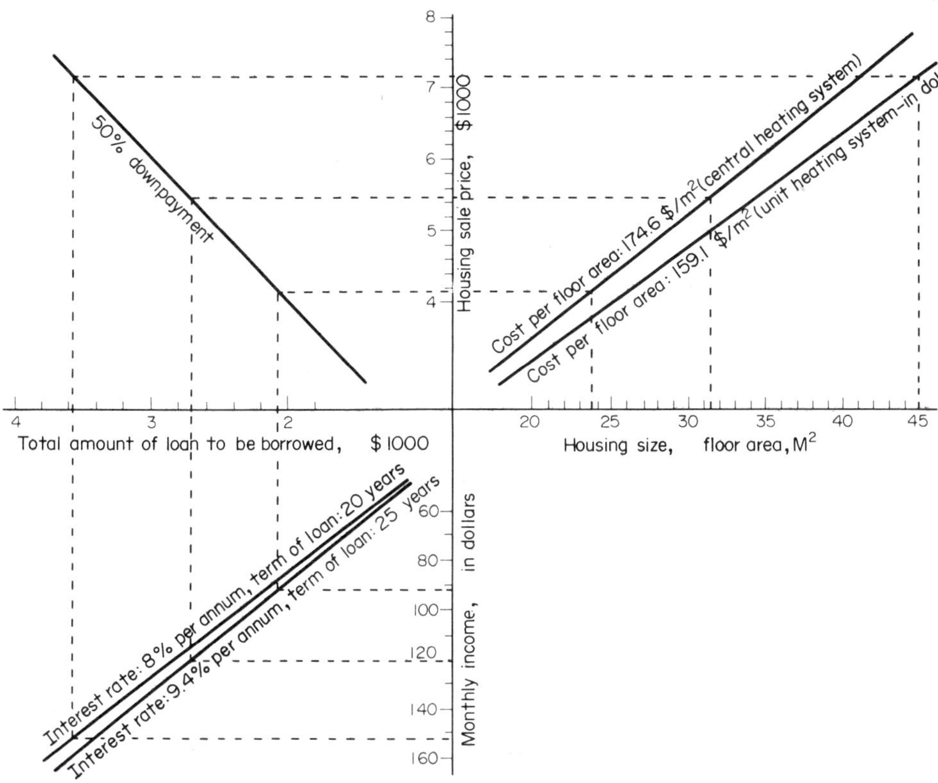

Fig. 5.3. Inter-relation between Income, Loan Conditions, Price and Housing Size.

On the basis of the results shown in Table 5.13 and Fig. 5.3 three types of 25, 33 and 43 m^2 were selected as the optimum sizes.

Of the scheduled housing units, 13%, or 1,100 units, consists of housing with floor area of 25 and 33 m^2, and the rest, or 7,610 units, with 43 m^2.

The estimated housing price per m^2 set at $159.30 for the average five-person family with a monthly income of $150.50 was intended to secure the maximum housing size by installing traditional Korean floor heating system, which is less expensive compared with central heating system.

The households with a monthly income of over $136 are provided with housings of 43 m^2 of floor area mostly on a rental basis since the prospective occupants do not bear the capacity to purchase the unit at their income levels. The rental housing units, thus, occupy a large proportion of 53% of the total 7,610 units.

Table 5.14. Renting Conditions of 43 m^2 Apartment Housing Units in Jamsil area.

Unit: dollars

Deposit			824.70	
Renting conditions	Monthly charges	Monthly rental	24.70	Includes depreciation costs, loan interests, and fire insurance premium
		Management fees	4.10	Includes expenses for cleaning common use areas and repairs
		Total	28.80	
Rent-Payable households must have their monthly income of above $136				Repayment capacity was set at 21% of the monthly income, including housing repair costs of the housing expenditure as shown in Table 5.12.

Housing design

Though the housing sizes are decided on the basis of income levels, they appear to be too small, for living in terms of physiological, social, and psychological, social, and psychological aspects or amenity.

That is, when estimating the number of persons of a household at three for the 25 and 33 m^2 units and five at maximum for the 43 m^2 units, the floor space per person is about 8 m^2, which is far short of the 16 m^2 recommended as the generally optimum unit floor space.

Therefore, the first concern in the floor planning was to design the unit in such a way as to make it expandable when necessary. Taking this into consideration in designing the apartment units for the low-income groups in Jamsil, the 25 or 33 m^2 units were so designed that two units could be united for combined use.

The second concern in the floor-planning was to reduce the area of common use as much as possible. Generally, most of the apartment housing units under constructi are so designed that one stairway is open to two households for common use, but, in the case of small units such as 25 and 33 m^2 units, when one stairway is to be used by two households the ratio of the common use area to the whole unit area is too large. Therefore, in this planning, stairways are so designed that one stairway can be commonly shared by four households.

Table 5.15. Comparison of Common Use Areas.

Unit: m^2

Type by Size	One stairway for two households			One stairway for four households		
	Total area (A)	Common use area (B)	B/A(%)	Total area (A)	Common use area (B)	B/A(%)
25 m^2	25.3	5.5	21.7	25.3	4.6	18.0
33 m^2	32.6	5.6	16.7	32.6	4.6	14.1
43 m^2	66.0	5.8	8.8	Not applicable		

The third attention was to try to use new types of structure. Table 5.16 shows a comparison of construction costs in terms of unit m^2 price for three types of structures; reinforced concrete structure, prefabricated concrete structure, and masonry structure using sand-lime bricks.

On the basis of this comparison, the 25 and 33 m^2 units were designed for sand-lime brick masonry structure and the 43 m^2 units were designed for prefabricated concrete structure. Figure 5.5 shows a comparison of the conventional construction method (reinforced concrete structure) with prefabricated concrete structure in terms of their construction periods and the manpower employed for construction. Figure 5.7 shows the assembling drawing for 40 m^2 apartment units of prefabricated concrete structure.

Especially, the masonry structure using sand-lime bricks which was the first attempt in practice, had to undergo research and study before its designing.

It can be seen from Table 5.16 that the Jamsil apartment housing units consisting of both the sand-lime brick-masonry structure and the prefabricated concrete structure represent reduction of construction costs by 2-3% as against the conventional reinforced concrete structure.

D. Project evaluation

The result of sales and the interview survey with homebuyers indicated that the 25 m^2 unit was just enough to accommodate a family size of 3 persons.

The typical plan firstly attempted to use one stairway for 4 units, consequently reduced the common use area but it gave more spacious opening of stairway to users.

The first attempt towards the 5-story-masonry-bearing wall system with sand-lime bricks saved the construction costs and also seemed successful in regard to its structure and construction as it was expected.

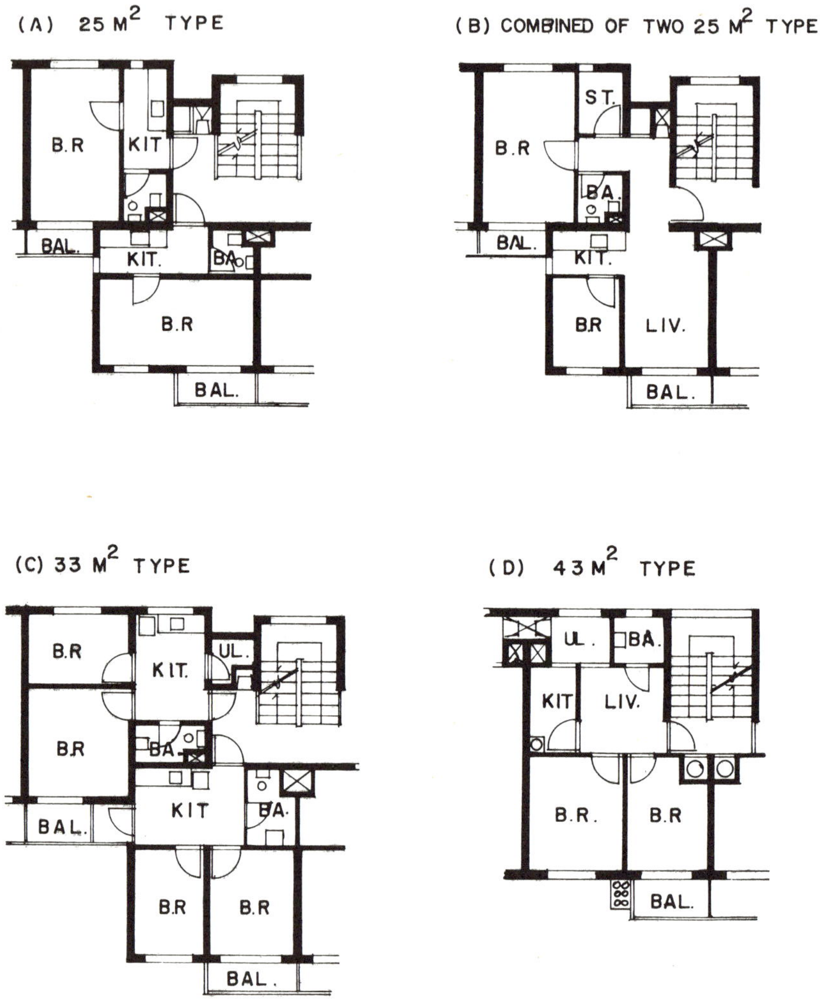

Fig. 5.4. Unit Plans of Low-Cost Housing in Jamsil Area

5.5. CURRENT R&D ON LOW-COST HOUSING AND BUILDING MATERIALS

R&D at KIST

Ceramic Materials Laboratory of KIST has long been concerned with research on utilization of agricultural industrial wastes and by-products for housing materials. Utilization of such industrial solid wastes and by-products contributes not only to conservation of natural resources but to reduction of material cost in industrial

Table 5.16. Comparison of Construction Costs.

Type by size	Structure	Building costs* ($/m^2)	Total construction costs ($/m^2)	% Saving (A-B)/A x 100
25 m^2	Reinforced concrete (A)	83.40	171.90	3.43
	Sand-lime brick masonry (B)	77.50	166	
33 m^2	Reinforced concrete (A)	76.40	Not available	
	Sand-lime brick masonry (B)	73.60	151	
40 m^2	Reinforced concrete (A)	80.50		
	Prefabricated concrete (B)	77	147.50	2.31

Note: *Architectural work only

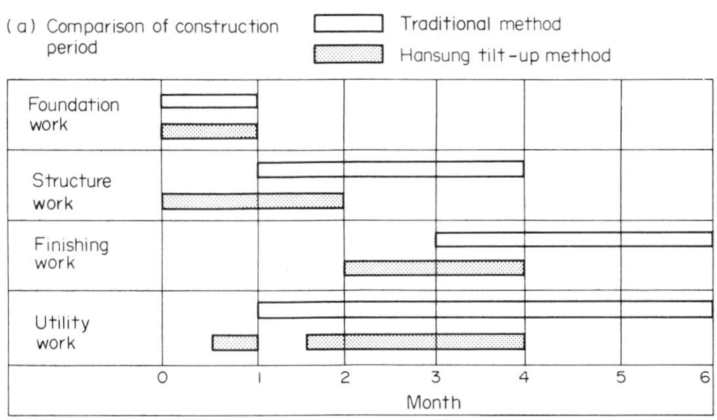

Fig. 5.5. Comparison of Construction Period and Men Power.

processing for manufacturing of building materials. Several R&D projects have been carried out to make use of waste materials for building materials. The waste materials concerned are blast-furnace slag, coaltail, fly-ash from power plants, chemical gypsum from fertilizer plant.

The following topics are the summaries of two most significant results obtained from the R&D projects conducted by the Ceramic Materials Laboratory.

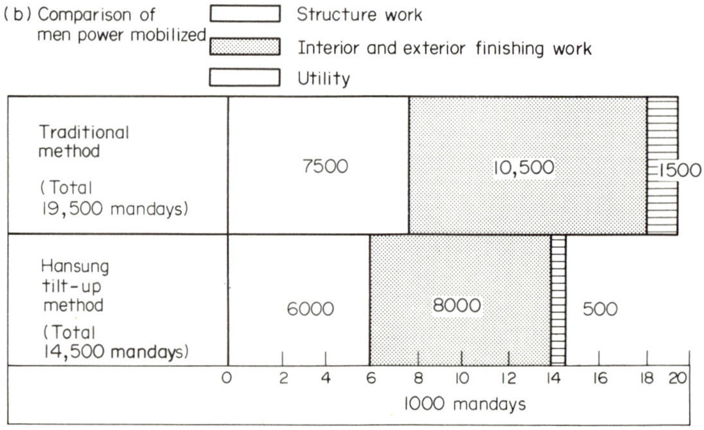

Fig. 5.6.

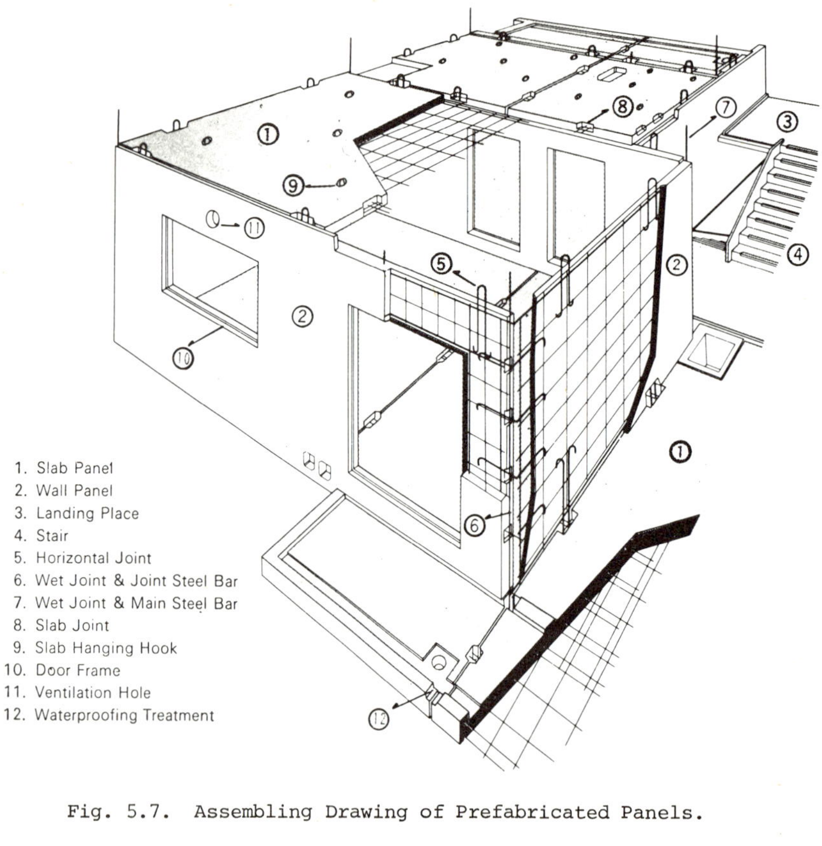

Fig. 5.7. Assembling Drawing of Prefabricated Panels.

A. *Improvement of sand-lime brick quality*

Calcium-silicate bricks have received wide acceptance in many European countries. A sand-lime brick manufacturing plant has been recently introduced by the KNHC to supply high quality, low-cost building component to the housing industry in Korea because of the availability of raw materials with low-fuel requirement. The Korea Silicate Brick Company has been in production operation since June, 1976. However, due to different physical and chemical characteristics of the starting materials, the end-product revealed inferior quality in its compressive strength. Both of the raw materials, sand and hydrated lime, created many problems in manufacturing process such as difficulty in control of moisture content of the mixture and lack of green strength of compacted pieces for mechanical handling. Particularly, the lack of fine part in particle-size distribution of the sand being used adversely affected the development of strength during autoclaving process. But neither pulverizing of sand nor transportation of fine sand from a remote deposit proved practical due to the cost increase of product.

Little work has been done to elucidate the effect of fine mass other than sand on the compressive strength of calcium silicate brick, although it has been recognized in general that the addition of fine sand to the lime-sand aggregate admixture increases the compressive strength.

The primary purpose of this investigation was to identify the problem areas in both the materials being used and the processing in the manufacturing flow. Then, the task was given to seek a substitute for fine sand to improve the strength of brick without increase of production cost. Many attempts have been made to employ various inorganic fine aggregates such as blast-furnace slag, diatomaceous silica, fly-ash, etc., with a view to assisting the manufacture to control the compressive strength at will depending on specific pose and usage.

Compressive strength alone in this R&D project, however, could not be viewed as the single improvement factor, since the quality of the bricks is to be optimized in terms of their general properties. Consequently, the effect of the addition of fines on water absorption, density and apparent porosity of specimens was also examined. This had an additional advantage of observing any correlation between these properties.

Fly-ash has shown a significant improvement in quality of brick specimens among various candidate additives through laboratory work. Fly-ash is the principal waste from burning finely ground coal in steam-generation power station and is collected as fine dust in cyclones. It has been known that this fine material possesses pozzolanic reactivities — to combine with lime in the presence of water to form compounds with cementing properties — suitable for substitution of a fraction of fine sand in aggregate particle size distribution. It has shown that small fraction of fly-ash addition to sand-lime mixture gave many advantages not only in improvement of the product quality but in reducing the production cost and pollution problem.

Five to 30% by weight of fly-ash was added in substitute for commensurate part of sand in the batch-mix containing lime as shown in Table 5.1. For instance, 10% addition of fly-ash to the sand-lime mix showed significant improvement in various properties of sample specimens having a size of 6 cm cube which have been compacted at a pressure of 210 kg/cm^2 in a steel mould. Then the green specimens were autoclaved at 16 kg/cm^2 of steam pressure for 6 hours. The result has shown that a small addition of fly-ash to the batch-mix greatly improved both compressive strength and water absorption of the brick specimens.

The cost of hydrated lime amounts to approximately one-half the total production cost of sand-lime brick. Thus, the amount of lime to be blended with sand is directly related to the production cost.

Fly-ash can be supplied from a power station within a city boundary. Therefore, the transportation cost of the fly-ash to the brick plant could be minimized and the price of fly-ash per unit weight is almost negligible as compared with that of hydrated lime supplied from remote points.

Assuming sand-lime brick requires a fixed compressive strength of 350 kg/cm^2, the batch mix should contain 14% of lime and 20% of finely ground sand if fly-ash is not used under the current manufacturing condition. However, a 5% of fly-ash addition to the raw-batch mix could reduce the necessary lime by 4% in order to maintain the same strength. Based on this configuration, the production cost would reflect a 10% reduction.

B. *Utilization of blast-furnace slag for building materials*

The Pohang Iron and Steel Company (POSCO) is one of the largest steel-making plants with present annual production capacity of 2.6 million tons which will be further expanded to 8.5 million tons by 1981. Blast-furnace slag is presently produced in a quantity of nearly 0.8 million tons each year from POSCO alone and it is expected to reach to 3 million tons per year in 1982. Therefore, the solid-waste treatment has emerged as a growing problem in the country. The object of this work was, therefore, to develop slag-lime brick as a building component. Slag-lime bricks are similar to sand-lime bricks in process and properties which are widely used in many countries.

Blast-furnace slag is a by-product of steel mill, and is obtained by cooling or quenching the molten mass consisting of the earthy constituents of iron-ore and limestone. The essential components of slag are the same oxides as in the portland cement, such as SiO_2, Al_2O_3, CaO, MgO, and Fe_2O_3 but in varying proportions.

The process of water granulation of slag yielded a material which, when mixed with lime, developed good cementing properties. Slag has already found considerable use in road construction and building industries mostly as an aggregate in concrete, slag-cement, and in the manufacture of slag wool.

The cause of the hydraulic activity of slag and its relation to the chemical composition and physical state of slag were studied extensively in the past.[10] It is found that hydrated lime reacts with sand and/or slag granules under the presence of high-pressure steam resulting in development of very high strength in the compacted mixture. For a given value of compressive strength, batch-mixes containing sand and slag required less amount of lime than in the conventional sand-lime bricks.

Several batches were prepared by varying the mixing ratio of lime/sand/water-quenched slag or of lime/air-quenched slag/water-quenched slag with wide variety of particle-size distributions. The batch-mixes were compacted into a cube having one side of 6 cm at pressure of approximately 200 kg/cm^2. The compacted pieces were, then, autoclaved at steam pressure of 16 atms for 6 hours.

Some of the experimental results showed 300-400 kg/cm^2 in compressive strength with 11-12% of water absorption.

It seems to be quite promising to utilize blast-furnace slag for industrialization of building components such as bricks and blocks with high strength characteristics and price competitiveness against other similar products.

The present results obtained by laboratory scale operation are still in the preliminary stage but the plant will be designed for mass production of these components in the near future.

R&D at KNHC and Other Institutes

A. Development of industrialized housing units

Housing cost can be reduced to some extent by rationalizing the whole production process from housing-planning, design, material delivery, and to construction. But there is still another aspect of housing cost reduction which is through an assembly-line operation from the plant to the construction site thereby saving time and labor in the interest of quality mass production. For the latter case, it requires the housing production to be industrialized with the system of prefabrication. Prior to this, there is a need to have standards; such as the minimum space standards, performance standards of building components and the basic or planning modules.

Studies on establishment of modules in Korea have been carried out mainly by four institutes: MOST (Ministry of Science and Technology), MOC (Ministry of Construction), AIK (Architectural Institute of Korea) and Housing Research Institute of KNHC (Korea National Housing Corporation).

The Korean Standards (KS) has already introduced the modules for industrial use, the basic of which is IM equivalent to 10 cm, and planning modules are 3M horizontally and 2M vertically. Studies on the minimum space standards for low-cost housing have been carried out by such institutes as MOC and Housing Research Institute of KNHC for the last several years.

It is expected that such minimum space standards will soon come out from MOC and AIK. One of the most urgent and important studies on standardization of opening size was already introduced as the Korean Standards by the efforts of MOD, AID and KNHC.

Basic dimensions for brick and block are standardized in accordance with 2M and 3M respectively. And it is hoped that various components with modular dimensions will be developed continuously. Studies on prefabricated housing have been mainly carried out by KNHC and other private companies. And as a result, tilt-up prefabricated construction method was applied to 5-story-walk-up apartments by KNHC up to present, and about 11,500 dwelling units have been constructed by this method. At the same time, numerous types of prefabricated single-detached houses for both urban and rural areas have been tried out by private companies.

B. Prefabricated single detached house

In 1976, MOC established the prefabricated rural-housing plan under which numerous studies are now in progress and some private firms have actually developed production schemes. The following are those typical examples made by private firms.

Table 5.17. Prefabricated rural house — Production scheme by Urim Concrete Co., Ltd.

(a) Type of house:

Size (in m^2)	No. of Rooms
40	2 - bedroom, 1-liv. room, 1-kit.
50	3 - bedroom, 1-liv. room, 1-kit.
60	4 - bedroom, 1-liv. room, 1-kit, 1-storage.

(b) Materials used for the structure:

Classification	Materials	Remarks
Foundation		To be prefabricated
wall	Precast concrete panel	thickness: 100-150 mm
Roof		Pitched roof

(c) Comparison of characteristics of systems:

System	Construction period (in days)	Labor requirement (in persons)	Compressive strength of wall components (in kg/cm^2)	Cost per m^2 for 60 m^2- floor area.* (in dollars)
Prefabricated	15-20	100-150	210	73.60
Conventional	50-60	200-250	40-60 (cement bricks or hollow blocks)	

* Cost based on the following evaluation:

Precast panel	$ 1,937
Transportation and erection	303
Others	2,140
TOTAL	$ 4,380/60 m^2

Elevation

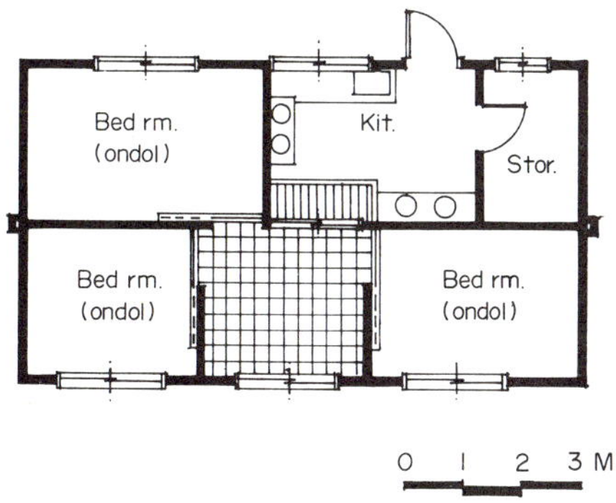

Plan (floor area 51.4 M^2)

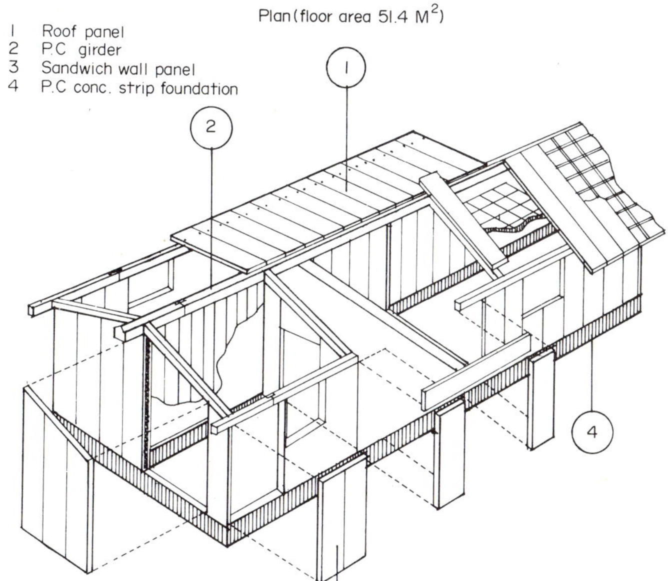

Isometric view of URIM prefab system

Fig. 5.8. URIM Housing System.

TABLE 5.18. Prefabricated Rural House — Production Scheme by Ssangyong Cement Industrial Co., Ltd.

(a) Type of house by size: 40 m^2, 50 m^2, and 60 m^2;

(b) Materials used for the structure:

Classification	*Materials*
foundation	*in situ* concrete
wall	light-weight precast concrete panel for curtain wall
roof	asbestos slate on light-weight precast concrete

(c) Physical characteristics of light-weight concrete:

Bulk density	1.20 + 0.15
Compressive strength at 4 weeks	60 kg/cm^2
Young's modulus	2.10×10^4
Thermal conductivity	0.245 kcal/mh°c

(d) Unit cost per m^2 for 60 m^2 — floor area: $90.50*

*cost based on the following evaluation:

precast panel and other components	$1,640
transport, erection and finishing	2,981
contingency (20%)	845
Total	$5,376/60 m^2

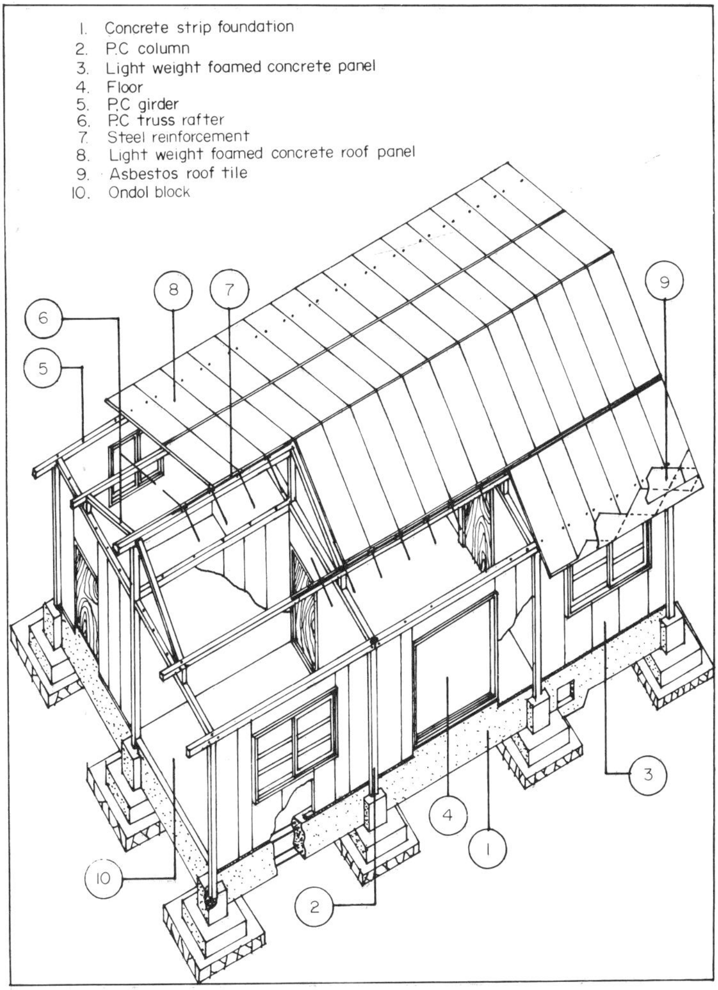

Fig. 5.9. Ssangyong Housing System.

5.6. FUTURE PROSPECTS

R&D on Housing Units and Relevant Projects for Future

A. R&D proposal – KNHC

The Korea National Housing Corporation has been identifying various problem areas through the past experience in housing plannings and implementations. The following R&D projects are directed towards finding early solutions to the problems prior to implementation of consecutive construction project in future:

1. *Prototype low-cost housing for rural villages.* The objective of this project, in a branch of the Saemaul Movement, is to develop low-cost housing units on a prevailing rural standard. This project is concerned with maximum utilization of local indigenous materials and self-help construction skill. The research will be collaborated with a KIST group.

R&D on Manufacturing of Glass-Ceramic Building Materials from Iron-Steel Slags

B. R&D proposal – KIST

The use of metallurgical slag as a raw material for the production of crystallized glass (or glass-ceramics) was intensively investigated by many Russian researchers during the past decades. Their research result has already been carried to a level of mass production of the glass-ceramics known as SLAG-SITALL.[11]

The proposed project is derived from the research work to the manufacturing technology for glass-ceramic materials similar to the slag-sitall developed by the Russians.

In-country Information Exchange System

It has been recognized for many years by various agencies, institutions and construction firms engaged in low-cost housing projects that a foundation of in-country information linkage system and an efficient operation of such a system is a vital necessity to help their activities. The Korea Institute of Science and Technology (KIST) has been participating in the "Low-Cost Housing Technology Project" sponsored by the Technology and Development Institute (TDI) of East-West Center and its cooperating network countries. Consequently, an in-country linkage was formed by collaborative effort of KIST and KNHC through the Roving Workshop II on Low-Cost Housing Technology in 1975. The necessity of expansion of this embryonic in-country linkage was realized by many other R&D institutes within the country.

In order to understand the complex nature of low-cost housing problems and in the interest of mutual exchange of relevant information, experiences and results of R&D, the in-country information network is planned to be organized with following institutions:

A. International exchange of information

The KNHC functions as the information center in cooperation with 48 institutions in 20 countries and international organizations.

The KNHC will present the research papers on low-cost housing at request from foreign countries.

Table 5.19. Network Institutes and Information Source

Network institutes	Relevant information source
Korea National Housing Corp. (Center)	Materials, technology, design, policy, statistics
Korea Institute of Science and Technology	Materials, energy, transportation, environment, computer application
Korea Scientific and Technology Information Center	(1) Current contents of domestic journals in science and technology;
	(2) Current contents of foreign journals;
	(3) Documentations and information; and
	(4) Technology information.
Korean Standard Research Institute	Korean standardization of building components and material
Korea National Construction Research Institute	Testing materials, standard design, structure and construction methods
Korea Development Institute	R&D papers, research papers, site plannings
Architectural Institute of Korea	R&D papers, rationalized materials, design
Institutions in Universities and Colleges, Government Agencies, Private Organizations and Others	Urbanization, population policy, financing, statistics, environment, etc.

B. Future planning of information center

In order to overcome the limitation of manual information processing and to reach the level of information service system in developed countries, KNHC will have a computer laboratory to introduce an automatic information processing and to develop it into a world-wide model of information linkage system.

Future Housing Situation in Korea

By the successive economic development plans, Korea is achieving a remarkable level of economic growth. It is expected that the GNP will expand at an average annual growth rate of 9% during the Fourth Five-Year Plan period (1977-81). Government convinces that *per capita* GNP will be raised from $532 in 1975 to $1,284 in 1981.

The total national investment will be increased by 52% from $23.6 billion of the Third Five-Year Plan period to $36.5 billion for the Fourth Five-Year Plan period, of which the social development sector; i.e., education and manpower development, health, housing, water and sewerage facilities shares 21%.

A total of $5.2 billion will be devoted to the housing component. This is more than three-quarters of the investment required for the social development sector.

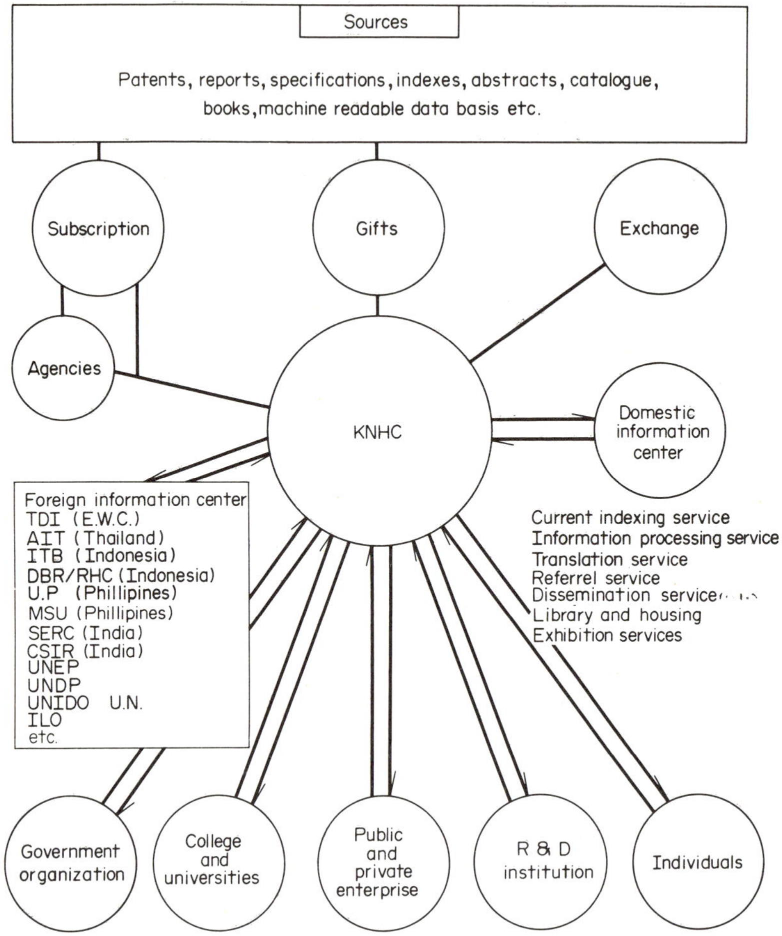

Fig. 5.10. Information Network System.

The Government has proposed some 1.3 million dwelling units to be built by both public and private sectors during the Fourth Five-Year Plan period to reduce the country's housing shortage from the present 26-20% in 1981.

Following Table 5.20 shows the total investment outlay for housing during the Fourth Five-Year Plan period. The investment in housing consists of 3.84% of the GNP.

It must be noted that the Government has now turned its development policy towards one more concentrating on housing not like those set out in the foregoing three Five-Year Economic Development Plan.

By this favorable situation, it is evident that the housing problem will be reduced to some extent as shown in the table, but the problem will remain still serious as long as a rapid urbanization is taking place at the present pace in this country.

Table 5.20. Investment in Housing, 1977-1981.

Unit: billion won at 1977 market price

Sector	Construction unit in 1,000	Outlay	
		In billion won	Percentage
Public	512	600.3	22.7
Private	818	2,040.2	77.3
Total	1,330	2,640.5	100.0

Table 5.21. Housing Situation between 1975 and 1981.

Classification	1975	1981
No. of housing stock (in thousand unit)	4,825	6,016
No. of dwelling units to be constructed yearly (in thousand unit)	180	320
Per capita floor area (in pyong)	3	3.22
No. of construction units (per 1,000 persons)	5.1	8.2

REFERENCES

1. Paik, Y. H., *Housing Situation and Development Policy in Korea*, presented at the Roving Workshop II (Korea Phase) on Low-Cost Housing Technology, Seoul, Korea, Oct. 1975.

2. The Korea Housing Bank, *9th Annual Report*, 1975, Seoul, Korea.

3. Study on Raising of Housing Funds and Housing Financing System (Summary), 1976, Korea Housing Corporation, Seoul, Korea.

4. *Housing Annual Report*, 1975, Korea National Housing Corporation, Seoul, Korea.

5. *Study on Housing Policy Formulation*, Dec. 1974, Korea Industrial Development Research Institute, Seoul, Korea.

6. *Korea Housing Policy and Long-Term Housing Construction Plan* (Draft), March 1975, Ministry of Construction, Seoul, Korea.

7. *Monthly Economic Statistics*, Vol. 30, Oct. 1976, The Bank of Korea.

8. *"Saemaul Undong" – from beginning to present*, 1976, Ministry of Home Affairs.

9. Kim, H. K., *Socio-Economic Considerations in Low-Cost Housing with References to Korea*, presented at the Roving Workshop II (Korea Phase) on Low-Cost Housing Technology, Seoul, Korea, Oct. 1975.

10. F. M. Lea, *The Chemistry of Cement and Concrete*, Third Edition, Chemical Publishing Company, 1971, pp. 454-489.

11. Pincus, A.G., *Soviet Building with Slag Sitall*, The Glass Industry, pp. 6-32, Jan. 1972.

6

LOW-COST HOUSING IN HAWAII

David C. Firth

6.1 INTRODUCTION

As of October, 1976, Hawaii's housing situation is rapidly evolving into a "have or have not" proposition. Those who cannot afford to purchase housing are frozen out of the market, and there is little hope that the situation will improve. The source of the problem is that as much as 85% of Hawaii's population is priced out of the housing market, i.e., during 1975 less than 15% of all families on Oahu earned enough to qualify for a conventional loan that would buy a median-priced ($60,000) condominium (and the figure is 10% for a single-family home).

Low-cost housing in Hawaii is virtually non-existent. In 1970 the MKGKY Study for the Hawaii Housing Authority concluded that 50,000 low and moderate income units were needed state-wide at that time. By January 1977, 80,000 total units had been built, mostly by private industry. So the availability and quality of housing had "improved significantly" according to official sources.

However, since 1970, housing prices (85%) have increased at more than twice the rate of family income (34%). Thus, while we have 80,000 more units we also have:

(a) Fewer people able to afford housing;

(b) Higher vacancy rates;

(c) No appreciable increase over the original estimate of needed low and moderate income housing units;

(d) A lesson learned — "more" is not necessarily "better".

Government housing agencies in Hawaii have yet to demonstrate the capability to provide housing in large volume, especially low-income housing. We should not expect them to. Among others, the public sector is faced with the following trends and constraints: costs and inflation spiral upward while public revenue does not increase concomitantly; existing housing appreciates in value rather than filtering down; housing is not delivered according to consumer need and preference; public programs cannot effectively subsidize low-cost housing in volume; and, government programs tend to squander available community resources that could be an asset to any housing program.

Government agencies have traditionally been the scape-goat for our housing woes. They are certainly an easy target for critics to hit. Yet government, with all of its constraints, probably produces housing at its optimum level.

Maybe we need to take an overdue look at a major, yet neglected factor in the housing problem — that of the public's tastes. We live in a highly consumption-oriented society where the "American Dream" has evolved into the equivalent of a $100,000 house. We have many families of four living in five and six bedroom houses. Our tastes and preferences are galloping ahead of our financial capability. Now we are going to have to adjust to the fact that the harsh economic reality of our time will no longer permit our reach to exceed our grasp, at least not if we really want to own housing.

The only example of low-cost housing ($8,000-$20,000) being developed is done by individuals on their own initiative, and by private non-profit agencies. During 1976, 1,326 units were built in this one-to-one manner, according to statistics from the Honolulu Building Department. While this number does not make a significant impact on Hawaii's total low-cost housing need, it does compare favorably with the total units of housing built by the public sector.

One major reason for this phenomenon is simply that it is cheaper to build housing than to buy housing, especially if part of the construction is self-help. Thus, there is low cost housing built, on a modest scale, by individuals on their own. There are housing organizations which provide one-to-one technical assistance in the development of housing, especially self-help. But these organizations are severely handicapped by a lack of funding, and by the absence of institutional support from the public sector.

A brief look at the recent evolution of housing would be instructive to determine how we have reached our current housing dilemma. Past policy has been one of rapid economic development, in that attempts were made to raise the level of population, employment, business activity, and construction. During the 60s this policy was beneficial — today, however, it is catastrophic. Our limits to growth are emerging — population, energy, natural resources and pollution are particular areas of concern. We behave as if the limits of our environment were endless.

Between now and the year 2000 we will be an island system which is going to be under great strain. Our present annual growth rate of 2.3% exceeds the environmental capacity of the state. As this trend continues, adjustments to our way of life are going to be forced upon us.

It is important to realize that nobody *makes* policy — the current situation dictates housing policy to us and we simply adjust to it. One of the adjustments that we are going to make is to realize that things are not going to improve unless we get innovative in a hurry. We need to develop new alternatives and prod the "system" with them. In particular, we need to examine housing with a new perception as a social problem, not as a physical one.

Nevertheless, we can see several trends and factors which are going to influence our future housing delivery system. Public outcry will dictate that we change our present unilateral housing delivery process. There will be more needs assessment and input from the local community. We will make more use of community resources, both physical and human, and we will emphasize what to build and for whom to build. We will see more "shell" structures built at low cost. We will see more innovative financing methods. We will see a dramatic reduction in the dollar level of housing demand — personal tastes will change, from highly-consumptive to more self-sufficient. Ultimately, in the final analysis, we are going to build less and use less.

6.2 OVERVIEW

The basic housing problem in Hawaii is no different from elsewhere in the world — the average person simply cannot afford to pay for basic housing needs. While the cause of this global problem may vary somewhat from country to country, the result is the same. People cannot afford, for whatever reason, to purchase housing that is adequate to meet their needs. Therefore, it is useful to analyze the experience of areas where industrialization and technology of housing are considerably advanced, such as Europe, Japan, and North America (especially Hawaii), in order to determine not only to what extent housing technology is applicable, but what the social, political, and economic constraints have been, how the market has been aggregated, and how the whole process has been organized and managed.

Housing delivery in Hawaii has always been a market function, i.e., private developers have built and marketed only what they could sell, while national and local government have made relatively little effort to enter the housing field. People with a need for housing have had one basic option: purchase housing at the "fair market value", or rent housing. Over the past 15 years, rising income, coupled with rapid population growth and spiraling inflation, have combined to triple and, in some cases, quadruple the cost of housing. New development has proceeded with the assumption that the market demand for new housing would continue unchecked. The recession, which began in earnest in 1974, caught everyone by surprise. Sources of conventional funding dried-up and interest rates soared. New house prices and interest rates froze an estimated 85% of Hawaii's homeowners out of the market. While owner-occupants were hurt, it was the investor who received the most damage from the recession.

Hawaii has always been heavily dependent upon the investor and speculator for housing construction. In the market function, the investor has been the guarantor of funding for new housing development (estimates are that over half of Hawaii's housing is owned by other than occupants), i.e., the investor purchases new housing before it is physically constructed and subsequently rents or offers the unit for resale. However, the recession made it difficult for the investor to get money, except at huge interest rates. The end result was that the investor got scared, and felt that the effective demand for housing was not as great as it had been earlier.

This role of the investor is pivotal in supporting Hawaii's real estate industry. Investors buy condominiums, and condominiums add many units at once to the housing stock. Housing demand decreased after the recession, i.e., in early 1976 vacant units (mostly condominiums) totalled 14,300 out of a total condominium inventory of approximately 210,000. Investors withdrew from condominiums which resulted in an overall slump in the total housing market. Developers of new housing refused to risk new construction until existing housing was sold, and banks and other financial institutions started to more carefully scrutinize new housing proposals.

It is interesting to note that between 1974 and 1975 building permits for single-family housing construction dropped 32% but multi-family units, especially condominiums, dropped 62%. For the first quarter of 1976, building permits for single-family housing are on the upswing while the multi-family trend continues (off 51%).[1] Thus, there is a rather large housing surplus at present, and it appears that the market is improving. Honolulu County officials estimate that within a year the housing surplus will be reduced to manageable proportions — three units are being sold for each unit that is being built.

While it is difficult to determine the new housing development pattern at this time, there are certain trends that are emerging. For example, while buyers and investors are entering the market, fringe and low-financed developers are being phased-out by lenders who got hurt during the recession. Currently, both the rental and resale markets are weak, i.e., investors with small amounts of capital tied-up in a project are simply withdrawing. In addition, with more mortgage money available and at lower interest rates, buyers are more particular about the type of housing that they want. When prices of condominiums and single-family units become comparable there is an overwhelming trend toward the purchase of single-family homes. The reasons for this phenomenon are unclear — however, it does appear that, at comparable prices, people of all income levels would rather have a yard than communal amenities.

For the low income people of Hawaii there is virtually no hope for adequate housing, other than rental units. Home ownership is an impossibility. The low income folks are simply priced out of the market.

Housing in Hawaii does not filter down — instead it appreciates. The sole source of subsidized housing programs for the low income is through federal and/or local government housing agencies. Chief among these is the Hawaii Housing Authority, the state housing agency. Since 1970, when Act 105 was adopted, the State has attempted to work with private developers to create home ownership and rental plans for low income people. As of June 1976, the HHA had completed over 4,300 units with an additional 2,000 under construction. At the start of 1976, almost 800 units of HHA-built housing were vacant, although the total is presently substantially less. While there are no doubt many reasons to account for this high vacancy rate, probably the chief reason is cost: the average unit of "low cost housing" for low income people sells at over $30,000. Low income people obviously cannot afford to pay this price. As a result, the State has quit trying to build low cost housing and has turned its attention to providing housing for the "gap group" (those people with a gross annual income between $12,000-$20,000).

6.3 DEMOGRAPHICS

The Hawaii Housing Authority estimated in 1970 that there was a need at that time for over 50,000 units of low and moderate income housing throughout the State. Between 1970 and the end of 1976 over 80,000 total housing units have been built,[2] very few of which have been low and moderate income units (largely due to the fact that the price of a single-family unit rose at more than twice the rate of average family income during this period). By and large, the inability to construct and market low and moderate-income housing has been the result of economic constraints, rather than a failure to recognize the need for low-cost housing.

While statistics which officially define income levels into "low", "moderate", and "upper" are open to interpretation, there are almost 89,000 people who receive some form of public assistance at a total cost of approximately $4,260,000 per month.[3] Thus, it would be reasonable to suggest that there are at least 89,000 low or moderate-income people out of a total state-wide population of over 750,000. Many of these people (and there are probably many more) live in dilapidated, overcrowded, costly, and otherwise substandard or inadequate housing. Furthermore, the cost of housing constitutes approximately 45% of all public assistance expenditures. The Office of Research and Statistics at the Department of Social Services and Housing estimates that a total of $32.3 million was allocated for housing costs during Fiscal Year 1973.

Housing is one of the most neglected social problems in the State of Hawaii. The demand for adequate low cost housing has outraced the supply and the shortage is increasing. In short, the housing problem is getting worse and not very much is being done about it. Population projections are alarming, especially in light of the fact that the State cannot keep pace with the present need. Recent projections by the United States Bureau of Economic Analysis anticipates that a state-wide growth (exclusive of Navy crews) to a population of 886,100 in 1980, 1,030,500 in 1990, and 1,178,000 in 2000 will take place. The Honolulu City and County General Plan projects a population of one million on Oahu by the year 2000. These projections carry significant implications for persons concerned with Hawaii's housing shortage. They indicate that new residential development will find it difficult just to keep up with the current growth, much less catch up with existing deficiencies.

According to the 1970 Census there are 730,095 people living in households throughout the State. Home ownership rates have always been low but are increasing significantly:

Percent of Housing Units Occupied by Owners

1910 - 13.0	1960 - 41.1
1940 - 25.4	1970 - 46.9

According to the Census, in 1970 only 5.6% of all units (approximately 10,000) lacked some or all plumbing facilities. It may be useful to note that the traditional definition of "substandard" housing is in terms of minimum quality based upon two aspects — the presence or absence of certain plumbing facilities such as hot and cold running water, tub/shower, and flush toilet — and the structural condition, which is either sound or dilapidated. The classic comment on the limited character of this definition is that "a nearly weathertight box with pipes in it" constitutes a "standard" house. My conclusion is that there are considerably more than 10,000 substandard unites throughout the State because the importance of plumbing in this definition has virtually assured the elimination of "substandard" housing from the statistics by making it identical with the spread of tubs and toilets.

Of this 5.6% figure, only 3.3% of the units on Oahu lacked some or all plumbing facilities; however, 15% of all housing on the more rural Neighbor Islands lacked some or all plumbing facilities. This figure is particularly alarming since, as noted above, there may very well be a much larger number of houses which are indeed "substandard".

Greater attention has been focused recently on "crowding" in housing. It has become accepted that an average of more than one person per room in a housing unit is undesirable and reflects need. According to the 1970 Census, approximately 20% of all occupied units in the State had 1.01 or more persons per room, with the median number of rooms in single-family structures being 4.6. Geographically, the areas with the highest percentage of overcrowding were: Nanakuli — 47% overcrowded; Laie — 44%; Waimanalo — 39.4%; and Hauula — 39.3%. All of these areas share a common high percentage of low income residents.

The cost of housing is extremely high. The median value of owner-occupied units in 1970 was $35,100 state-wide. Of this total, Honolulu showed a median value of $43,200; Oahu — $38,100; and the Neighbor Islands — $24,600. Vacancy rates were low. Over the last ten years housing costs have increased more rapidly than incomes. At the end of 1971 the construction cost per square foot in Hawaii was $21.88 whereas the U.S. national average was $15.27. Furthermore, new homes insured under FHA Section 203 loans, as of December 1971, averaged $45,048 for Hawaii as compared to a national average of $24,749.

The minimum standard continues, however, to be a limited one and to ignore the major aspect of the problem — the relationship of housing costs to ability to pay. Federal housing programs recognize, if the statistics do not always, that a household paying an unusually high (over 25%) percentage of its income for housing which may meet other minimum standards still represents a need for assistance. In general, the ratio of housing costs to ability to pay exhibits a regressive pattern. FHA borrowers with incomes of $7,000 or less are likely to pay 30% or more of that income for housing, while those with incomes of $14,000 and above are likely to have housing costs which do not exceed 20% of their income.[4] By this definition, indications are that the number of households which cannot afford housing of average quality without devoting too large a share of their income to its cost may be 3-5 times as high as the number currently living in substandard or overcrowded units.

As noted earlier in this section, a study by Daly & Associates revealed that since 1970, single-family housing prices have increased by 85% while average household income has gone up by only 34%. While 80,000 units have been constructed state-wide during the past seven years, much of this housing has been priced beyond the means of the average person. At present, as much as 85% of the total population of the State of Hawaii cannot afford the type of housing that they want, where they want it, without devoting an excessive amount of their income toward obtaining it. Statistics suggest that a substantial number of people in Hawaii — particularly those with low or moderate income — are affected by the housing situation.

Changes in Hawaii's Housing Supply. 1975-76

A survey of changes in Hawaii's housing inventory during the twelve month period ended on March 31, 1976 reveals the following highlights:

(a) The number of housing units in the State increased by 9,449, bringing the total to 270,646;

(b) Net increase on Oahu amounted to 6,241 housing units, raising the inventory to 214,078 units by April 1, 1976;

(c) Net additions to the inventory were well below the record numbers (10,500 for the State) reported for 1974-75;

(d) Private housing increased at a somewhat more rapid rate than housing units under government control.[5]

The net changes indicated by the above data were reduced by adjustment factors to allow for structures authorized but never built, units intended for non-resident occupancy, and unreported demolitions. Data on authorized additions to and deletions from the housing stock tend to overstate net growth to a significant degree. Past experience indicates that demolitions are generally under-reported and that many building permits are left unused. Actual net increase amounted to only 64.5% of authorized net growth between 1960 and 1970.[6]

Statistics compiled by the Hawaii Housing Authority, while limited to lower-income families, show that vacancy rates are higher and turnover is increasing slightly. The number of families applying for housing on Oahu dropped from 5,655 in 1974 to 3,970 in 1975 on the HHA Listing. There were 4,217 on the HHA Oahu waiting list at the end of 1974 and only 3,291 a year later. Annual turnover on Oahu amounted to 665 (out of 4,532) in 1975. Thus, housing is becoming more readily available, to at least some extent — although not necessarily at prices easily paid by most Island households. (More extensive coverage of housing costs can be found in the Appendix.)

One of the more striking features of recent residential construction in Hawaii has been the high proportion of units in high-rise structures. Hawaii is increasingly growing upward, rather than outward, as land use becomes more intense. As a result, elevators have become one of the most common forms of transportation in the Islands. The number of Honolulu housing units in structures with four or more floors rose from 1,728 (or 2.1% of the inventory) in 1960 to 15,413 (or 15%) in 1970. For all

Oahu, 9.3% of the housing inventory in 1970 was in structures with four or more floors. This total has no doubt increased dramatically between 1970 and 1977, due to the recent proliferation of condominium construction.

6.4 FACTORS

The Hawaii Housing Authority estimates that there is a need for at least 82,000 total housing units throughout the State between 1977 and 1985. At present, the City and County of Honolulu indicates a housing need, at all income levels, of 5,300-9,200 dwelling units per year through 1985. There are other estimates of Hawaii's total housing need — suffice it to say that a total of 82,000 units may be a bare minimum of housing units needed at this time. The factors involved in this production of housing can be reduced, at the risk of oversimplification, to the following: land, labor, technology, and financing.

Land and construction costs are the major drawbacks preventing private agencies and firms from having a significant impact on the housing problem. No State agency, or other public organization, exists which could comprehensively acquire land, hire architects and planners, choose builders and developers, arrange subsidies, issue building permits, override zoning and building codes, make mortgage loans, conduct closings, and perform related services including advancing front money for these purposes. Furthermore, there is no public agency which works with, and is responsive to, the low income people who are most adversely affected by the housing shortage.

While Act 105 gives the Hawaii Housing Authority the power to go into joint ventures with private home builders, the cost is prohibitive (HHA estimates that the cost of their typical three bedroom house is approximately $31,000). Private industry is not eager to enter into joint ventures with the State because private industry will build and market only what it can sell. Private developers have a justified fear of a low profit margin and government red tape. In short, the profit factors discourage private developers.

Families, especially those of low and moderate income, cannot obtain mortgage money from banks and savings and loans because lenders insist that the borrower be a "good" credit risk. Low and moderate income people in particular are considered to be poor credit risks unless there is some agency available to guarantee the loan. Therefore, a valid sentiment is expressed by many local housing experts who concede that there is really no such thing as "low-cost housing".

Land

Between 1950 and 1975 land on the island of Oahu increased in price approximately 16 times while residents' personal incomes went up only 4.2 times. On a Statewide basis, land prices increased 17 times. The end result is that people are increasingly less able to pay for the land on which they live. Consequently, housing development is largely stymied due to the high cost of fee simple land.

Nor is leasehold land exempt from high cost. There has been a great deal of controversy recently over fair return on the value of leasehold land that is paid in rent by residential leaseholders. This controversy has created a furor that in some quarters has resulted in a cry for land reform, specifically, the conversion of leasehold land held by Hawaii's estates and trusts into fee simple land at a fair market value. The basis of this controversy is the fact that 34% of the living units on Oahu are on leasehold land. Furthermore, the number of units on leasehold land has increased five times since 1960 while the number of units on fee simple land has increased only 1.6 times. In addition, the 1975 tax appraisal increased the value of the land and lowered the value of the houses, which adversely affected the resale market. Therefore, an owner pays more when buying the land while getting less when selling the house.

While land reform on a massive scale appears to be somewhere in the future (if it happens at all), some of the larger trusts have agreed to sell certain parcels of land in fee simple. It remains to be seen what effect this action is going to have on future land values. It may increase speculation and sales. Or it may result in owners not being able to afford to sell their homes.

Yet the fact remains that the high cost of land is the single crucial deterrent to the construction of housing. Land is not yet readily available from Hawaii's estates and trusts, and escalating land values lead current landowners to hold what they own on the assumption that their land will steadily appreciate in value. While it may be cheaper to build new housing, rather than buy new housing, the average resident still cannot afford to purchase land. The price that one must pay for fee simple land is simply beyond the means of the average resident.

Labor

The Hawaii Community Design Center, in its self-help enterprises, has estimated that the elimination of labor (including contractors and developers) can save a home-builder up to 40% of the construction cost of a house. Through collective bargaining and other means, Hawaii's labor unions have escalated the cost of labor in the housing industry beyond the ability of the average consumer to pay. As a result, the construction industry in Hawaii faces record unemployment numbers, with little hope that the depressed market will ever again reach the employment levels of the late 1960s. Although land has had the primary effect on depressing the housing market, the high cost of labor has also contributed to the total price of a new dwelling unit. It is somehow ironic that past demands for higher wages and better working conditions has resulted in Hawaii's labor costing itself out of the housing market.

Technology

It has become accepted that construction materials and methods technology has reached its zenith, i.e., there are no new materials nor methods which could substantially reduce housing costs. Costs are fixed for materials, substantially stable for labor, and beyond the average person's reach for land — other than self-help, the cost of a housing unit is limited within certain predetermined boundaries. None of these factors are likely to be reduced significantly, at least not by technology or design. Increasing attention has been focused by housing experts on innovative financing as the sole method of pricing housing within the range of the consumer.

A major hindrance to the use of innovative technology has been the refusal of the average consumer to readily accept bold new housing designs and materials which deviate from convention. An example is the dome — numerous attempts to market this concept have fallen by the wayside due to the steadfast refusal of the consumer to accept this and other new designs, regardless of the merits or inadequacies of the product.

Increasing attention has been devoted in recent months to minor technological concepts such as the minimal structure, similar to the old "shell house" concept, which can be built at a relatively low cost, short time, and small dimension, and then expanded and improved upon as the occupant family and its finances grow. In particular, attention has been focused on the urban house which is designed for expansion upward, rather than outward. As yet, however, the concept of expandable housing has not been implemented in Hawaii.

Financing

Increasing attention has been devoted to innovative financing methods which could, theoretically, reduce housing costs to a level where the average consumer could afford to purchase a dwelling unit. On Oahu during 1975 less than 15% of all families earned enough to qualify for a conventional loan that would buy a median-priced $60,000 condominium. Furthermore, less than 10% on Oahu earned enough during 1975 to qualify for a conventional loan which would purchase a median-priced $83,000 single-family home.

As a result, the real estate market during 1975 "softened", especially in condominiums where there was a surplus of units. Mortgage foreclosures increased noticeably. Most important, during 1975 the percentage of owner-occupied housing decreased from 46% to 44.3% — the first time in over 50 years that the percentage did not rise.

The combination of escalating housing costs coupled with the reluctance of lenders to take risks has resulted in Hawaii's worst housing crisis in history. Developers with a surplus of housing, especially condominiums, have resorted to requiring no down payment and other innovative financing concepts in the hope that buyers, particularly speculators, can be attracted. The public housing agencies have done little to discover new financing methods for home ownership other than the limited number of traditional housing subsidies and "buy back" provisions for a few moderate income people. Thus, "tight money" has combined with high land prices, expensive labor, and inflation, to price a potential 85% of consumers out of the housing market.

6.5 PRIVATE NON-PROFIT HOUSING DEVELOPMENT

There is an interesting, widespread phenomenon known as the private non-profit housing corporation. It is unique, at least in scope, to North America. While it may take the form of a "community design center" or a "community development corporation", its primary function is to provide housing for low income people. In addition, this type of corporation takes on other roles, such as:

(a) Provides technical assistance (planning, design, construction, and financing) to those individuals and organizations which cannot afford these services commercially. In effect, it provides free services to the poor.

(b) Accepts an advocacy role for the clients that it serves. It involves citizen input into planning and housing-related matters in the community, and acts as liaison between the community and the bureaucrats. It assumes that people who live in any given community have skills and knowledge that are useful to planning and housing, and endeavors to refine and promote these skills.

(c) Attempts to provide housing at cost, usually through self-help. Elimination of profit enables housing to be built at reduced costs to the consumer.

(d) Acts as enablers of innovative housing ideas to government. It can suggest, but not force, government agencies to change housing policies, where appropriate.

There are several other features which these private corporations share. First, they constantly struggle for survival due to inadequate and sporadic funding. Since the cost of services cannot be passed along to the consumer, these organizations must rely upon government grant funding and on the generosity of private trusts and foundations. Money is not always available. Next, they are forced to muddle through many problems in which no precedent has been set. Frequently, innovative solutions are required where traditional approaches have failed, and the element of risk is always implicit. Projects fail, or sometimes are not completed. Finally, they tend to project a "renegade" image for several reasons: the traits that make them acceptable to the low income community tend to make more staid people nervous; they are by and large of a younger generation; they do not always follow the rules; and under-capitalization is not conducive to high fashion. But they have been known to provide at least partial solutions to some of our housing problems, as some examples will demonstrate.

The Hawaii Community Design Center, Ltd.

The Hawaii Community Design Center provides architectural, housing, planning, and educational services to low income individuals and non-profit organizations throughout the State. The preeminent objective of the Design Center is to provide the technical assistance necessary to translate community goals into reality.

In order to be eligible for Design Center services, the individual or community group must demonstrate that they cannot afford to pay for commercial services, and that the project's use or its aesthetics are for the common good. The client pays a nominal fee when the application is accepted, and may be assessed for the cost of materials beyond that point. Otherwise, the client does not pay for the services.

The Design Center was established in 1969 by the Hawaii Chapter of the American Institute of Architects and the Department of Architecture at the University of Hawaii. Impetus was provided by other AIA chapters across the United States which organized other community design centers. Today there are 83 CDCs nationwide.

Originally located at the University, the Hawaii CDC received its first full-time staff with the arrival of three VISTA Volunteers in March 1973. Shortly thereafter, the Design Center incorporated and moved to the Waimanalo Human Services Center in order to better serve the needs of the low income people of the area. Construction of the $5,500 "Minimim House" followed in April 1974. This project, for which HCDC is most widely renown, became a springboard for Design Center involvement in low-cost housing throughout the State. Other self-help houses were constructed at low cost, and the Design Center subsequently expanded state-wide, opening offices on the Big Island and on Kauai.

Architectural, planning and housing projects grew in number from 20 in 1973, to 31 in 1974, to over 59 in 1975. HCDC has recently participated in the planning phase of a 47-acre Wailua park; provided a land use plan for 244 units of self-help housing in Kahuku; remodeling of the Koa pre-school and other day care center renovation projects; assisted community groups in opposition to publically-planned projects which would have an adverse effect on their communities; and, worked with community groups interested in constructing their own housing as an alternative to public housing or relocation. Other projects include housing rehabilitation, community building designs, playground designs, graphics, and other planning activities.

The Design Center is able to perform these services at a fraction of the cost of a private consultant due to maximum utilization of volunteers, both professional architects and planners, as well as VISTA Volunteers from the federal ACTION agency. The Design Center is able to develop programs and expand its staff of professionals without an increase in operating costs. Thus, HCDC is uniquely suited to coordinate community objectives with government and other public programs.

The Minimim House

Self-help housing provides the only viable alternative for the people of Hawaii who cannot afford adequate housing at existing prices. The Design Center was contacted by a local organization known as "The Hawaiians" in March 1973 to determine the nature of housing services which HCDC could provide to the Hawaiian people. This organization wanted to know if the Design Center could develop a low-cost housing system which could be utilized by low income Hawaiian people. The basic problem was three-fold: residents could not afford existing housing prices; they had no input into the design of their own houses; and, the residents wanted to build the houses themselves, to save costs.

Fig. 6.1. How To Build Your Own House.

Realizing that the basic problem was actually state-wide in scope, the Design Center set out to implement a housing system that would have a positive impact on the housing problems of low-income residents throughout the State. HCDC investigated various housing alternatives — none were satisfactory, so the Minimim House was created and plans were developed for implementation. It was decided that the experimental/demonstration three bedroom house would be built on the grounds of the Waimanalo Human Service Center with $5,500 grant from the Governor's Task Force on Progressive Neighborhoods and the Hawaii Housing Authority.

Fig. 6.2. Minimim House.

Fig. 6.3. Variation of Minimim House.

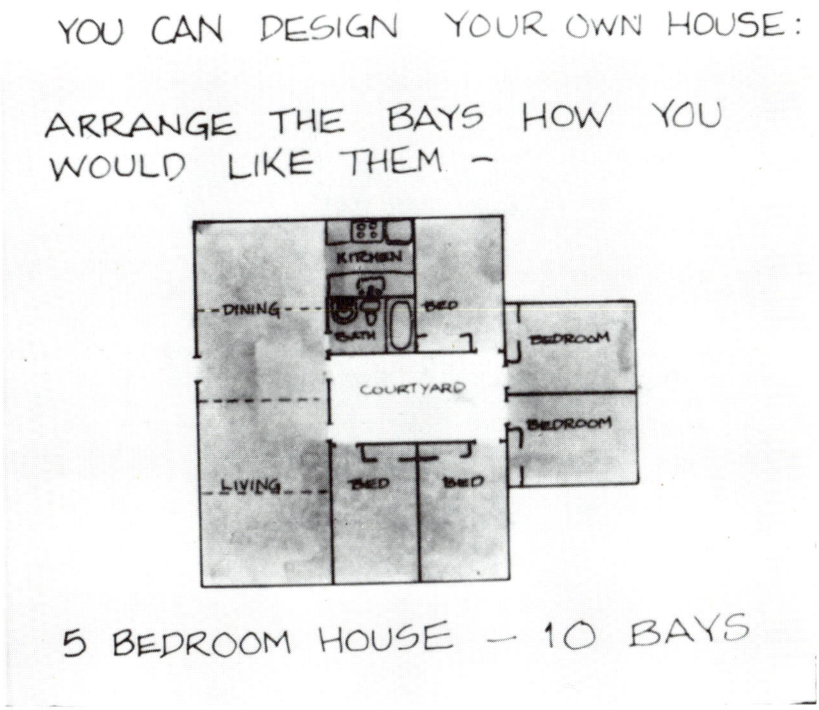

Fig. 6.4. Minimim House Floor Plan.

The Minimim House is a single-wall post and beam wood house, designed to be easy to construct and maintain. The three bedroom house, which meets all building codes, is available to anyone. Design of the house is based upon the 8' x 12' modular system — the three bedroom house is 8 modules, or 768 ft^2 (roughly 71 m^2). The utility of the modular system is that it enables each family to choose its own floor plan and make other design changes to suit their own needs. Furthermore, the house is expandable, i.e., a family can build a small house now and expand it later when finances permit. A Building Manual showing the step-by-step process of construction has been compiled, allowing a person with no construction experience to build his own home by following a simplified process. Construction and work teams are organized in the "ohana" style (extended family and friends helping friends), a cultural situation indigeneous to the Hawaiian Islands.

The low-cost of $5,500 was due to a simplified process of construction which eliminates the needless over-reinforcement of conventional housing. Costs are further cut by the reduction of material waste. Since lumber comes in stock lengths of 8', 12' and 16', and since the modular system is 8' by 12', there is less cutting of lumber involved. This also reduces labor time, as well. Total construction time for the project was 994 man-hours, or 25 days of part-time effort.

Since completion of the Minimim House prototype other models have been built. Each varies according to the needs and finances of the owner, but the basic floor plan and construction process remains substantially the same. The Design Center offers a standard package of individualized floor plans and a Building Manual to low-income people for a nominal charge. In those cases where a client feels that the system does not meet his needs, individual house design is arranged.

Completion of the Minimim House was followed by a mixed response from the island community. For some, the Minimim House met their needs and finances, and was the first such house to do so. For others, the aesthetics of the house were, to say the least, unappealing. Somehow lost in the shuffle was the project goal of creating a demonstration house that would clearly show the minimal structural requirements of Oahu's building codes. While the intent was that the house could be (and should be) upgraded, many people could not visualize the aesthetic change that a different roof, glass windows, and wood stain would make. The lesson that was learned for future experimental designs was to "cheesecake" the final product, in order that the lay person could more readily envision all of the design options, and their cost.

Following completion of the Minimim House, the Design Center assumed that interest in low-cost housing from the public sector would be keen. Instead, the merits and impact of the house were apparently lost on Hawaii Housing Authority and others. Ironically, three years after completion of the Minimim House, there have been policy rumblings from HHA to the effect that maybe what is needed in housing is a "shell house" concept, where contractors would provide a basically complete structure which the occupant family would finish according to their own needs and tastes. Should this be adopted as policy, there are many in the State who feel that the public sector will have reached a milestone in its efforts to construct low-cost housing.

Labor Union Housing

There are several other organizations throughout the State which actively deliver housing to low and moderate-income people. The International Longshoreman's and Warehouseman's Union (ILWU) has been especially effective at assisting its members in organizing their own low-cost housing programs. In general, ILWU members are former plantation workers (sugar or pineapple) whose jobs have been phased out or reduced. Most are older people with fixed incomes who are desperately in need of new housing. They place great emphasis on maintaining their life-style and their community, i.e., they want to ensure that they have maximum input into the type and cost of housing that they will receive. In most cases, the landowners recognize the contribution that these people have made over the years, and are reasonably amenable to seeing that the people receive financial and technical assistance in the replacement of their housing.

In general, the housing goal of the ILWU has been the replacement of plantation housing with consolidated subdivision housing (usually single-family units). The ILWU housing program has been successful, but the replacement of plantation housing with consolidated subdivision housing has not been as successful elsewhere as it has been in Hawaii. There is no parallel phenomenon — not with housing for migrant workers on the mainland, not with housing for rural workers in the South, nor the lumber camps, nor the coal mining areas. The ILWU housing success story is unique.

Organization of the program has illustrated several principles which could serve as a model for the delivery of low cost housing. First, an equitable agreement has been made between all parties on the scheduled phase-out of plantation camps. Second, there are no forced moves — each family has a choice as to what type of housing and its location, and can move when they want. Third, there is great effort made to guarantee some form of financial subsidy in order that the residents will be able to meet the new higher housing costs. In fact, the program is made as financially attractive as possible, and in most cases the type and method of financing usually determines the nature of the housing design during the planning stages of the project.

Furthermore, there is a comprehensive educational package offered by the ILWU, in which the managers of successful housing programs share their experiences with groups in other parts of the State who are embarking on their own housing programs. If a community group is starting a new housing project, they are able to go to other ILWU groups for assistance. All of these community groups affiliated with the ILWU demand maximum resident input into all phases of the housing program. The people themselves plan the nature of their future housing and community. The ILWU provides counseling services and physical models in which people can visualize the type of housing that they need. This type of housing program has been enormously successful because it makes maximum utilization of the expertise of the people affected — the residents themselves.

One of the lessons of the ILWU housing program is that those who have adequate housing tend to speak in terms of "quality of life", while those who have inadequate housing look for the basics. Thus, the concept of smaller lots and modest homes is being accepted with increasing frequency. Furthermore, great attention is devoted to the community as a whole — most replacement housing is available only for the current residents, and not for outsiders. The net result is simple but durable housing designs at relative low cost for low and moderate income people who wish to maintain their rural and ethnic life-styles. They have been successful largely due to their cooperative efforts and organization.

A typical example of this type of housing is the Kahuku Housing Corporation. Kahuku housing is a group of former sugar plantation workers who have a need for 244 units of replacement housing, located on a 40 acre site on Oahu's North Shore. Campbell Estate, the landowner, has agreed to negotiate a new 40-year lease for the property. The Housing Corporation contacted the Hawaii Community Design Center for a land use plan for the 244 units. At present, the Corporation and Campbell Estate are ironing out the fine points of the new lease. The Corporation is considering a self-help or modified self-help (partially subcontracted) program for the replacement housing. The residents of the area are consulted and must agree on all matters concerning their future. With a new lease the Corporation will be able to arrange financing for its members. General agreement has been reached that the replacement housing will conform in design to existing units, that there will be adequate open space on each lot for gardening and chicken-raising, and that the entire community will have recreational and open space. In addition, special housing is being built for senior citizens.

This type of housing program is not unique in the State of Hawaii. There are other community groups, such as Kahaluu and Ota Camp, who are taking affirmative action in cooperation with public agencies to generate housing programs which meet the needs of their respective communities. The trait that these programs share is that they work — and they work because they involve maximum participation and cooperative effort from community residents. One lesson is clear: if low-cost housing for low income people is to be a reality in Hawaii, it will depend upon well-organized, knowledgeable community groups who can pressure public agencies and private landowners into allowing them to develop their own housing programs

Mokauea Fisherman's Village

Another organization, separate from the ILWU, which also aims at developing its own housing to preserve its cultural life-style is the Mokauea Fisherman's Association. This organization, the last subsistence fishing village on Oahu, has existed for years. The people are essentially squatters, as title to the property has never been established. In 1974, during the construction of the Honolulu Airport reef runway, the State tried to evict the people. Legal action proved unsatisfactory, and during this time 5 of the 14 houses on the island somehow burned down. The fishermen threatened the State with a lawsuit unless the housing was replaced. The State government, through the Governor's Office, agreed to provide funds for replacement for approximately 40 people.

The community organization contacted the Design Center to provide assistance in the design of the houses. The fishermen distrusted the State's track record in the provision of housing, at least housing that was sensitive to their unique needs. The community wanted input into the design of their houses, and wanted to maintain their life-style. While the people of Mokauea Island had no electricity, water, or sewage, the State indicated that it would be willing to install a water line out to the island.

The design that HCDC provided for the fishermen included five single-family units, estimated at approximately $15,000 per unit, built on a self-help basis with State funds. The houses would be situated on the water, with a pier as an integral part of each house (a large part of the cost per unit). In addition, each house would be designed to function as a work space. Each house would have a low-pitched roof for the drying of nets and fish. Also proposed were independent energy-efficient utility systems, such as solar water heaters and a Clivus waste disposal system. Thus, five houses (of four bedrooms each) will be built at a total cost, including utility systems, of roughly $75,000.

Apparently, low cost housing in Hawaii is a reality only when community groups take matters into their own hands and use innovative construction methods, design, and financing (and do most of the work themselves) to guarantee housing that meets their particular needs and life-style. It remains a mystery why State and City and County housing agencies do not make use of the resources within each community to cut costs and produce housing which meets the needs of the people. Experiments have shown that the self-help concept is viable — it is up to the public sector to make optimum use of this housing concept.

The Hawaiian Energy House

Increasing attention at cutting costs of housing has been devoted to simplified design and alternative energy systems (particularly the latter) on the part of public agencies. The Hawaiian Energy House, a good simple house modified to the local climate and life-style, has been developed by Professor Jim Pearson at the University of Hawaii Department of Architecture. Based upon standard Hawaiian architecture, this house is a single-wall post and beam simply-framed wood house which can be developed at low cost. Total size is 1,260 ft^2 based upon a 30' x 42' modular system. Interior size is 3 bedrooms, 1 1/2 baths, easily expandable for an additional bedroom beneath the roof. Features of this unique house include:

(a) Facing south for maximum exposure for solar collectors;

(b) Narrow east-west exposure to low sun for better heat control;

(c) Carport on the west to block the sun;

(d) Two-storey design to compact the roof area (for less heat gain) and to use less land area;

(e) Longer roof overhangs to keep the sun off of the interior;

(f) Roof is reflective (white enameled aluminum), light weight, and well-ventilated (including a hole in the roof, with a flat rain lid, to eliminate hot air);

(g) Sliding walls with mosquito screening to take advantage of the gentle climate;

(h) House itself is off the ground which adds cooling while eliminating ground termites and dampness;

(i) Bedroom jalousies are low at bed level for cross-ventilation;

(j) Solar water heater which provides 100% of the hot water, and which saves as much as 50% of the cost of electricity;

(k) Wind generator which powers the lights (although it is expensive for the average home at $3,300);

(l) All appliances, lights, plumbing and fixtures are designed to save electricity and water;

(m) The landscaping cools, provides cover, and produces edible food.

This innovative house demonstrates realistic methods for cutting costs, both through its simplified design as well as its alternative energy systems. It is this latter factor that is receiving increasing attention in Hawaii, as a means to cut costs.

6.6 NEEDS

Government housing programs have generated many more failures than successes in the construction of housing. The major cause for this phenomenon is that we simply do not have adequate information on housing needs. There is no foolproof manner in which to obtain data that could be used to match public housing programs with housing demand. It seems that there are five areas in particular which need to be examined by housing policy-makers:

(a) Why people live where they do;

(b) Changing attitudes and life-styles which will produce effective demand for future housing;

(c) A Price Index on housing amenities (i.e., how much people are likely to pay for an additional bedroom, carport, pool, etc., based upon their particular values);

(d) Trends that influence potential home owners (i.e., analyzing people throughout the home buying process to establish motivation and preference);

(e) Ways and means to determine how to make multi-family housing demand comparable to single-family demand.

While we are inundated with various statistical and demographic information on housing, nobody has successfully completed a "bottom-up" study, measuring at the community level attitudes, preferences, and needs as perceived by the consumer. Furthermore, little attempt has been made to look at successful neighborhoods (those with little in-out migration) to determine reasons for this phenomenon. Nor has adequate attention been focused on changes in life-style that will affect future housing choice, such as: the tendency for the young and the elderly to live apart from their families; "ohana" (extended family and calabash) living by choice, not necessity; motivation for investment property; dissatisfaction with older or changing neighborhoods; trends and factors in ethnic areas and racial mix; and, estimates of future number and types of households that will form (i.e., divorced singles, doubles, age differentials, etc.).

Certain lessons have been learned over the years which point out potential areas of future study. Neglect of this information, or failure to thoroughly analyze these trends, is one of the primary reasons why housing that is inadequate for people's needs continues to be built. More information is needed, especially concerning the following:

1. Local attitudes determine the success or failure of housing programs — we cannot measure how much to build until we know *what* and *for whom* to build;

2. Information from neighborhood preferences must be evaluated separately from that of housing-type;

3. Traditional definitions of "housing-poor households" (substandard, over-crowded, ability to pay) neglect other factors, i.e., a resident may be "housing poor" if he perceives himself as such;

4. The source of the housing problem is the ability to pay rather than the lack of technology;

5. Housing programs must concentrate on meeting the needs of households, not on a magical number of units;

6. Perhaps housing is over-produced, e.g., there may be no housing "shortage", only fewer single-family units than demanded;

7. Analysis is needed as to why single-family housing is preferred;

8. Instability of income may be a cause of the failure of people to purchase affordable housing, i.e., maybe people *prefer* to rent rather than buy;

9. The issue of housing "standards" must be evaluated in the context of the necessity of building codes and financing requirements;

10. There is a public fear of government subsidized housing, and a social stigma attached to subdivisions built at public expense.

This emphasis on data related to housing needs cannot be overemphasized. Perhaps in no other phase of existence does the human being have deeper emotions than those related to his home. He has spent his life in the home(s), become attached and habituated to it, and is not easily diverted from the way of life associated with it. He is generally strongly conservative in his views concerning what constitutes a home, especially if he is an owner. Custom, preference and prejudice are all strong constraints in the selection of a home. There is no other aspect of housing design that is as difficult and as important as this one. Yet virtually no attention is devoted to this factor. Research is desperately needed to determine what constitutes socially acceptable housing.

The City and County of Honolulu's Department of General Planning has specified the following "Housing Needs and Problems":[7]

1. 5,300-9,200 new dwelling units per year through 1985;

2. Units should be matched by price and size with needs of the whole range of households (although how these "needs" will be determined is not mentioned);

3. More housing should be delivered at minimum cost:

   (a) Direct private development (with land use approvals requiring some units at minimum cost with limited profits);

   (b) Assist private development (with land acquisition, site development, financing, and reduced risks with reduced profit);

   (c) Public development (with public agency as the developer, thus no taxes or profit).

In addition, there are other factors which the public sector could examine to create housing which meets the needs of the people. One method is the standard subsidized housing package, where local government reconciles the ability to pay by the consumer with the fair market value of the housing unit. Another method is the density bonus, where economic incentives are provided for families to live in high density multi-family units. Yet another method is the tax incentive for owner-occupants, or lower taxes for those people who live in their own homes.

There are other efforts which the public sector could make in an attempt to match housing consumption with needs. One method is to assist housing consumers by making low interest mortgage loans available from government bonds, and by providing subsidies to those who cannot afford the minimum cost. Another method is to control units built under public housing programs by: purchase or lease of some units; selection of occupants by specification of income criteria; imposition of owner-occupant and buy-back rules on individual units; and, restriction of future rent increases on rental projects.

It seems realistic to conclude that an analysis of housing needs should be completed before additional housing is constructed, at least at public expense. It is the public sector which is uniquely suited to compile and analyze this data. It would provide an invaluable public service and allow us to collectively address the housing problem in a more realistic manner.

6.7 TRENDS

Although it does not focus on housing alone, the environmental movement in the United States had had an important effect on shaping our attitudes toward housing and related issues. The "environmental movement" itself has been the most effective social movement of our time, especially with regard to the institutional change which it has engendered. More and more Americans are becoming increasingly concerned with the quality of life, with the relationship of man to the air, water, land, flora and fauna. The importance of ecology, and the citizen concern associated with it, can best be illustrated by the following excerpt on the importance of the environmental problem: "Though the word ecology was not coined until the 1870s, the central ecological principle had in fact been operative forever. It is that all aspects of planetary life, from one-celled plants to many-celled mammals, are ultimately connected to one another by a web of cause-and-effect relationships. If one strand is severed — or even jiggled a bit — the disturbance will inevitably be felt and will cause changes throughout the web.... Even today we are only minimally able to predict what the ecological consequences of very simple environmental actions will be."[8]

In the housing field today, much attention has been devoted to the development of low cost ecological building systems which can be adapted to the modern world, and which will provide inexpensive, ecologically-efficient systems in the future. Some of these systems are basically simple, such as building with renewable resources (such as timber or vegetable fibers) which do not consume the earth's stock of non-reusable materials; building with materials which would otherwise cause pollution when discarded as industrial waste, such as sulphur and garbage; and building with indigeneous materials which can be recycled at the end of the building's life. Other methods are being experimented with which involve the use of innovative water and waste disposal systems.

Energy-efficient, self-contained systems are in the experimental stages at this time which, once developed for the mass market, could result in substantial monetary savings as well as providing abundant energy at no ecological cost. Such systems involve simple technology like using wind energy to generate electricity, and making use of solar energy to heat water, to cook, and to purify polluted or sea water. In addition, there are currently smaller methods of cutting costs and eliminating waste which are employed now, such as avoiding waste by relating the size of all building components through modular coordination.

All of these energy-efficient/ecological innovations have implications for future housing development. This concern for ecology has evolved directly from our environmental problems, which in turn have evolved from our common life-style which is based upon high technology and high rates of consumption. Thus, there are questions as to how compatible our traditional economic, social and political systems are with overall environmental stability. In short, our consumptive energies have outrun our ability to make and pay. The upshot is that future housing development, in addition to being affordable, will have to be ecologically sound. We will no longer be able to look at housing cost, technology, and our environment as independent entities. Furthermore, we have no choice in the matter — it is now an economic necessity, in that we either have to pay more or use less. Now that we are putting a price on such things as clean air, pure water, and open land, we also need to put a price on non-development, and poor development of housing, in order to meet the needs of an increasingly less-consumptive population.

It has been estimated that at our present rate of consumption we will deplete our oil resources within the next 35-65 years. Thus, the limits to growth of a petroleum culture have obviously been reached. When we run out of oil we will have an opportunity to make use of new and cleaner energy sources, including solar, geothermal and wind power. Yet development of "exotic" energy sources has been hindered by the lack of research and planning funds, the opposition of conventional power producers, and the shortsightedness of our government bureaucracy. Apparently, problems must reach the danger zone before any effective action is taken. At present, most of our public funding is being poured into employment programs with the result that needed funding for housing and energy programs is simply "not available at this time".

Land use and controls, like energy use and controls, have a direct relationship to housing problems. Land has been inundated with contaminants and toxins, eroded, flooded, covered-over with concrete — any cleanup or restoration will take longer and use more dollars than similar action with water or air. Land use, especially on an island, is an area where it is imperative that we change our pattern of use and abuse. However, changing land use habits may prove even more difficult than changing our oil-consumptive habits. Attempts in Hawaii to impose land controls, restrict private property rights, and transfer estate-held land to the public domain have been controversial, to say the least.

Hawaii can feed and shelter just so many of us, provide us with just so many useful resources, absorb just so much of our waste and poisons, tolerate just so much growth and development. What is disputable is just how close we are to these finite environmental limits. On Oahu, the trade-offs are starting to emerge. We are rapidly reaching our population limit that can be supported by existing water and land use patterns. Our annual growth rate of 2.3% exceeds the environmental capacity of the island. Already the trend in dwelling units is to build upward, rather than outward, and to consolidate multi-family living into a smaller area. Dire predictions have been made concerning when the island will run out of water. Transportation is increasingly becoming a more difficult problem, as are employment, education, and social services. Yet time-controlled growth and other population growth inhibiting policies, successful elsewhere, have not been implemented in the new General Plan which will shape the destiny of Oahu between now and the year 2000.

Regardless of the shape that future housing programs in Hawaii will take, several trends are clear. These will evolve by necessity, whether we like it or not. First, government agencies will reduce housing production efforts, other than provision of a limited amount of housing for the moderate income (the gap) group. While housing production is a role for which the public sector has historically been ill-suited, only now is this reality becoming clear. Second, the public sector will step-up the programs for which it is best-suited — that of providing incentives for private industry and the public to become more active in housing development. This would include tax incentive plans for private developers, provision of public land for private development, and subsidies, density and tax bonus plans for the consumer. Third, we will ultimately use (and build) less, though not for a while yet. Fewer available housing units will act as an unintentional, though viable, limit to population growth — it appears that it will be later, rather than sooner, that we will develop an orderly time-controlled growth policy. Fourth, the limited new housing that is built will gradually begin to incorporate energy-efficient utility and waste disposal systems. Fifth, increasing use of existing housing stock to prolong its life will be instituted through rehabilitation and renovation. Sixth, housing will eventually be placed into its proper perspective as a social problem, and not as a physical or economic one. Finally, the housing *needs* of the consumer will be determined and implemented, to ensure that new housing will not become useless housing.

REFERENCES

1. Housing in Hawaii, *Hawaii Business*, June 1976.

2. *Housing for Hawaii's People*, Daly & Assocs., January 1977.

3. Office of Research & Statistics, Department of Social Services and Housing, January 1975.

4. *Hud Statistical Yearbook, 1970*, pp.214-15.

5. A Statistical and Analytical Report on Housing and Community Development in Honolulu, *Housing and Community Development Research*, Department of Housing and Community Development, City & County of Honolulu, July 1976, p.1.

6. *Redevelopment and Housing Research*, No. 32, July 1972, p.8.

7. *Alternatives for Residential Development*, Department of General Planning, City & County of Honolulu, 1975, p.4.

8. GILBERT, BILL, "My Country, 'Tis of Thee'", *Sports Illustrated*, Vol. 45, No. 25, December 20-27, 1976.

Part II

LOW-COST HOUSING: BY TOPIC

7

MATERIALS

Albert G. H. Dietz

7.1 MATERIALS

Building materials obviously play an important part in any consideration of housing for low to middle-income people in developing areas. The approach to this enormous and complex subject depends upon the objective to be attained. The point of view adopted here is that indigenous materials either now or potentially available must be employed to the fullest possible extent in order to keep costs down, avoid draining limited foreign exchange, and provide employment locally. Furthermore, the construction methods adopted should use these materials in such a way that unskilled and semi-skilled labor, much of it on a self-help basis, can be used to construct the needed housing, thereby reducing costs, providing employment, and at the same time imparting at least some training in construction skills.

In this brief treatment, materials are described that meet these requirements. They are mostly locally available and in use now, or are available now but not widely used, or are potentially available. Among the first group are soil, wood, bamboo, naturally-occurring cementitious materials, and combinations based upon them. In the second category are such materials as sulfur and portland cement (plentiful or scarce, depending upon the region). In the third category are the plastics and related polymeric materials, as well as the composites based upon them. While these are at present largely high-technology products not readily available in most developing areas, many of these same areas have extensive petroleum resources and are building plastics-producing capacity.

The approach taken here is to attempt to explain, simply and briefly, the basic properties of these materials as they stem from their internal structure, without becoming entangled in highly technical considerations. From these properties the principal advantages and limitations of the materials are derived, and some of the typical uses are described. No attempt at detailed completeness is made, rather, it is hoped that with some insight into the properties of the materials, new ideas will be generated respecting ways to use them that will lead to the provision of satisfactory housing.

7.2 SOIL

Soil is probably the most widespread and abundant material of construction. It is in many ways both the easiest and the most difficult to employ.

Soil is widely used in its natural state and, to a much lesser but growing extent, modified by the addition of various other materials. Natural soil is extremely variable, and by no means all of it is suitable for construction, even of small dwellings. The properties needed depend upon the construction method or methods employed, the requirements of the dwelling, and the local climatic conditions.

Methods

Before considering soils and their principal pertinent characteristics for house construction, it is desirable to review briefly the principal construction methods.[1,2] They differ in degrees of advance preparation and methods of placement.

Wattle and daub

A framework of posts and poles supports a matting or other assemblage of woven or otherwise intertwined reeds or sticks. Soil mixed with water to the proper consistency for easy placement is plastered to both sides. A variant consists of a double wall of poles and sticks, with the interspace filled with mud. In either case, as the soil dries it is likely to shrink and cause cracks which must be filled and repaired. The surface mud is also likely to fall away from time to time. Because rain causes erosion, the walls should be protected by wide overhanging roofs and by an applied surface coating more resistant to erosion than the soil. Fairly constant maintenance is to be expected.

Cob

This method, like other techniques employing soil, goes by a variety of names, such as swish in Ghana.[3] In this method, the soil, either used as dug from a pit near the construction site, or consisting of a mixture of soils to obtain the desired characteristics, is kneaded and trodden with water to provide the proper consistency. It is then formed into balls about the size of a football, small and light enough to be tossed by a man on the ground to a second man on the wall. The balls are placed and pounded into a solid mass. The soil is applied in courses, usually about 30-45 cm high, each course being left to dry about one to three days before the next course is placed. The mud is trimmed to shape and desired thickness and the surfaces smoothed as the wall is built. As the wall becomes higher, the workman sits astride it, so no scaffolding is required.

While drying, the soil shrinks, and cracks are likely to occur. They are repaired with more mud. When the wall is dry, a surface layer of mud is applied.

As is true of wattle and daub, the material is subject to erosion by rain and should be protected by wide overhanging roofs and by a surface layer more resistant to erosion than the wall proper.

Poured adobe

By much the same method as portland cement concrete, adobe may be cast monolithically between forms. The soil is mixed with water to a softer and more plastic consistency than for cob or swish, and may have straw or other binder added. The mixture is poured or shovelled into the forms. As is true of portland cement concrete, it must neither be too dry and stiff to be placed readily, nor so soft and wet as to segregate or leak out of the form. Evidently, the wetter the mix, the greater the tendency to shrink and crack upon drying. If the material is placed in courses, straw reinforcing may not penetrate and intermingle with previously-placed courses if the latter have begun to dry and harden.

While this method makes placement of the material easier than cob, and the forms provide alignment and control of thickness, the forms themselves add to the cost. They are most practical, as is true of forms generally, when they can be re-used, with minimum alteration and repair, a sufficient number of times to justify the cost.

Adobe brick

Adobe bricks or blocks have long been widely employed. Sizes, composition, and methods of manufacture and construction vary from place to place.

Bricks are molded from a clay-based mixture with a high-enough water content to produce a plastic or workable consistency that allows the material to be formed in simple molds. Optimum combinations of sand, silt, and clay have to be developed on the basis of local materials, and no hard and fast rules respecting combinations can be given. Experience and the "feel" of the mixture are the criteria.

The mud is usually placed by hand in simple wood single-cavity or multi-cavity molds. The mold and brick are placed on a suitable bed and the mold removed, leaving the brick to dry in the sun.

Adobe bricks have traditionally been reinforced with straw or other fibers, but opinions differ as to their value.

The dried bricks are laid up in the wall in much the usual fashion, with horizontal courses and with vertical joints in successive courses staggered or "broken" to avoid long continuous vertical joints. Mortars are commonly compositions similar to the brick, and an effort is made to minimize shrinkage in the mortar, because the dried bricks shrink relatively little.

Minimum shrinkage is one of the chief advantages of adobe brick. Shrinkage occurs in drying, before the bricks are built into the wall. Furthermore, their uniform size permits walls to be laid accurately. There is no long drying period.

A major disadvantage is the double handling of the heavy material. It is first dug, mixed, formed into bricks and dried. The bricks must then be rehandled and transported to the building site, entailing a good deal of manual labor. Furthermore, during drying and hardening, the bricks must be protected from rain which can easily severely damage or wash them away. Once they are dry and hard such damage is minimal.

Rammed Earth (Pisé)

Cast adobe and rammed earth are similar in that both use forms, but in rammed earth the soil is rammed into place, whereas in cast adobe it is merely poured or shovelled.

Rammed earth requires heavy movable forms that can be shifted along the wall as each course is formed in successive steps, and can be raised for successively higher courses. Commonly two forms are used, one for corners, and one for the straight portions of the wall.

Moist sandy loam is usually employed, placed in thin layers within the form and tamped or rammed down hard with a hand-driven or pneumatic tamper. Composition and moisture content must be carefully controlled to obtain the proper consistency.

As is true of other soil-based walls, the surface should be protected against rain erosion by employing wide over-hanging roofs and an erosion-resistant surfacing coat.

Because it is relatively dry and heavily compacted, rammed earth tends to shrink and crack less and is more resistant to erosion than many other soil walls.

Cajon and nogging

The open spaces in a structural wall frame of timber are filled with wall panels of soil, either packed in place, or fitted in the form of brick. The procedures for mixing and placing the soil are the same as for cast-in-place soil or for adobe brick. The soil forms a base for surface coating such as stucco or plaster.

Material

A thorough discussion of soil and the mechanics of soil is beyond the scope of this brief exposition. The following points appear to be the most important from the standpoint of the simple construction employed for small dwellings.

Types

Soil consists not only of the surface layer of earth or loam capable of growing things, and usually called top soil, but, more importantly, of the material further down, with little or no organic matter. This may well vary with depth, as in lateritic soils, formed in place by the gradual breakdown of the underlying bedrock, brought about by the natural processes of weathering, altered and augmented by leaching of some constituents and infiltration of others. Although such soils obviously vary greatly from place to place, in a given bed the topmost layer is characteristically a growing layer containing much organic material, underlain by friable mineral material, loose at the top, but becoming gradually more compact with depth, merging into hardpan and heavily fissured rock, the fissuring becoming less as the essentially solid bedrock is reached.

Other soils result from particles transported by water or, to a less extent, wind, deposited in beds, often of considerable extent and depth, ranging from exceptionally homogeneous to extremely variable. Constituents obviously depend on the source or sources of the transported particles.

Constituents

Soils specialists generally grade soils on the basis of particle size from the smallest or colloidal particles through clay, silt, sand, gravel to cobbles or boulders. Several such classifications are shown in Fig. 7.1.

For earth houses, soil is often simply classified as clay and sand, the dividing line being between silt and fine sand in Fig. 7.1. Since "clay" in this case includes silt, and "sand" includes gravel, a somewhat better terminology is "fine" and "coarse". This simple two-part classification is often good enough, but if competent engineering talent is available, the greater subdivision of Fig. 7.1 is better for the best selection of soil fractions for a given application.

Moisture content

Moisture content for engineering purposes is usually expressed in percent as the ratio of the weight of water in a given volume of earth to its dry weight.

Atterberg limits

As the amount of water increases, many soils pass through four states defined as solid, semi-solid, plastic, and liquid. The boundary moisture contents between

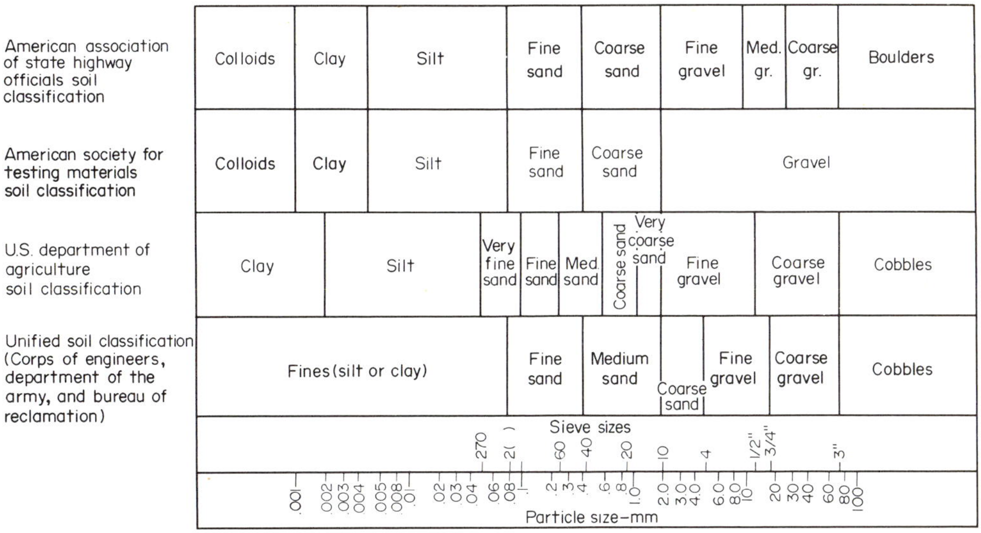

Fig. 7.1 Soil Classification.

these four states, devised by Atterberg, the Swedish soil scientist, are known as the shrinkage limit (solid to semi-solid), plastic limit (semi-solid to plastic), and liquid limit (plastic to liquid).

Shrinkage limit. The shrinkage limit is the moisture content below which shrinkage ceases. Because clay is the most important constituent causing shrinkage, this limit is a rough indication of the amount of clay present.

Plastic limit. This is the moisture content below which the soil begins to break or crumble when rolled into a 1/8-inch diameter thread. Clay content largely controls the position of this limit. Some silty and sandy soils that cannot be rolled into 1/8-inch threads have no plastic limit and are called non-plastic.

Liquid limit. The arbitrary definition is the moisture content at which the sides of a 2mm-wide groove cut in a soil sample flow together a distance of one-half inch under the impact of 25 blows in a testing device. Sandy soils have low liquid limits, such as 20 or less, whereas silty clays and clays may range above 80.

Plasticity index. The numerical difference between the liquid limit and the plastic limit is called the plasticity index. It has been used as an excellent index of the performance of soil for construction. Desirable soils for earth house building have low plasticity indices (e.g., 10-15), whereas soils with high plasticity indices (e.g., higher than 20-25) exhibit high shrinkage and swelling unsuitable for earth house construction.

Optimum moisture content

Dry soil is difficult to compact to maximum density, whereas too wet soil has so much water that it is overly porous when it dries, and does not attain maximum density. For a given effort at compaction, the optimum moisture content is that content at which maximum density is achieved after the soil has dried.

Absorption

Penetration of moisture into earth walls is evidently important in determining their resistance to deterioration. Increased sand content results in larger pore spaces and increased permeability, but too much fine material can result in increased capillarity. The optimum combination for a given soil must be determined by test.

Strength

For earth houses with typically thick walls, the commonly-specified compressive strengths of 20-25 kgf/cm^2 (300-350 psi) are usually ample to carry the super-imposed loads. Some tensile stresses may be developed by lateral wind loads, but tensile strengths of 5 kgf/cm^2 (50 psi) are normally sufficient. Earthquakes, of course, may develop significantly higher stresses, including shear, and call for stronger construction.

Soluble salts

Large quantities of soluble salts, as often found in arid regions, may make a soil unfit for earth construction. Their detection may require laboratory test.

Weathering

As is true of many materials of construction, weathering is difficult to define and to measure satisfactorily enough by laboratory tests to predict accurately how a given material will behave under actual exposure. For earth construction, four types of weathering tests are common: wetting and drying, freezing and thawing, erosion by rain, and abrasion, occurring separately or simultaneously.

Just how any given soil will react to these tests, or in a structure, is difficult to predict and must be ascertained by test. Sometimes increasing the clay content is reported to be beneficial, in other instances it is not.

Example

An example of the application of some of these concepts may be drawn from the work of the Building and Road Research Institute of the Council for Scientific and Industrial Research, Kumasi, Ghana.[3,4] The lateritic soils commonly found and used in Ghana, primarily for swish or cob construction, are defined as the "hardened or soft (reddish to brownish) product of tropical and sub-tropical weathering which is leached of bases, but enriched in sesqui-oxides of iron and alumina in the form of clay minerals, especially kaolinite and secondary iron minerals such as goethite, limonite and hematite. There is little or no combined silica in laterite, but depending on the parent rock, laterite may have high quantities of quartz."

The laterites are classified as: (1) rock and boulders (particles $>$ 60 mm), (2) gravelly soils, and (3) fine-grained (gravel fractions less than 10%). Specific gravity varies between 2.5-3.5. Atterberg liquid limits vary from 65 to non-plastic, and plasticity limits between 40 and non-plastic. Under standard compaction, maximum dry densities vary between 1500-2100 Kg/m^3 (92 and 128 cfm) for optimum moisture contents varying between 9-22% during compaction.

It has been found that soils whose range of constituents from fine to coarse falls within the envelopes of Fig. 7.2 are suitable for earth wall construction. These soils are generally sandy clays and clayey loams. Liquid limits are 30-40%, and plasticity indices 12-20%. Volumetric shrinkage is not more than 20%.

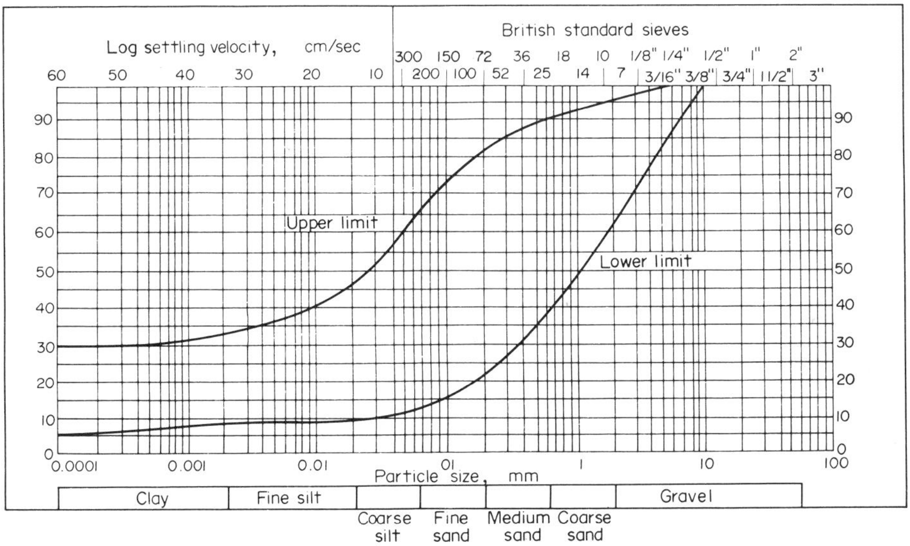

Fig. 7.2 Grading Envelope for Selecting Lateritic Soils for Building

The example of the Ghanaian laterites should be considered only an example and is not necessarily indicative of soils elsewhere. Soil in each locality must be examined on its own merits. Different authorities differ, depending upon their particular experiences.

Figure 7.3 summarizes the recommendations of a number of investigators respecting grading of soils for rammed earth, adobe, and machine-made blocks. Wide variations exist. A few extremely general observations follow.

Rammed earth. The monolithic wall must shrink as little as possible to prevent excessive cracking. Clay is kept to a minimum, and a narrow range of 60-75% sand is usually specified. Atterberg liquid limit is recommended at 35 or less, plasticity index 2 to 15, and shrinkage limit less than 25, although for tropical laterites which usually exhibit less shrinkage and swelling than temperate-zone laterites, these values may be low.

Adobe blocks. Because these are dried before use, shrinkage is largely eliminated, and the limits for grading can be broader than for rammed earth. Some authorities say that any earth containing at least 50% sand is suitable. There must be enough clay to provide good cohesion to the dried block. Another authority says any soil within the limits 80% clay-20% sand is suitable.

Cob and wattle and daub. These permit the widest variations. In wattle and daub the soil is held in place by the wattles and frame, and cracks are repaired with more mud under more or less continuous maintenance. In cob, shrinkage cracks typically occur around the individual balls of earth and are therefore numerous, scattered, and fine — seldom occurring as single large serious cracks. The numerous fine cracks may afford bonding points for stucco and plaster.

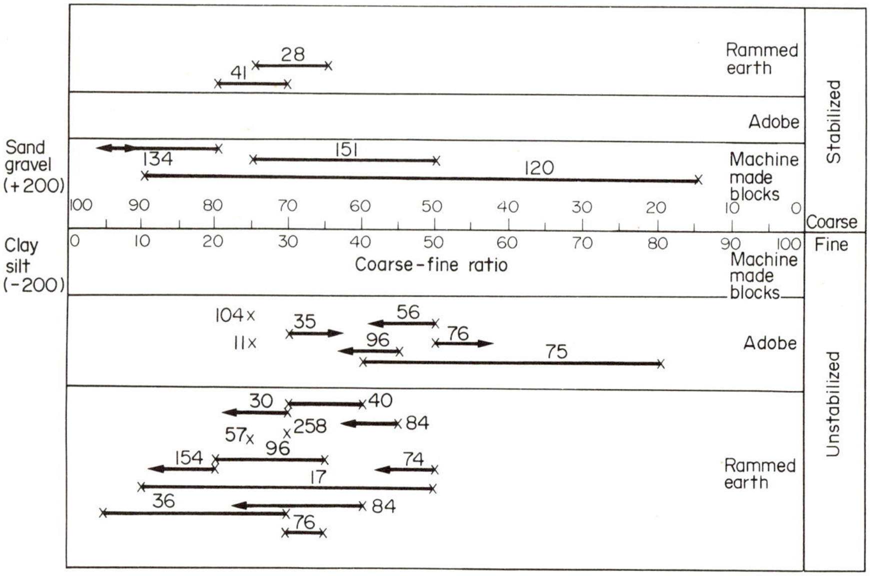

Fig. 7.3 Recommended Grading.

Soil Stabilization

Numerous attempts have been made to stabilize soil by admixtures, usually to increase its resistance to deterioration by water, but also, in some instances, to increase the strength, especially the wet strength.

As indicated above, excessive shrinkage can often be reduced by changing the ratio of sand to clay (coarse to fine). If the clay content is too high, sand and gravel can be added, provided this does not result in too great porosity and loss of cohesion. Fibrous materials such as sawdust, wood shavings and various available fibers help, presumably by increasing the tensile strength, thereby reducing the tendency to crack, and dividing such cracking as does occur into numerous small cracks instead of a few large ones.

Stabilization may be achieved by improving the adhesion among the grains of soil by various admixtures that adhere well or change the surface tension characteristics of the binding material, or in other ways change the chemical nature of the mix, such as waterproofing the grains, or changing the ionic character of the soil.

Among the many indigenous materials that have been employed, usually on a completely empirical basis, the following are said to have been successful in varying degree: cattle dung and urine, grass, flax straw, oat straw, jute fiber, sawdust, wood shavings, tannic acid, molasses, various plant juices, pulverized clay brick and tile, gum arabic, and hardwood ashes. Some analyses of vegetable ash seem to indicate constituents that have hydraulic characteristics similar to fly ash. Ashes have, of course, long been used to modify agricultural soil.

Probably the most extensive work in stabilizing earth has been done with portland cement, lime and lime admixtures, and emulsified asphalt.

Portland cement

The extensive work done in many parts of the world has resulted in considerable variation in results depending upon composition of the soil and the method of preparing the soil-cement mixture. In general, soils with relatively high sand content can be stabilized more readily and with less cement than can high-clay soils, particularly when the sand is well graded. This is similar to the experience in making satisfactory mortars and concrete. In general less cement is needed with moist mixes than with wet, but thorough mixing is necessary, and is easier to achieve with wet mixes.

What proportion of cement to soil should be used depends upon the soil, the degree of stabilization wanted, and whether increased strength is also necessary. Some significant reduction in absorption of water has been reported with as little as approximately 3% cement, but other research has reported significant shrinkage occurring in optimally-compacted high-clay soils with 20% cement. Contents of 6-10% appear to be common, but variations are large. In one example in the Philippines ratios of 1 to 12 appeared to be satisfactory, but with two different soils this ratio could be varied by more than 2 to 1. Any soil must be tested with various mixes until the best one for the purpose is found.

Lime and lime-admixtures

In many parts of the world portland cement is scarce and expensive, whereas lime is available and has long been used. Variations in mixes occur because of variations in soil and in the degree of stabilization needed. Data from controlled experiments appear to be meager, but contents from 6-14 or 20% seem to be employed. In some areas such as Indonesia, both lime and pozzolans are readily available, and combinations of lime and pozzolan in ratios 5 to 1 appear to be useful. Portland cement may also be added, and the mix may include essentially all sand, or various combinations of sand and fines. Where fly ash is available it may be used instead of or in combination with pozzolan. For any given soil and application trial mixes must be made to find the optimum.

Emulsified asphalt

Asphaltic emulsions "break" or settle at various rates. For earth construction, slowly-breaking compositions are usually best. Unlike portland cement, asphaltic emulsions do not normally increase the strength of a given soil, they may decrease it, but they do retain the dry strength in the presence of moisture. The minimum amount of emulsion to achieve the objective is the optimum amount, in turn depending upon the soil and the objective. The Housing and Home Finance Agency has suggested the following broad range: sandy loam, 4-6%; clay loam, 7-12%; heavy clay, 13-20%. These are merely approximations, and for any given situation test batches must be made to determine the best mix.

Sulphite liquor

This has been found to be effective in imparting cohesive strength to soil and in making it relatively impervious to water. Where such liquor is available, often as a waste material from pulp and paper manufacture, it can provide a useful stabilizer at low cost.

Other chemical stabilizers

Many chemical materials have been used experimentally, as well as in practice where conditions such as cost and availability have been favorable. A few may be listed: polyvinyl alcohol, sodium tetraphosphate, calcium acrylate and polyacrylate, polyacrylamide, vinsol, aniline-furfural, silicate of soda, and many others. Although often relatively costly, percentages are usually low, ranging from 0.1 to 2% in many cases. Sometimes they are used alone, sometimes in combination with soil-cement.

Surface Coatings

Surface coatings are commonly applied to limit water absorption, erosion, and general weathering. In moderate climates such coatings may be sufficient protection for unstabilized earth, but in severe climates earth may have to be stabilized even with surface coatings.

Many different plasters and stuccoes or other thick coatings are employed. Where cost is of overriding importance and better materials are either too expensive or unavailable, mud plaster is widely employed. This may simply be more of the same material as is used in the wall, especially if cob or wattle and daub, with water content modified to make it workable enough to apply to the wall. Frequently, however, the clay and sand (or fine and coarse) components are varied to obtain the locally optimum combination of plasticity, minimum shrinkage (sand) and water tightness (clay) that can be attained. Where kaolinite clay is available its combination of minimum shrinkage, cohesiveness, and adhesiveness can be used to advantage in the formulation.

Various admixtures, such as cow dung, may be employed to improve the characteristics.

Where lime is available, lime stucco may be employed, or the mud plaster may be enhanced by adding lime. When lime plasters or stuccoes are employed, care must be taken to obtain the optimum combination of lime and sand. Too much lime leads to excessive shrinkage and cracking, whereas too much sand results in softness and porosity. Once the stucco has undergone the slow process of hardening by reaction with carbon dioxide in the air, a durable weather-resistant coating results.

Portland cement, if available and not too costly, can provide a hard durable protective plaster or stucco when properly formulated with a good grade of clean sand, using standard mixes. It can be extended with lime: as a matter of fact, a lime plaster can often be made good enough by adding modest amounts of cement. If volcanic materials such as pozzolan are available, usually at lower cost than portland cement, lime plaster can be improved, and portland cement plasters reduced in cost, by adding these materials.

Mud plasters are sometimes improved by the addition of asphalt, when available, but the resulting color is often objectionable.

References

1. Housing and Home Finance Agency, *Earth for Homes*, HHFA Ideas and Methods Exchange No. 22, Housing and Home Finance Agency, Washington, D.C., 1956.

2. Wolfskill, L. A., Dunlap, W. A. and Gallaway, B. M., *Earthen Home Construction*, Bulletin No. 18 E 18-62, Texas Transportation Institute, A and M College of Texas, College Station, Texas, 1962.

3. Opeka, Y. A., *Housing the Rural Communities*, Rural Housing Research Series (No. RH-1), Building and Road Research Institute, Council for Scientific and Industrial Research, Kumasi-Ghana, 1970.

4. Hammond, A. A., *Lateritic Soils for Rural Housing*, Current Paper No. 1, Building and Road Research Institute, Council for Scientific and Industrial Research, University, P. O. Box 40, Kumasi-Ghana, 1972.

7.3 WOOD

Wood is a familiar material found in virtually all parts of the world, versatile, and relatively easy to fabricate. Its high strength and toughness relative to weight make it attractive, but its susceptibility to destruction by a number of agencies militate against it. The problem, as with all materials, is to utilize its advantages and circumvent its limitations.

Classifications

Wood is classified in various ways, the most common probably being hardwood and softwood. These terms correspond roughly to deciduous or broad-leafed trees and to conifers or needle-bearing "evergreen" trees. This classification leaves out the large class of palms and similar trees that belong to a different major division.[1,2]

Growth

The palms grow by adding new wood fibers more or less throughout the trunk, whereas the hardwoods and softwoods add new wood fibers to the woody core directly under the bark. Growth is often seasonal, resulting in distinct rings of new wood, but growth may also be more or less continuous.

Structure

The structure and properties of wood follow from its pattern of growth. It consists mostly of long hollow tapered cells oriented parallel to the axis of the stem. These hollow cells, closed at their ends, are typically 100-200 times as long as their diameters. A few short blocklike cells grow radially in the tree and form wood rays. Some softwoods have resin-filled specialized cells called resin ducts. Hardwoods have large-diameter specialized cells called pores or vessels for primary transportation of sap (see, also, Bamboo). The orientation of the long longitudinal cells parallel to the axis gives wood its distinctive directional properties, as reflected in the familiar terms "along the grain" and "across the grain".

Because fibers are generally large in diameter and thin-walled early in the growing season ("spring wood") and thin and heavy-walled late in the season ("summer wood"), annual rings often exhibit distinct changes in color from light to dark.

The walls of the wood cells consist of strands of cellulose wound in a long spiral parallel to the axis, or long direction of the cell, somewhat like the fibers in a rope. The cellulose strands or fibrils are extremely strong along their lengths, but only loosely held together laterally. Cells are cemented together by lignin.

If wood were truly "solid" its specific gravity would be approximately 1.50, and it would rapidly sink in water. All wood is porous because the wood cells are hollow. The apparent specific gravity of oven-dry wood ranges from figures as low as 0.10 for such exceptionally light woods as balsa, to well over 1.00 for heavy tropical woods such as lignum vitae and greenheart. The apparent specific gravity of wood is a rough indicator of strength and hardness; the heavier, the harder and stronger. Consequently, many "softwood" species are harder than many "hardwood".

Moisture Effects, Shrinkage

In freshly-cut wood ("green") the cells are filled with water, some held freely in the hollow cavities, other held more firmly or "imbibed" on the cellulose strands in the cell walls. When the wood dries by evaporation, the free water leaves first. The imbibed water then leaves and the cellulose strands move close together, causing the cell to shrink laterally, but practically not at all longitudinally. As all the longitudinal cells in a block of wood shrink laterally, the block also shrinks laterally, that is, "across the grain". There is very little or no shrinkage along the grain, a fact of great importance in the use of wood.

The degree to which the imbibed moisture evaporates depends upon the relative humidity of the surrounding air. At 100% relative humidity there is no loss. This is called the fiber-saturation point. At zero the wood becomes completely dry. At 50% relative humidity, the average moisture content is about 12% of the dry weight of wood. This is often called "air-dry". At 100% relative humidity, when the walls are saturated but there is no free water, the average moisture content is about 25-30% of the dry weight.

Evidently, as the relative humidity of the air surrounding a piece of wood changes, the moisture content of the cell walls also changes, and the wood either shrinks or swells across the grain. Because the change is not instantaneous, daily variations in boards and thicker pieces are minor, but seasonal variations can cause noticeable shrinkage and swelling.

The extent of shrinkage and swelling depends upon the apparent specific gravity, or amount of wood present in a given volume; the higher the apparent specific gravity, the greater the shrinkage. It is greatest in the direction tangential to the growth rings, less in a radial direction, and least longitudinally, i.e. parallel to the grain. Tangential shrinkage from green to completely dry ranges typically from 4 to 10%. Radial is about one-third less.

Strength

Strength of wood is closely related to the structure of the wood cell. In a single cell, strength is great in tension along its axis because of the cellulose strands, but tensile strength at right angles is low, much as a rope is strong in tension along its length, but fibers can be pulled apart laterally. In compression, a better analogy is a soda straw. Push on a single straw and it buckles, push on a bundle of straws and although they still buckle eventually, they sustain each other. Compress a cell laterally and it flattens, but does not actually break. A block of wood is similar, it has extremely high tensile strength along or parallel to the grain, poor tensile strength across the grain (it splits easily), good compression along the grain (about 40% of tensile) and moderately good compression perpendicular (about 25% of parallel). In shear, wood fibers can be slid past each other longitudinally or along the grain fairly easily against only the strength of the lignin cement, but only with difficulty laterally; in fact, the cells crush before they slide past each other laterally. Wood is accordingly moderately strong in shear parallel, but much stronger in shear perpendicular to the grain.

The bending strength parallel to the grain is intermediate between tensile and compressive strengths parallel.

All strength properties are affected by moisture; the drier the wood the stronger, but compression and bending are more markedly affected than tension. Compressive and bending strengths may vary 25-40% or more in changing from dry to a condition of saturated fiber walls. Free water has no effect.

When test values of strength of wood are not available, the apparent specific gravity can be used as a rough indicator of strength. The apparent specific gravity can be determined relatively easily by measuring and weighing a block of wood. The moisture content can be determined by drying in an oven. The oven-dry and air-dry volumes, of course, are less than the green volume.

Table 7.1. gives formulas for the calculation of approximate strength and hardness values.

Structural Timber

In timber used for construction the orientation of the grain significantly affects strength. In a perfect timber free of all blemishes, in which the cells are oriented parallel to the edges of the timber -straight grained- the full strength can be achieved. If the grain is not parallel to the edges the strength rapidly falls off so that at an angle of only 1 in 10 with respect to the edges, strength may be off as much as 50%, and such a timber is best not used for carrying loads, although it may be perfectly acceptable for such things as simple enclosure (sheathing boards) where no great loads are encountered.

Knots cause local deviations of grain direction and therefore loss in strength, although not in shear, where they can act as keys. Splits, however, reduce the shear strength.

Commercial wood members must therefore take into consideration diagonal grain, knots, and splits such as checks (caused by shrinkage in drying) and shakes (splits occurring naturally in the tree). Engineering timbers are graded on this basis, plus an allowance for dimensions, duration of load (impact, wind, snow, long-term load), and a suitable factor of safety.

Destroyers

As an organic material, providing food for some types of organisms, and subject to oxidation, wood can succumb to a number of destroyers. Protective measures are then needed.

Table 7.1. Relations Between Strength and Apparent Specific Gravity.

Property	Unit	Green	Air-dry (12% moisture content)
		Moisture Condition	
Static Bending			
Fiber stress at proportional limit	psi	10,200$G^{1.25}$	16,700$G^{1.25}$
Modulus of rupture	psi	17,600$G^{1.25}$	25,700$G^{1.25}$
Modulus of elasticity	1,000 psi	2,360G	2,800G
Impact Bending			
Fiber stress at proportional limit	psi	23,700$G^{1.25}$	31,000$G^{1.25}$
Modulus of elasticity	1,000 psi	2,940G	3,380G
Compression parallel to grain			
Fiber stress at proportional limit	psi	5,250G	8,750G
Maximum crushing strength	psi	6,730G	12,200G
Modulus of elasticity	psi	2,910G	3,380G
Compression perpendicular to grain			
Fiber stress at proportional limit	psi	3,000$G^{2.25}$	4,630$G^{2.25}$
Hardness (load to embed 0.444" ball one half its diameter)			
End	pounds	3,740$G^{2.25}$	4,800$G^{2.25}$
Radial	pounds	3,380$G^{2.25}$	3,720$G^{2.25}$
Tangential	pounds	3,460$G^{2.25}$	3,820$G^{2.25}$

G = apparent specific gravity of oven-dry wood based on volume at moisture content indicated.

Source: Wood Handbook, U.S. Forest Products Laboratory.

Living organisms

Fungi. Decay or rot is caused by fungi, primitive plants unable to synthesize their own food. These begin with spores or seeds carried from parent bracket fungi to a favorable place on a suitable piece of wood. The spore germinates and sends out long threads or hyphae capable of dissolving wood for its own life processes. As the wood becomes riddled it gradually weakens, softens, and crumbles.

Being living plants, fungi need food, water, air, and suitable temperature conditions. Wood provides the food, but if poisoned by the proper preservatives, can be protected. Temperatures comfortable for humans are conducive to decay. Freezing merely makes the fungi dormant. Heating, especially in moist heat, to 65° C (150° F) or higher can kill fungi, but will not stop reinfestation at normal temperatures.

From a practical standpoint, the most promising points are the need for water and air.

Dry wood does not rot. Furthermore, it need only be air-dry, i.e., well below the fiber-saturation point. If a structure is so built as to be well ventilated, with no pockets to hold water, and not in contact with damp spots such as earth, it can last for centuries or thousands of years, as attested by ancient Egyptian artifacts, and by centuries-old wood structures in all parts of the world.

At the other extreme, wood continuously immersed in water does not rot. Piles under water, or driven below the ground water level in soil, have lasted for centuries. Logs of species extinct for many thousands of years have been found in deep excavations.

Keeping wood dry or immersed, therefore, keeps it from decaying.

Molds. Molds are similar to fungi, but do not destroy wood. Sapwood is more susceptible to attack than heartwood. The chief deleterious effect usually is discoloration and stain. Mold, however, indicates that conditions may be favorable to decay.

Termites. These highly-organized efficient insects use wood as food. Their destructiveness is well known.

Subterranean termites have their headquarters in the ground and fan out through subterranean galleries in the ground, or in covered runways when they must come to the surface. Other termites, such as the dry-wood and damp-wood types, do not need contact with the soil, but make their abodes directly in the wood.

In either event, the colony revolves about a queen and king who produce the eggs from which termites are hatched. A special class of sexual insects leaves the colony at maturity, and an occasional successful pair will start a new colony. The warrior types guard the colony, especially if galleries are broken and subject to attack. The most numerous class is the workers, blind, asexual, whose function is to find wood, digest it, and bring it back to the colony for use.

Workers do their work under the surface, often riddling the wood without revealing their presence, so that damage may be well advanced and failure may even occur before they are detected. Testing for hollow sounds, or digging into the suspected wood is often necessary to find them. Keeping the wood dry does not help, although some termites prefer damp wood. Immersion, of course, is a deterrent. Poisoning the wood with preservatives, or periodic fumigation of a building, are frequently the only answers.

Carpenter ants, grubs, beetles. These all burrow into wood. Carpenter ants use wood for habitation, but not for food. In any case, infestation may be inconspicuous, but usually small holes in the surface, or small piles of wood dust are indicators.

Teredo, Limnoria. These are both found in salt water. Teredo, or shipworms, are mollusks. They attach themselves to the surface of a wood member in salt water, and using a pair of shells, bore long tunnels, about the diameter of a lead pencil as they grow and elongate. They do not come to the surface. Infestation under favorable conditions can be rapid, and wood marine structures have been known to fail in as little as six months.

Limnoria are small crustaceans that swim on the surface of the water and attack the surfaces of wood members, burrowing into the surface, and weakening it so that erosion occurs. A pile, for example, has an "hour-glass" figure, deepest at mean tide, and tapering to high and low tide. Unlike teredo, their presence is readily evident.

Preservatives (See also, Bamboo)

Many materials have been and are utilized as wood preservatives. Their function is to poison the wood against living organisms that attack it. One convenient classification is into: oily preservatives, water-soluble salts, and solvent-soluble organic preservatives.

Oily preservatives

The most widely-used is creosote, the fraction of coal tar that distils in the approximate range 50-120°C (125-250°F). It is a mixture of organic materials found effective in combatting wood-destroying organisms. Where wood tar creosote is available, it too is highly effective. Creosote is sometimes blended with petroleum. Creosote is highly effective, but its odor and black appearance make it unusable in many locations, especially indoors. It can generally not be painted effectively. For the most demanding applications, however, creosote is often the best answer.

Water-soluble salts

For applications in which odor, appearance, or both are important, certain water-soluble salts have been found effective, particularly where dampness is involved, but not running water. Zinc chloride, plain and chromated; sodium fluoride, and others are in this category. Borax and boric acid were used in Roman times and are still employed.

Solvent-soluble preservatives

Various water-insoluble chlorinated phenols, such as pentachlorophenol, are effective. They are impregnated into the wood with solvents which are then evaporated leaving the preservative behind. Once the solution is gone, there is no residual odor and little discoloration.

Fire

Preservatives against fire are commonly snuffing salts such as the ammonium phosphates. While wood so treated can be destroyed in hot enough fires, the wood does not support combustion and is self-quenching. Intumescent paints contain a constituent that volatilizes at high temperatures, and gives off an inert gas at the same time that the coating becomes viscous. The resulting gas bubbles form a thick insulating blanket that retards the penetration of heat from the surrounding fire into the wood.

The most effective treating methods are not readily available in many areas, and some raise costs, already too high for many people, to prohibitive levels. Badly needed are indigenous low-cost readily-available materials that can be effectively applied by simple means such as steeping or brushing.

Adhesively-Bound Wood Products

Although plywood was known to the ancient Egyptians, and some of their glues are still used today, it is only recently, with the advent of strong, reliable, waterproof adhesives, that bonded wood materials have leaped ahead. Among the most important are plywood, laminated wood, and particle board.

Plywood

In plywood, thin sheets of wood called veneers are glued together, with adjacent sheets at right angles to each other, the total number of plies being three, five, seven or some other odd number to achieve a balanced structure not prone to warp or twist, as would occur with an even number of plies, or plies not symmetrically arranged with respect to the central plane (Fig. 7.4). For most general purposes, veneers are peeled off the log in long sheets and cut into the desired sizes. For fine figures and thinner sheets, veneers are sliced off square logs or flitches. Refractory woods are sometimes sawn into veneers.

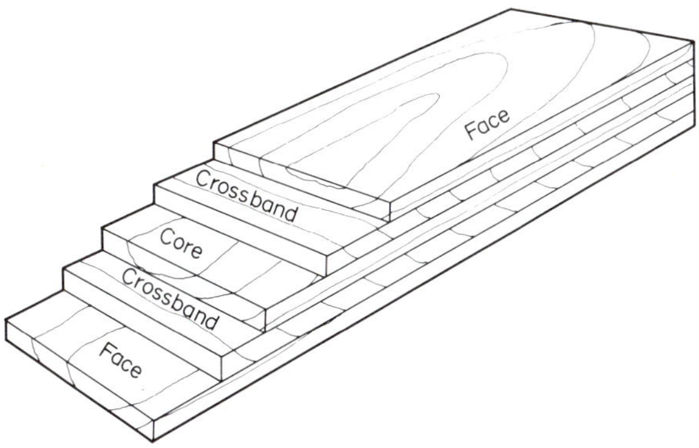

Fig. 7.4. Essentials of Plywood Structure Showing Faces, Core, and One Pair of Cross-Bands.

To achieve balance, corresponding plies must have corresponding properties, i.e., same shrinkage, moisture content, thickness. The structure of the plywood must be symmetrical, i.e., a mirror image about the central plane.

Because of the cross-plied construction, plywood shrinks and swells very little. Similarly, splitting is virtually eliminated, and nails, for example, can be driven close to the edges without splitting the board. Strength and stiffness are more nearly equalized in the longitudinal and cross directions of the sheet than is normally true of wood. Finally, plywood is available in large, thin sheets, generally not possible to obtain with plain wood.

Because of their lightness, good strength, stiffness, and size, plywood sheets are extensively employed as sheathing, roof boards, and subflooring or underlayment for finished floors. The large sheets act as diaphragms with framing members to provide stiffness and rigidity to the structure. If waterproof adhesives and suitable surface veneers are used, the plywood can provide exposed interior and exterior surfaces.

Laminated wood

In laminated wood, as contrasted with plywood, layers of wood are glued together with all layers or laminae parallel. Usually the laminations are horizontal, but they may be vertical (Fig. 7.5). The resulting member, therefore, has many of the same attributes as "solid" wood; strength, stiffness, shrinking and swelling, and so forth. There are important differences. In laminated wood the individual layers — customarily boards or thin planks — are easily dried, and must be dried for successful gluing. The resulting timber is dry, stable, and stronger than an equivalent green solid timber. Large timbers are difficult to season, take a long time, shrink, and tend to check or otherwise become degraded. In laminated timber, top-grade material need be used only where stresses are maximum, the rest can be lower grade. More important, however, is the ability to obtain large sizes — larger than feasible in saw timber — and, most important, efficient curved shapes such as arches (Fig. 7.5). These are easily made by bending and gluing thin layers. Lightweight efficient structural shapes can therefore be obtained with laminated wood or combinations of laminated wood and plywood.

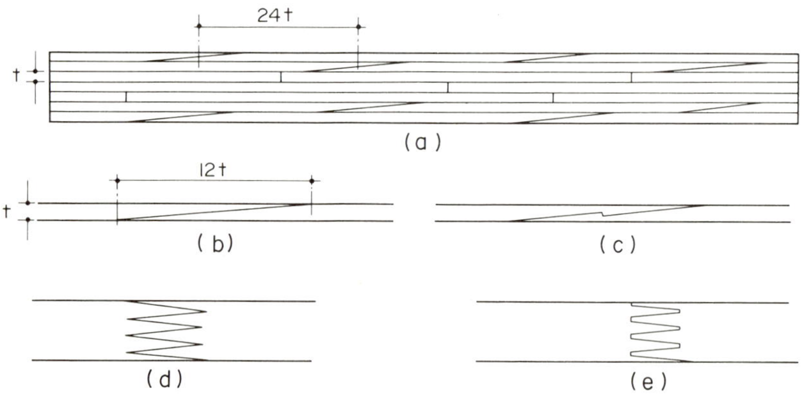

Fig. 7.5. Types of Scarf Joints and Positions of Scarf and Butt Joints in Laminated Beams.

The principal drawbacks are cost and waste. Individual laminations have to be cut and surfaced smooth for gluing, resulting in sawdust and planing waste. To achieve continuity in a single lamination, long tapered joints (scarfed) must be used where stresses are appreciable (Fig. 7.5). Glue must be spread, the assembly must be put together and clamped until the glue hardens, and almost always the surfaces must be re-planed, resulting in additional waste. All these operations increase the cost. The advantages of shape, size, dryness, and use of mixed grades must be sufficient to offset the increased unit cost per board foot. Frequently they do, and laminated wood has found widespread use.

Particleboard

Waste wood from lumbering and milling operations can be chipped into small particles, irregular or flat, and bonded together with a suitable adhesive into large boards. Because of the random orientation of the particles, the material has essentially equal strength properties in all directions, and for the same reason, shrinkage and swelling are equalized as well as greatly reduced because in any one direction the strong longitudinal chips oppose the shrinkage or swelling of transverse chips. Sometimes flat flake chips are arranged on the surfaces to provide a sandwich-like structure with enhanced surface properties and increased resistance to bending.

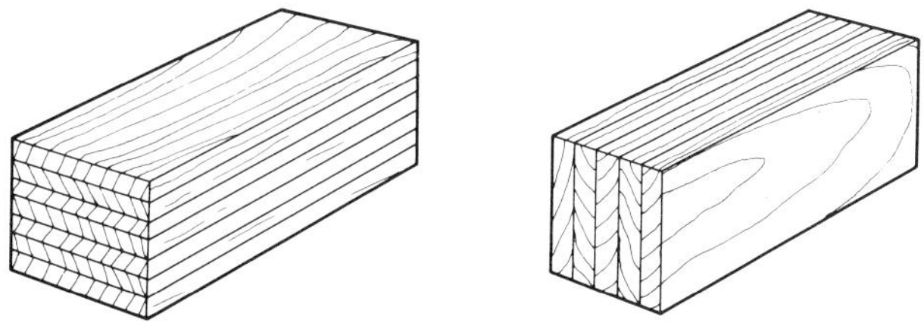

Fig. 7.6. Two Arrangements of Laminations in Laminated Timber.

Depending upon the adhesive employed, the particle board may be waterproof and suitable for exterior use or intended only for interior use. Not infrequently it provides the cores for sandwiches with facings of wood veneer or other sheet materials.

Hardboard

If wood chips are subjected to high-pressure steam in an autoclave until they are completely penetrated by the steam, the thermoplastic lignin softens, and when the pressure is suddenly released by opening a fast-acting valve, the sudden explosive expansion of the steam tears the wood cells apart into a fluffy random mass. This can be reconstituted by pressing in a hot-plate press into a board whose density and hardness depend upon the pressure employed. Waterproofing agents, such as waxes, and preservative materials can be incorporated into the mass before pressing to provide moisture resistance (tempered hardboard) and resistance to destructive organisms.

Hardboards are extensively employed as surface layers of sandwich structures, doors, and furniture, as underlayments for finish flooring, as sheathing and surface finishes for walls and ceilings, and many other applications in building and other industries.

References

1. *Wood Handbook*, Forest Products Laboratory, U.S. Department of Agriculture, Madison, Wis., Published by Superintendent of Documents, Washington, D.C.

2. Dietz, A. G. H., *Materials of Construction: Wood, Plastics, Fabrics*, D. Van Nostrand Company, N.Y., 1949.

3. Gurfinkel, G., *Wood Engineering*, Southern Forest Products Association, New Orleans, La., 1973.

7.4 BAMBOO

Distribution

Bamboo and the related family of reeds are among the oldest and most widespread building materials in the tropical and subtropical regions of the world, used widely for house construction, especially in the rural areas and villages. According to F. A. McClure[1]:

"Bamboos occur as more or less prominent elements in the natural vegetation of many parts of the tropical, subtropical, and mild temperate regions of the world, from sea level to altitudes of more than 13,000 feet, wherever a suitable combination of ecological factors prevails. Their natural distribution is very uneven, both as to abundance and variety of kinds in a given area. Through the agency of man, the distribution of many species of bamboo has been greatly widened. This process probably has been going on for a long time, and the actual extent of it has not been surveyed. However, some of the most valuable species have not been distributed to any important extent and much remains to be done to make these more generally known and available."

"The greatest concentration of bamboos and the highest development of their use are to be found on the southeastern borders of Asia and on adjacent islands. This area extends from India to China on the mainland, and from Japan to Java among the islands. Some 20-odd species of bamboo have been reported from the little-known flora of Africa, and many of these are used by the native peoples for house construction. The Island of Madagascar, whose flora is more fully known, has been found to have more native species of bamboo than are known to occur in the whole of Africa. Australia has perhaps a half-dozen native species; Europe none. In the Western Hemisphere, the natural distribution of bamboo extends from southern United States to Argentina and Chile. Some 200 species are native to this area, but they are very unevenly distributed. Many of the recorded species are very imperfectly known, and some kinds have been recorded under more than one name, but the known bamboo flora of the world probably totals more than 700 species, classified in about 50 distinct genera."

Growth

Although most bamboo grows in forests and is naturally propagated, commercial plantations are now common in many countries. Species not native to some regions have been introduced to improve the stock or to meet requirements better than is possible with indigenous species.

Bamboo are grasslike woody plants. New trunks, called culms, are formed annually in clumps growing out of the spreading rhizomes or roots. Generally, the culms grow to their full height in four to six months of the first growing season.

Culms vary markedly in height and diameter. Some grow to 35-36 m heights (approximately 110 ft) while others are mere shrubs. Diameters may vary from 1 to 30 cm (0.4-12 in.). Once the full height has been reached, lignification of the walls of the culm takes place during the subsequent two to three years. Full maturity is usually reached in five to six years, but may take longer.

Structure

The culm is a hollow cylinder divided at intervals by raised nodes from which any branches arise (Fig. 7.7). At each node a transverse wall called a septum or diaphragm separates the cavity of one internode from that of the next. Cavities are extremely variable and may be large in diameter when culm walls are thin, or practically non-existent when culms are virtually solid.

Tissue of the culm wall consists of parenchyma cells and vascular bundles made up of thick-walled fibers, vessels, and sieve tubes (see section on Wood). All cells including parenchyma are parallel to the axis of the culm, unlike the parenchyma in wood rays. Fibers constitute some 60-70% of the total weight. Vessels constitute about 15%. In the internodes vessels are parallel to the axis, but at the nodes they are strongly branched into the nodal wall or diaphragm. The vessels are the primary transporters of water. The outside and inside of the culm are covered by hard waxy cuticles resistant to the absorption of water.

Strength

Strength properties, as is true of wood, vary with species, growing conditions, position within the culm, moisture content, and other variables. Generally, properties vary more among species than within a species. Table 7.2 gives a few properties reported by various investigators.[2] Modulus of rupture varies from 623 to 1,926 kgf/cm^2 (8,800 to 27,000 psi), modulus of elasticity from 73,000 to 183,000 kgf/cm^2 (1,022,000 to 2,6000,000 psi), and compressive strength parallel to culm from 320 to 703 kgf/cm^2 (4,500 to 10,000 psi). These values are comparable to those of wood parallel to the grain; if anything, they tend to run somewhat higher than the customarily employed structural species of wood.

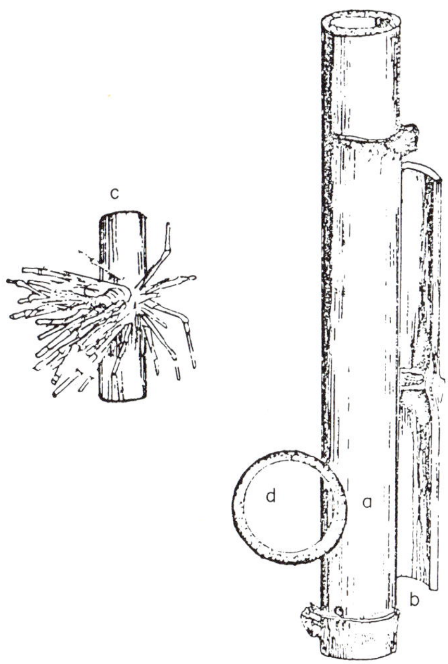

Fig. 7.7. Gross Features of a Bamboo Culm

a. The culm
b. Vertical section of the culm
c. Branches of a node
d. Cross section of the culm

Because of the alignment of cells parallel to the axis of the culm, bamboo is notably weak in tension perpendicular to the "grain" or longitudinal direction, and splits easily.

TABLE 7.2. Strength of Bamboos Used in Building Construction.*

Species	Location	Age (years)	Condition and moisture content (%)	Splint or full round	Specific gravity	Static bending: Fibre stress at elastic limit kgf/cm^2	Static bending: Modulus of rupture kgf/cm^2	Static bending: Modulus of elasticity 1000kgf/cm^2	Crushing strength parallel to grain kgf/cm^2	Author
Dendrocalamus strictus	Dehra Dun, India	Average of 1/2 2 1/2	Green 72.2	Full	0.575	394	696	124	320	Limaye
Dendrocalamus strictus	Dehra Dun, India	Average of 1/2 2 1/2	Kiln-dry 10.6	Full	0.732	668	1139	159	578	Limaye
Dendrocalamus strictus	Dehra Dun, India	Average of 1/2 2 1/2	Dry 5.8	Splint	0.666	1219	1926	175		Limaye
Dendrocalamus str. (solid)	Dehra Dun, India	2 1/2	Dry 13.0	Full			1190	183	619	Limaye
Bambusa arundinacea	Puerto Rico		Dry 9.4	Splint	0.50[a]	363	682	80	320 (results	Heck averaged
Bambusa balboa	U. Pradesh, India		Green 42.0	Full	0.783		651	73	466	Limaye
Bambusa nuntans	Dehra Dun, India	Average 1-4	Green 95.1	Full	0.631	334	623	98	453	Sekhar
Bambusa spinosa	Philippines		Air-dry	Splint			1180		535	Espinosa
Bambusa tuldoides	South America		Air-dry 11.1	Full	0.83	844	1547	162		Heck
Bambusa vulgaris	Puerto Rico Puerto Rico		10.0 9.5	Full Splint	0.70[a] 0.61	653 659	1224 1171	159 562	584 562	Heck Heck
Bambusa tulda	Puerto Rico		10.0	Splint	0.79[a]	931	1247	196	703	Heck
Phyllostachys edulis	China		Green	Splint	0.666	490	995	81 (results	424 averaged)	Chinese publication

*The information contained in this Table has been obtained from various sources. (2)

[a]Based on oven-dry weight and volume.

In any given application, the properties of the particular bamboo to be employed under the conditions of use should be ascertained, rather than using average figures. This is true also of wood and other materials, especially those of natural origin.

Deterioration and Preservation[2]

One of the most serious limitations in the use of bamboo for housing is its susceptibility to decay, caused by fungi, and to attack by insects, including termites and an assortment of beetles. Frequently, the life of a bamboo house is as little as 2-3 years, especially the parts in contact with the ground[1] With preservative treatment the life can be extended to 15 years or considerably more.[2] The problems with preservative treatments are cost and the ready availability of simple processes adaptable to village and rural situations.

Processes used for preservative treatment include leaching, whitewash, brushing, spraying, swabbing, dipping, hot and cold bath, Boucherie method, and pressure treatment.[1,2]

Many preservatives are employed, depending upon the conditions to be combatted. Among them are familiar wood preservatives: coal-tar creosote; various combinations of copper, chrome, arsenic, boric compounds, and zinc; zinc chloride; copper and zinc naphthenate; borax and boric acid; Dieldrin; and pentachlorophenol. Some are more readily available at the village level than others.

Methods

Some of the available methods are summarized below and in Table 7.3.[2]

Leaching. Immersing the culms in water for a few days to several weeks to remove the starches and sugars greatly increases resistance to attack by certain beetles that use the starch and sugar for food. Running water is preferable.

Whitewash, similar coatings. Lime, tar and combined tar and lime, often sprinkled with sand, are frequently used as protective coatings. Like all coatings, they are effective only as long as they are continuous and intact.

Brushing, spraying, swabbing, dipping. These treatments are best for temporary protection during storage, or before impregnation. Various chemicals, such as Dieldrin, Aldrin, DDT, borax, and pentachlorophenol have been found useful. Dipping is probably the most effective and least wasteful. Spraying can often be accomplished with hand-operated sprayers. Brushing and swabbing are effective, but slow.

Steeping. This is similar to leaching, except that the culms or strips are immersed in the preservative instead of in water. Steeping is usually the cheapest and simplest method, and can give very good results. The principal drawback is the long time required, usually about five weeks for moderately good impregnation, and longer to meet more severe conditions such as contact with the ground, or for large thick-walled culms. If the bamboo is split, soaking times can often be reduced by one-third or one-half. Culms should preferably be green, because the pits in the walls of vessels and other cells are open and allow the preservative to penetrate. In dry bamboo the pits are often closed or blocked by dried sap.

Boucherie process. This process is particularly effective for green culms in the round. A flexible tube is clamped to the base of the green culm with branches and leaves left on, and the other end of the tube goes to an elevated tank of preservative (usually about 10 m or 30 ft above ground). If a closed tank is available, air pressure inside the tank, easily applied by a bicycle pump, can be used instead

TABLE 7.3. Recommended Preservatives for Different End Uses of Bamboo.

End use of bamboo	Recommended *preservatives*[a] Dry Bamboo	Green Bamboo	Concentration of preservative (%)	Loading of dry chemical in bamboo (kg/m^3)	Proposed treatment	Expected service life (years)
Use in the open and in contact with the ground (e.g., posts, pale-fencing, etc.)	a			80-128	-Open tank for pressure process	15
	b,c		6-8	8-12	-Pressure process	
		b,c	6-8	5-8	-Modified Boucherie for 6-8 hours or steeping for 35-40 days.	
Use in the open but not in contact with the ground (e.g., bridges, scaffoldings, ladders, etc.)	a			48-80	-Hot dipping or open tank or pressure process	15
	b,c		5	5-8	-Pressure process	
		b,c	5-6	5-8	-Modified Boucherie for 4-6 hours or steeping for 20-25 days.	
Use under cover:						
a) House building, walls, trusses, purlins, rafters, tent poles, etc.	a			32-48	-Hot dipping or open tank or pressure process	20-30
	b,c		4	4	-Pressure process	
	d,e,f		6	8	-Pressure process	
		b,c	4	4	-Modified Boucherie for 4 hours or steeping for 15-20 days.	20-30
		d,e,f	6	8	-Modified Boucherie for 4 hours or steeping for 15-20 days.	
b) Screens, ceilings, doors and door panelling, furniture, etc.	b,c		3	3	-Pressure process	10
	d,e,f,g,h		5	5	-Pressure process	10
		b,c	3	3	-Modified Boucherie for 2-3 hours or steeping for 8-12 days.	10
		d,e,f,g	5	5	-Modified Boucherie for 2-3 hours or steeping for 8-12 days.	10
Prophylactic treatment: green bamboos including splints and round bamboos for eventual full-fledged treatment after air drying.		i,j			Dipping for 5 minutes.	

a The letters in these columns refer to entries in the following list of preservatives:

TABLE 7.3. (cont.) List of Preservatives

(a) Coal tar creosote and fuel oil, 50:50 by weight.
In highly termite infested areas it is preferable to add 1% dieldrin, and in highly decaying areas 1% pentachlorophenol.

(b) Copper-chrome-arsenic composition (Ascu).
A typical composition of this preservative comprises copper sulphate ($CuSO_4 . 5H_2O$), arsenic pentoxide (AS_2O_5) and sodium or potassium dichromate ($Na_2Cr_2O_7 . 2H_2O$ or $K_2Cr_2O_7$) in the proportion of 3:1:4.

(c) Acid-cupric-chromate composition (Celcure).
A typical composition of this preservative comprises 1.68 parts of chromium sesque oxide (Cr_2O_3) (equivalent to 2.5 parts of sodium dichromate), 50 parts of copper sulphate and 47.5 parts of sodium dichrcmate.

(d) Copper-chrome-boric composition.
This consists of boric acid (H_3BO_3), copper sulphate and sodium or potassium dichromate in the proportion of 1.5:3:4.

(e) Copper-chrome-zinc-arsenic composition.
A typical composition of this preservative comprises 28 parts of arsenic acid ($H_2AsO_4 . 1/2\ H_2O$), 25 parts of sodium arsenate ($Na_2HAsO_4 . 12H_2O$), 17 parts of sodium dichromate and 30 parts of zinc sulphate ($ZnSO_4 . 7H_2O$).

(f) Chromated zinc chloride.
This consists of zinc chloride ($ZnCl_2$) and sodium or potassium dichromate in the ratio of 1:1.

(g) Boric acid-borax, 2.5% each.

(h) Copper naphthenate and zinc naphthenate.
These are salts of naphthenic acid and should contain 0.5% of copper and 3% of zinc by weight, respectively.

(i) Dieldrin pentachlorophenol emulsion.
Dieldrin 18% emulsifiable concentrate, 1 part; PCP 12% emulsifiable concentrate, 4 parts; water, 75 parts by weight. To this may be added copper naphthenate (1% copper) in emulsifiable form, 1 part by weight.

(j) Water solution containing borax 2%, sodium pentachlorophenate 1%, and gammexane (water dispersible), 1% by weight.

of elevating the tank. The preservative, preferably water-soluble, diffuses under pressure into and along the culm and into the septa or diaphragms, displacing the sap and forcing it out the open end. Leaves and branches aid the process.

The process is most effective with green bamboos felled when full of sap. It is ineffective with dry over-mature bamboos.

A variant of the Boucherie process, called the steeping process, consists of standing the freshly cut culms, with branches and leaves left on, in a container of preservative to a depth of 30-60 cm (12-24 in). As moisture is transpired through the leaves, the preservative is drawn up the stem. One to two weeks may be needed for the process.

Hot and cold bath. In the simplest process, the culms (which can be dry as well as green) are heated in a tank of hot preservative to a temperature of about 90°C (194°F), and allowed to cool. Good penetration is promoted if the septa of the nodes are first bored through or otherwise broken to give the preservative access to the cavities. A variant of the method, used with wood, is to place the material in preservative at a temperature just above the boiling point of water. Moisture is converted to steam and most of it leaves. When the hot material is then transferred to a cold bath, the remaining steam condenses, creating a partial vacuum to draw in the preservative.

If the preservative cannot be heated, the culms may first be heated in water or in a liquid compatible with the preservative, and then the hot culms immersed in a bath of cold preservative.

Fire retardant treatment. The chemicals used in treatment of wood may be employed with bamboo. A recommended combination preservative for protection against fire, fungus, and insect attack consists of 3 parts ammonium phosphate, 3 parts boric acid, 1 part copper sulfate, 5 parts zinc chloride, 3 parts sodium dichromate, plus water to 100 parts. Such combinations are relatively costly and not always readily available.

Uses in House Building[1]

Frequently a rural village house is entirely built of bamboo, including foundation, frame, floors, walls, partitions, ceilings, doors, windows, and roof. Pipes and troughs are also commonly made of bamboo. Perhaps more frequently, the supporting frame of a house is of wood, usually a locally-available hardwood, and the enclosure of bamboo, either by itself or combined with other materials.

For these purposes, bamboo may be employed in the round as full culm, split in half, split radially into smaller segments or withes, shaved or split tangentially into thin strips, flattened into boards, and woven into mats.

Full culm

This is the commonest shape. To provide a nearly constant diameter, the tapered upper end is often cut off and used for such purposes as roofing or infill. If culms are to be closely fitted, noticeably enlarged nodes are trimmed to the same diameter as the rest of the culm. Curved or crooked culms may sometimes be straightened by heating and placing in forked pegs, or by pushing through holes in a post.

(In Japan, culms are sometimes constrained to grow up through a rectangular form, which forces the culm to take a rectangular shape with rounded corners, often more useful than the normal round shape. It has been speculated that at least the principal posts of a house could be grown this way.)

Half culm

Because bamboo splits readily and straight longitudinally, half or quarter culms are easily made with a heavy knife or by starting splits and then pushing the culm against a wedge or bar. Quarter and finer splits are made the same way. Machines to perform essentially the same splitting action have been developed (Fig. 7.8).

Fig. 7.8. Splitting of Bamboo by Machine.
Courtesy of Building Research Institute, Indonesia.

Splits

This term is used for shapes smaller than quarters. They may be radial or tangential, and be produced by hand or simple slitting machine. The hard outer tangential strip may be used for special purposes where a smooth hard surface is wanted. Radial strips with hard smooth edges are generally used as they are.

Boards

Boards are fabricated by making a series of cuts in a culm at each node, e.g., with an axe, then making one long split, and finally spreading and flattening the culm. The diaphragms or septa at the nodes are removed. Boards are stacked, face side to face side and inside to inside, weighted to prevent curling, and dried. If large sheets are to be made, boards may be plaited, or laid up in cross-plied arrangement as in plywood and glued or otherwise firmly fastened together.

Mats

Mats are made by plaiting splits by hand or by machine (Fig. 7.9.). A variety of patterns may be employed, similar to the patterns available in woven fabrics.

Fig. 7.9. Weaving Bamboo Mats by Machine. Bamboo mats, the traditional building material, have been improved by machine weaving and laminating the resulting panels to form stronger wall units. Courtesy of Building Research Institute, Indonesia.

Foundations[1]

Bamboo is commonly used as supporting posts. Usually large-diameter thick-walled culms are favored to carry the loads, but if such culms are not available, clusters of smaller culms may be lashed together. Posts may be continuous to the roof.

Untreated posts in contact with soil may last, on an average 2-3 years, possibly 5 if conditions are favorable. Here, treatment with a preservative such as creosote or pentachlorophenol, if available, can be especially effective against decay and soil-based insects. More durable foundations consist of naturally durable or treated wood, or masonry such as stone or concrete blocks.

Frames

Frames are often entirely of bamboo, but may be combined with wood. Wood usually provides greater stiffness, and stiffer joints. In earthquake regions, however, the resilience of bamboo frames often provides greater resistance to collapse than wood, and much greater resistance than masonry or earth.

Only whole culms are used for the principal parts of the frame. These include posts, beams, joists, rafters, trusses, studs, and braces, similar to the parts of a standard wood frame. The sizes and spacing of the culms must be selected not only on the basis of load-carrying capacity, but in light of the greater springiness or "bounciness" of the bamboo which often results in closer spacing than for corresponding wood members.

Few species of bamboo can be nailed without splitting. Holes must be drilled for nails and similar fastening such as bolts and pins. For this reason, as well as the scarcity of metal fastenings (formerly at least), the traditional method of fastening is by lashing. Withes are commonly split bamboo, less commonly rattan, also tough vines and bark, but soft iron wire, usually galvanized, is often favored when available. Some typical joints and fastenings are shown in Fig. 7.10.[1]

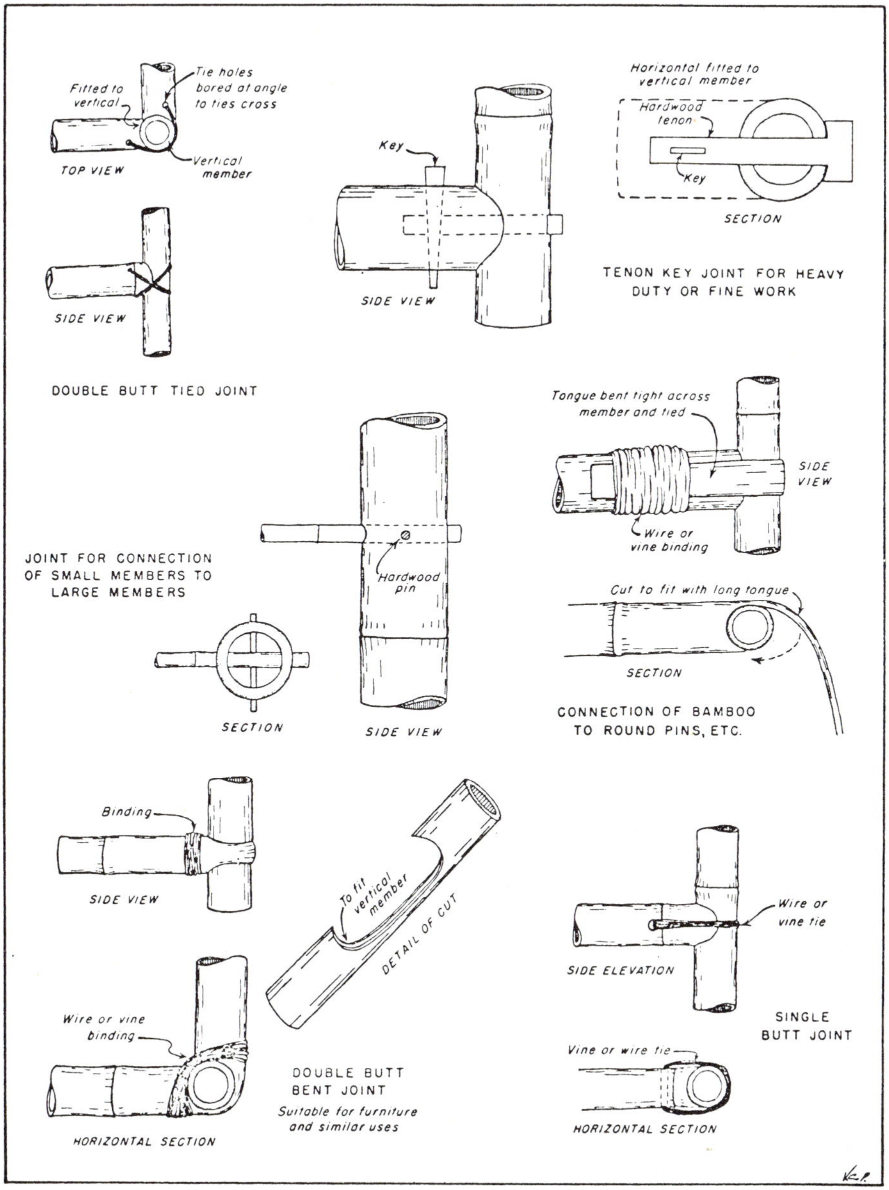

Fig. 7.10. Joints Used in Building with Bamboo.

Floors

Floors may be tamped earth, preferably raised to avoid inundation, and best with a high clay content, stabilized if possible (see section on Soil). Bamboo boards are sometimes pounded to the earth floor after it has been levelled, but before tamping.

It is often preferable to raise the floor for sanitary reasons and for improved ventilation. The floor may be high enough to allow lifestock underneath. In such instances,[1] the floor frame — beams, joists — may be bamboo culms of sufficient size and suitable spacing, and the floor itself closely spaced culms, strips or boards, fastened by lashing, nails, thongs or wire to the floor frame.

Walls, partitions, ceilings

Where substantial walls are wanted, bamboo lends itself well to wattle and daub, or to a double layer of wattles with earth fill between.

Walls may be built of whole or half bamboo culms closely spaced either vertically or horizontally. Vertical placement allows quicker shedding of rainwater with less susceptibility to decay because of water pockets than does horizontal. Thin walls and partitions are commonly made of woven matting, usually with strips horizontal and vertical, but sometimes inclined at 45 degrees, the latter being stiffer against lateral motion such as is caused by winds and earthquakes.

Where ceilings are employed, they are commonly bamboo matting, sometimes with a ventilating opening around the periphery and at partitions. Still better ventilation may be achieved by omitting the ceiling.

Roof

The strength of bamboo recommends it for roof framing and it is widely employed either for the entire frame or for supporting purlins for such roofing materials as thatch, tile, corrugated metal and cement-asbestos, shingles, and bamboo itself.

Halved culms are often laid with alternately convex and concave sides overlapping in much the same manner as Mission tile.

Troughs and pipes

Half culms, with diaphragms at the nodes removed, make good eaves troughs or other troughs to carry away water.

Pipes are sometimes made by forcing a long rod of the proper diameter along the cavity of a culm, knocking out the diaphragms at the nodes. They may also be made by halving a culm, knocking out the diaphragm on one or both halves, and refitting the halves, fastening them together by lashing or otherwise. Drainage pipes to be buried are sometimes made by this means, first cutting notches in one half to allow water to enter, and treating the halves against decay and insect attack before fastening them together.

Reinforcement for Concrete[3]

Because of bamboo's attractive high strength and ready availability in many regions where reinforcing steel is scarce and expensive, the idea of using bamboo as reinforcement for concrete has received widespread attention and trial.

Problems

Two principal problems must be met. The first is the relatively low stiffness or elastic modulus of bamboo (see section on Strength, above). Although its strength is good, it must stretch an appreciable amount to attain that strength, and in so doing it allows the adjacent concrete to crack. Although such cracks are not necessarily disastrous, they are at the least unsightly, and they may lead to deterioration of the concrete. Furthermore, sagging of concrete beams and slabs so reinforced may be excessive because of the low overall stiffness.

The second problem is that of obtaining a good bond between concrete and bamboo. If the two are to act together, the concrete must grip the bamboo strongly enough to prevent slippage between the two when the structure is loaded.

Both problems can be approached by using enough bamboo to keep the stresses so low that the problems do not arise, on the assumption that bamboo is plentiful and cheap. While this does help to overcome cracking and slippage it is wasteful. Tests have shown that if cracking and deflection are not limiting factors, the full strength of the bamboo can be developed if slippage does not occur. The latter can be ameliorated by using compositions of concrete that bond well to the bamboo, or by roughening the surface of the bamboo, particularly the smooth waxy casing. Heavy nodes in the bamboo also help and are frequently sufficient; they are similar to the deformed surfaces of steel reinforcing rods. Small metal wedges driven into the surface of the bamboo act as mechanical keys or shear connectors. Various adhesives applied to the surfaces of the bamboo have been investigated. For a given cross-section of bamboo, multiple thin strips provide more bonding surface than single culms, and the split surfaces bond better than the smooth waxy outer surfaces of the culms.

Moisture and temperature

Other important considerations respecting bamboo reinforcement are moisture sensitivity, and differential thermal expansion.

If too dry bamboo is embedded in concrete, it absorbs moisture, swells, and may cause longitudinal cracks in the concrete, leading to loss of bond and deterioration. If too green bamboo is used, it subsequently shrinks in the dry concrete, causing a loss of bond. To overcome this, various coatings have been tried in an effort to reduce migration of moisture and simultaneously to improve the bond. Brush coats of asphalt and coal tar, if not too thick, have helped reduce moisture problems and improve bond. Coatings of rosin in alcohol, subsequently coated with white lead, have been reported to be helpful. Coatings of epoxies and other adhesives have been investigated.

Bond strengths for untreated bamboo have been reported between 0 and 13 kg/cm^2 (0-182 psi), and for treated bamboo between 4 and 24 kg/cm^2 (56-336 psi).

The thermal coefficients of expansion of concrete and bamboo are different, with the result that marked changes in temperature can cause significant bond stresses capable of leading to bond failure.

Design

The following recommendations are made respecting design[3]:

The amount of reinforcement should not exceed 4-5% of the cross section. Higher percentages tend to result in no increase and possible decrease in strength of the reinforced beam or slab.

If whole culms are used, they should not exceed 1.8 cm (3/4 in.) in diameter. Distal and basal ends should alternate in adjacent culms to provide reasonably uniform reinforcement throughout the length of the member. If seasoned culms or strips are used in important structures, some form of waterproofing should be employed.

If excessive deflection is to be avoided (conforming to the usual limit of 1/360 span), stresses in the bamboo should be limited to 214-285 kgf/cm^2 (3,000-4,000 psi).

Vertical culms or strips are employed in much the same way as stirrups in steel-reinforced concrete. Inclined culms or strips may be employed for diagonal reinforcement.

It should be evident that while bamboo can be employed for reinforcement of concrete, and is an attractive material, it must be used with its properties and peculiarities clearly in mind.

References

1. McClure, F. A., *Bamboo as a Building Material*, United States Department of Agriculture, Foreign Agricultural Service, Washington, D.C.

2. United Nations, Department of Economic and Social Affairs, *The Use of Bamboo and Reeds in Building Construction*, study prepared by D. Narayanamurti and Dinesh Mohan, United Nations, N.Y., 1972.

3. Glenn, H. E., *Bamboo Reinforcement of Portland Cement Concrete Structures*, Clemson College Engineering Experiment Station, Bul. 4, Clemson, S. C., May, 1950.

7.5 SULFUR

Elemental or raw sulfur is found in all volcanic areas of the world and in extensive deposits elsewhere. Sulfur has long been recovered from pyrites and mined ores. More recently, large supplies, now estimated at approximately 30 percent of total production, are coming from the de-sulfurization of natural gas, smelter gas, and petroleum. As high-sulfur coal undergoes similar processing, or as stack gases are cleaned, additional supplies will become available.

For use in building, sulfur's melting point of 113-120°C (235-250°F) is important; high enough to avoid melting in most applications, but low enough to be easily melted with simple equipment. The heat of fusion is 30 cal/kg, a relatively small amount. Ignition temperature is 245°C (475°F) and boiling point is 440°C (830°F).

Among the properties of sulfur favorable to construction[1] is its relatively easy utilization, requiring generally simple handicraft techniques. It is waterproof, adheres strongly to many construction materials, has little or no residual taste or smell (except when burning, when extremely noxious sulfur dioxide is given off), does not unduly affect the skin of most people, and is a poor conductor of heat and electricity. It hardens quickly as it cools, has high resistance to acids and salt solutions, and can be stored indefinitely.

As is true of other materials, sulfur has its drawbacks. Foremost is flammability, although ways of reducing flammability are being developed by the use of relatively inexpensive fillers.[2] It is brittle, although the addition of hydrocarbons can reduce brittleness and increase plasticity. The incorporation of fibers likewise not only increases strength but resistance to impact.

Where sulfur is available at relatively low cost, such as in some developing areas with petroleum production, interest in its use as a binder in building units such as concrete block and tile, in coatings, and as impregnants, is growing.

Concrete and Masonry

Sulfur may be used in place of portland cement or other hydraulic cements in making concrete, or it may be used as an impregnant to improve the properties of masonry units such as concrete block and brick.

Sulfur concrete

Sulfur binds well to mineral and other aggregates used in concrete. Strength properties comparable to portland cement concrete have been reported. For example, compressive strengths of 190 kgf/cm^2 (2,700 psi) have been achieved with 16% sulfur and standard high-strength aggregates. Increasing the sulfur content to 30% and adding 50% sand and 20% other mineral aggregates has resulted in compressive strengths of 440 kgf/cm^2 (6,280 psi).

Such freezing and thawing tests as have been made appear to show that with proper mixes, sulfur concrete with good resistance can be achieved. In many developing areas, freezing and thawing are not important. Because sulfur is insoluble in water, resistance to rain is good.

Mixing of concrete evidently calls for modifications of standard equipment. It is reported that the best and quickest approach is to use a heated mixer in which the aggregate alone is first heated to approximately 115-121°C (240-250°F). The sulfur is separately melted and heated to approximately 121°C (250°F) and added in the mixer. A short time suffices to achieve thorough mixing. The hot mix must be deposited quickly enough to prevent cooling and hardening before the mold or form is properly filled. Segregation must be avoided, as in any concrete. Gloves help to protect hands from hot materials, as well as from possible irritation of the skin. Safety glasses or face masks also provide protection against spattering.

Once the concrete has cooled and hardened, it is ready to use. There is no curing time. Forms or molds can be stripped at once and reused. Waste or reject material can be remelted and reused. In theory, at least, if a building is demolished it need only be remelted and reused with little if any waste.

Most experimentation has been with concrete blocks and tile. Because the hot molten concrete fills molds well, depending on the composition and fineness of the aggregate, fairly intricate interlocking blocks have been produced. A problem is shrinkage, and this must be allowed for in the design and filling of molds. Figure 7.11 shows a house built of sulfur concrete blocks.[4]

Mass concrete awaits the development of suitable simple methods of handling large quantities of the hot mix and getting it into place before it hardens.

Impregnated masonry units

Marked improvements in strength of masonry units, including cinder concrete blocks and various types of brick, have been achieved by impregnating with sulfur.[5] Table 7.4 gives a summary of some test results, including impregnation with acrylic polymers (see Plastics, below). It is seen that compressive strength is approximately doubled by impregnation with sulfur.

Fig. 7.11. House Built of Sulfur-concrete Block.

Table 7.4. Results of Compression Tests on Individual Blocks and Bricks.

Type	Loading (%)	Ultimate load (lb)	Ultimate stress (psi)
Type A concrete blocks, control Average		253,300	4,161
Type A concrete blocks, PMMA treated. Average	12.38	477,200	7,840
Cinder block, control. Average		83,130	2,185
Cinder block, sulfur treated. Average	25.52	169,000	4,446
Solid brick, control. Average		322,670	11,293
Solid brick, sulfur treated. Average	13.25	664,850	23,653
Dartmouth brick, control. Average		257,000	9,430
Dartmouth brick, sulfur treated. Average	16.73	483,500	17,690

In water immersion tests, the same investigators report a 90% reduction of water absorption in brick, and 95% in cinder concrete block.

Impregnation is simply achieved by melting the sulfur in any suitable container such as a pail or drum, with any suitable heat source, immersing the oven-dried units in the molten sulfur and, after withdrawal from the bath, immediately cooling in water to prevent draining the molten sulfur from the large pores.

Surface-bonded masonry

Sulfur reinforced with fiber and treated with a plasticizer that also acts as a thixotropic agent is being used to construct masonry walls by a surface-bonding technique instead of the usual mortar joint.[6]

The procedure is to bond the first course of block to the concrete floor slab by pouring enough of the molten sulfur mix into the cavities of the block to effect a bond to the floor, and allowing it to cool. Subsequent courses of block are laid up dry. The sulfur-fiber mix is either brushed over the joints (2-4 mm thick), or is brushed or sprayed completely over the surface including joints. It must be done quickly enough to avoid cooling and hardening before it is in place (Fig. 7.12).

Fig. 7.12. Applying Sulfur-bonding Mixture to Surface of Dry-stacked Blocks (Southwest Research Institute).

Tests indicate that the wall is stronger than standard masonry.

Similar wall construction is obtained by brushing or spraying a mixture of glass fiber and portland cement paste on dry-stacked blocks. Glass fiber must be alkali-resistant.

Lintels can be made by bonding blocks with sulfur in the form of a beam, and lifting into place.

In fire tests of a small wall exposed to burning cooking oil, the sulfur surface bonded areas caught fire but often after the oil burned away the sulfur surface

Fig. 7.13. Dry-stacked Wedged Blocks with Sulfur-bonding Mixture Applied to Joints Only (Southwest Research Institute)

went out after periods of approximately one hour, leaving an intumescent charred surface.

Fire resistance remains an important consideration, depending upon the application.

Pavements

Sulfur combined with asphalt can provide superior roadbeds,[7] especially where aggregates are scarce or poorly graded, resulting in large voids among the particles. The sulfur fills those voids. It is reported that increased strength and hardness allow reductions in thickness of pavements of up to 20%. Depending upon relative availability of asphalt and sulfur, costs may also be reduced as much as 30%.

References

1. National Academy of Sciences, National Research Council, Washington, D. C., *Roofing in Developing Countries – Research for New Technologies*, 1974.

2. Dale, John M. and Ludwig, Allen C., "Fire-Retarding Elemental Sulfur", *Journal of Materials*, Vol. II, No. 1, March 1967, pp. 131-145.

3. McGill University School of Architecture, *"Sulfur", in the Problem Is*, Part II, Minimum Cost Housing Studies Report, McGill University, Montreal, Canada.

4. Rybczynski, Witoid, *Sulphur Building*, School of Architecture, McGill University, Montreal, Canada.

5. Chen, W. F., Menta, H. C. and Slutter, R. G., "Sulfur and Polymer-Impregnated Brick and Block Prisms", *Journal of Testing and Evaluation*, American Society for Testing and Materials, Philadelphia, Pennsylvania, pp. 292-293.

6. Southwest Research Institute, *Techniques for Sulfur Surface Bonding for Low-Cost Housing*, San Antonio, Texas, 1975.

7. Engineering News Record, *Cost Cutting Sulfur Pavement Making Inroads*, September 30, 1976, pp. 18-19.

7.6 PLASTICS AND COMPOSITES

Although plastics and other synthetic polymers are largely high-technology materials, relatively costly and not generally available in many areas for housing for low-income people, in other regions with plentiful petroleum and a growing petrochemicals industry plastics offer attractive possibilities both by themselves and in combination with other materials. It is therefore pertinent to look into their advantages, limitations, and applications, both existing and potential.

No attempt will be made here to discuss the chemistry or molecular structure of plastics and other polymers. It is sufficient to recognize that there are some 20-30 commercially important classes of plastics with thousands of different formulations modified by copolymerization, plasticizers, fillers, stabilizers, dyes, pigments, and other modifiers. Some are hard and brittle, others are soft and flexible. Some are transparent or translucent, others are opaque. Some stand up well to the weather and sunlight, others deteriorate quickly. Some burn readily, others do not support their own combustion, but all can be destroyed by hot enough fires.

There are two great classes: thermoplastics and thermosets. Thermoplastics soften upon heating, harden upon cooling, and may become brittle at low enough temperatures. They can be temperature-softened and hardened any number of times. Thermosets usually are soft and plastic -even liquid- to begin with but harden irreversibly; that is, once hardened will not soften appreciably. This may or may not require heat. Both types find use in housing.[1,2,3]

Properties

Important properties of plastics in building are mainly strength, stiffness, toughness, heat transmission, thermal expansion, light transmission, durability, and resistance to fire.

Strength

Strength of most unmodified plastics is good. Tensile, bending, and compressive strength are in the same general range as wood parallel to the grain, with some plastics running higher and others lower. Compressive strength is generally equal to, or higher than unreinforced concrete. Moisture generally has little effect, but thermoplastics are markedly affected by temperature, all of them becoming soft and plastic at some temperature, in almost all cases well above normal use temperature. Thermosets are much less affected by temperature, some moderately, others scarcely at all. Comparisons with some other materials are shown in Fig. 7.14.

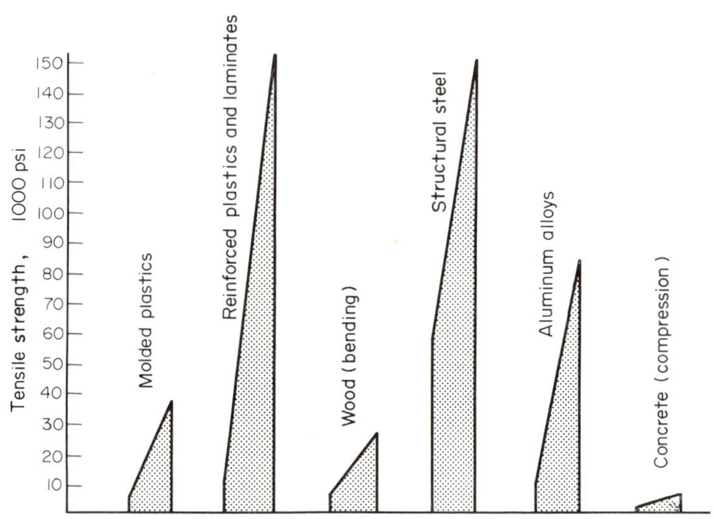

(a)

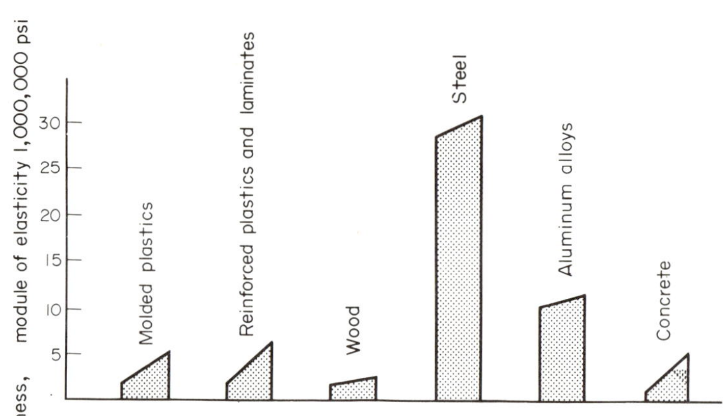

(b)

Fig. 7.14. (a) Comparison, Strength of Plastics and Composites with Other Materials; (b) Comparisons of Stiffness.

Stiffness

Compared with most other construction materials, the stiffness of unmodified plastics, as measured by elastic modulus, is low, substantially lower than most structural wood species parallel to the grain. On an equal-weight basis, however, because of their greater volume, plastics may compare favorably even with steel. As is true of strength, stiffness is markedly affected by temperature, especially among thermoplastics (see Fig. 7.14).

Toughness

This is hard to define and even harder to measure. Depending upon the use, some plastics such as polyvinal butyral, the interlayer in safety glass, are extremely tough. Other plastics are extremely brittle and fracture easily. Hardness and wear resistance, likewise, are hard to define. No plastics are as hard as steel and glass, but some are much harder than others. Melamine, for example, is harder than the usual hard finishes such as varnish, but polyethylene is about as hard as hard wax, and feels waxy. Wear resistance, depending upon the kind of wear, may be good, as in vinyl flooring and nylon bearings.

Heat transmission

All plastics when compared with metals are insulators, similar to wood parallel to the grain (Fig. 7.15). When made into foams, however, plastics provide some of the best thermal insulators available, and rank at least with the best of the mats, fiberboards, woods, cork, and similar materials.

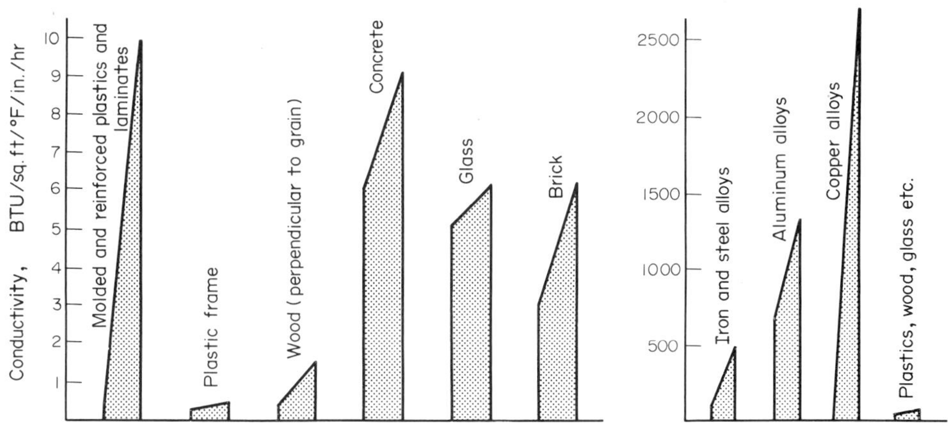

Fig. 7.15. Thermal Conductivity of Plastics Compared with Other Materials

Thermal Expansion

Compared with most construction materials, thermal expansion of unmodified plastics is high. Compared with glass at about 0.0000 8°C, iron at about 0.000012°C and aluminum at about 0.000025°C, plastics range from about 6 to 105 x 10^{-6}/°C. This must be allowed for in design, but can usually be accommodated with little difficulty. Figure 7.16 gives comparisons.

Light transmission

This property may be extremely high and be as good as or better than the clearest glass, or it may be moderate, or the plastic may be completely opaque, depending upon the composition, fillers, colors, and other ingredients.

Durability

For building purposes this term usually refers to weather resistance. Because of the short history of plastics in building (perhaps 50 years at the most, and 25-30

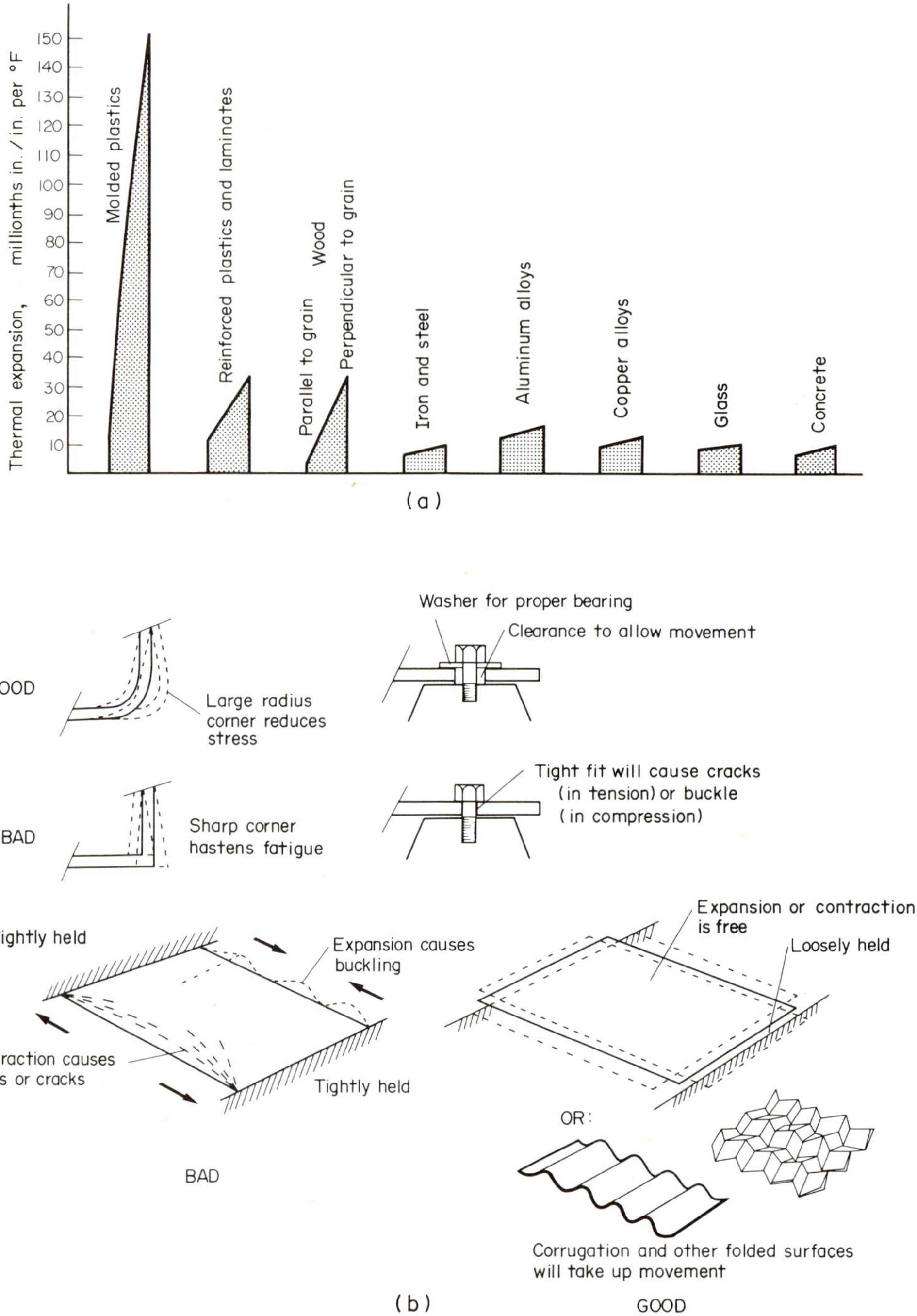

Fig. 7.16. (a) Comparison of Thermal Expansion; (b) Good and Bad Practices.

for any large-scale uses) there is no large body of data respecting actual exposure. Accelerated tests are uncertain, at best. A few materials, such as some acrylics, have had a good history of use outdoors for periods of perhaps 25 years. Others have been known to fail badly in a year or less. With every passing year, more experience is being accumulated and confidence is gradually growing with respect to what plastics are or are not reliable. Any given use should be considered in the light of what experience has accumulated.

Indoor uses, by and large, are much less extreme, and many plastics have given a good account of themselves, although here, as in outdoor uses, any given application should draw upon accumulated experience.

Because plastics do not corrode in the ordinary sense, they can often be used where metals would be unsuitable, and they are used for many applications such as piping, not only for water, but for corrosive chemicals. Many are, however, susceptible to certain solvents and this must be considered. Most are immune to decay and to most insects.

Fire

Being organic materials, all plastics can be destroyed by hot enough fires. Some burn readily, others with difficulty, still others do not support their own combustion and are rated as non-combustible by standard tests.

Like other organic building materials such as wood and fabrics, plastics may, depending upon composition and burning conditions, burn with a clear flame and give off harmless carbon dioxide and water vapor, or, again like other organics, they may give off large quantities of smoke, carbon monoxide, and other toxic or noxious gases. As is true of durability, each given case must be considered and such experience as is available used as a guide.

Plastics Used in Building

Among the important plastics employed in building are the following:

Thermoplastics

Vinyls. This term as usually used refers to polyvinyl chloride. In its unmodified state it is hard and fairly brittle and is used for pipe, plumbing fittings, siding, gutters, downspouts, hardware parts, cases, and similar applications. When modified with fillers or plasticizers or both it finds extensive use as floor sheets or tiles and wall covering.

Vinylidene chloride and copolymers. These provide higher temperature tolerances than plain vinyls and are used for hot water piping and similar uses. They also provide flashing.

Acrylics. These are among the clearest and most transparent of all materials and among the most weather-resistant plastics. They are accordingly much used for skylights, lighting fixtures, and glazing, especially where resistance to breakage is important. They can easily be colored with dyes or pigments and therefore find much use for illuminated signs, furniture, and similar uses.

Polyethylene. In building, among the major uses are piping such as for irrigation, electrical insulation, and vapor barriers, e.g., under concrete slabs.

Polystyrene. One major use in building is as foam for thermal insulation. Others include lighting fixtures and accessories. Normally brittle, when polystyrene is copolymerized with acrylonitrile and butadiene it forms a tough material, ABS, widely used for piping.

Polycarbonate. This is tough, transparent, weather-resistant, and often used for lighting fixtures and other light-transmitting applications where breakage hazard is high.

Nylon. This provides many tough wear-resistant parts for hardware, plumbing fixtures, and similar applications.

Polypropylene. This is similar to polyethylene but somewhat stiffer, harder, and more resistant to elevated temperatures. It is used for hot water lines.

Cellulosics. These are modifications of cellulose, a natural high polymer. Cellulose nitrate, highly flammable as film, provides tough tool handles and commercial movie film; cellulose acetate and acetate-butyrate, much less flammable, provide safety film and other film materials, also tool handles, extruded moldings, and molded small parts such as knobs and handles.

Thermosetting plastics

Phenolics. Phenol formaldehyde provides a great variety of molded parts such as knobs, handles and other hardware parts, components for electrical fixtures, and numerous other items, dark in color.

Urea formaldehyde. This light-colored material can be formulated for white and other light colors, unlike the phenolics, and provides many light-colored molded parts.

Resistance to moist, warm exposures is less than the phenolics. Foams are used for thermal insulation.

Melamine formaldehyde. Like the ureas, these materials provide light-colored moldings, but their resistance to moisture and elevated temperatures is superior. One large use for melamine formaldehyde is in decorative high-pressure laminates. Another is in dishes and cups.

Polyesters. For building applications, polyesters are mainly used in fibrous reinforced plastics, as described under COMPOSITES. The electrical industries use them as potting compounds, to embed delicate electrical parts. The *epoxies* are most used in building as high-strength adhesives and as coatings, but there is some use for cast components such as sinks in chemical laboratories. Both polyesters and epoxies are formulated with marble dust and other fillers to provide cast "marble" lavatories. In polyester concrete, portland cement is replaced by polyester (see COMPOSITES, below).

Polyurethane. Principal use in building is as foam, often foamed in place, especially in odd-shaped spaces difficult to fill otherwise. Soft resilient upholstery foam is also made. Polyurethane provides tough coatings.

Silicones. These, like the fluorocarbons have outstanding resistance to high and low temperatures, and exposure to sunlight, as well as to ultraviolet and infrared radiation. They are hydrophobic, i.e., non-wetting and are therefore employed as water-repellent treatments for masonry. Another important use is as rubbery caulking. Silicone coatings are resistant to extreme conditions.

Plastic-Based Composites

Frequently, plain unmodified materials do not provide properties needed for a given purpose. In such instances, composites may be the answer. Composites may be defined as combinations of materials whose combined behavior transcends or is different from the behavior of the constituents acting alone. For example, many plastics are moderately strong but may not be strong enough for a given application. Furthermore, their stiffness may be too low. On the other hand, many fibers are exceedingly strong and inherently stiff, but, unless somehow held in position, fall into a heap. When fibers and plastics are combined, the fibers are stabilized and kept from buckling by the plastic, thereby making their great strength and good stiffness effective. The combination has strength, stiffness, and stability not possessed by either constituent alone. The steel in reinforced concrete acts in an analogous manner. Wood is a natural composite of cells composed of strong cellulose strands, held together or stabilized by lignin as a matrix or binder.

Composites are frequently divided into three principal categories: particulate, fibrous, and laminar.

Particulate

Particles are embedded in a continuous matrix or binder. The best known such composite is concrete, consisting of fine and coarse particles of stone embedded in a matrix, usually of portland cement paste. If asphalt is the binder it is asphalt concrete. Recently, plastics such as polyester have been employed, providing a plastic or polyester concrete. Another important particulate composite consists of wood particles such as chips, flakes, or both, bonded with a plastics-based binder.

Polyester concrete, although new compared with portland cement concrete or the much older pozzolanic concretes used by the Romans, has exhibited good durability over the 15-20 years it has been used. Examples have shown less cracking than standard reinforced concrete.

Fibrous

Fibers, continuous or cut into short to long lengths, are embedded in a matrix. As indicated above, the matrix stabilizes or supports the fibers so their strength can be realized. Probably the best known such composite consists of extremely strong glass fibers embedded in a polyester resin. These reinforced plastics have found numerous structural and semi-structural as well as non-structural applications in building. They have certain advantages as well as limitations. Among the advantages are:

Lightness, strength, and toughness. Because the fibers and resin are both light in weight, reinforced plastics often provide extremely high strength-to-weight ratios. Because they are also tough, they can be used in thin sections, often as little as 1.5 mm (1/16 in), unlike standard concrete whose brittleness sets limits on thinness.

Formability. These materials have no inherent shape. They must be molded to shape. Consequently, inherently strong, stiff, and efficient configurations can be chosen.

Light transmission. In the thin sections often employed, glass fiber reinforced polyesters can have a high degree of light transmission. They can therefore provide a combination of structure, enclosure, and illumination. Among the limitations are:

Limited stiffness. Even with the addition of glass fiber (about the same elastic modulus as aluminum) the overall stiffness of fiber-reinforced plastic is only moderately high, ranging from less than two to three times that of most structural wood species parallel to the grain. Consequently, it is essential to use the inherently stiff shapes made possible by the formability of the material.

Cost. The unit weight cost is not low compared with other structural materials. It is therefore necessary to utilize the strength, lightness, toughness, and formability to the utmost.

Durability and fire. The observations already made respecting plastics are applicable. Durability can be enhanced, for example, by using acrylic-modified polyesters, and by using surface protective coatings of fluorocarbons. Deteriorated surfaces can also be renewed by field-applied coatings. When inorganic fibers such as glass, or other inorganic fillers, are employed, resistance to fire is enhanced, but susceptibility to fire is not eliminated.

Laminar

Sheets or films of various materials are bonded together in layers by a binder which may at the same time interpenetrate or impregnate them. The thin layers are stabilized and stiffened by being bonded together. Among the most common building applications are the decorative high-pressure laminates used for counters, furniture, doors, and similar uses.

Sandwiches (Fig. 7.17.) Sandwiches are a special case of laminar composites in which two relatively thin, hard, strong, stiff, dense facings are combined with a relatively thick, light-weight, lower-strength, lower-density, lower-stiffness core to provide a combination possessing strength and stiffness combined with low weight. The behavior is much the same as that of an I-beam which has high strength and stiffness because of its geometry. In addition, in sandwiches for building the facings provide weather and wear resistance plus aesthetic qualities, the core provides thermal insulation, and the combination provides acoustical isolation and resistance to fire.

Many materials can be and have been employed for facings. Among them are aluminum, steel, plywood, cement-asbestos board, glass, hardboard, high-pressure laminates, reinforced plastics, rigid plastics, reinforced concrete, polyester concrete, paper, fiberboard, particle board, and others.

Similarly, many materials find use as cores, including foamed plastics, foamed concrete, foamed glass, foamed silicates, fiberboard, gypsum, hardboard, particle board, and numerous cellular forms such as resin-impregnated paper honeycomb, egg-crate constructions and a variety of other grillages.

References

1. Dietz, A. G. H., *Plastics for Architects and Builders*, M. I. T. Press, Cambridge, Mass., 1969.

2. Dietz, A. G. H. (Ed.), *Composite Engineering Laminates*, M. I. T. Press, Cambridge, Mass., 1969.

3. Skeist, Irving (Ed.), *Plastics in Building*, Reinhold Publishing Company, New York, 1966.

4. *Modern Plastics Encyclopedia*, Modern Plastics, McGraw-Hill Publishing Company, New York.

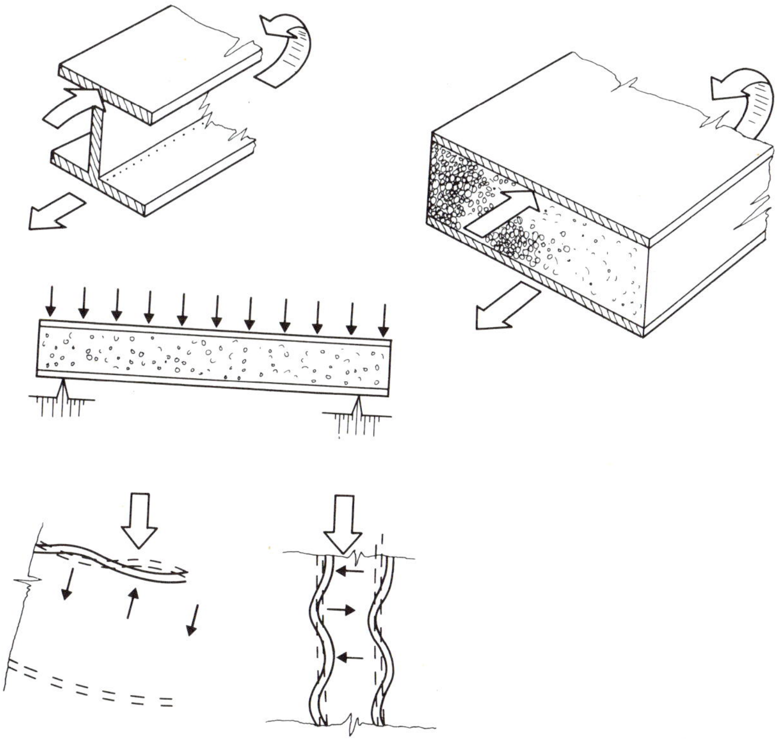

Fig. 7.17. Structural Sandwich, Showing Comparison with I-Beam and Stabilizing Effect of Core against Wrinkling of Facing when Compressed in Beam or Column.

7.7 CONCRETE

As usually used, the term "concrete" refers to a composite of mineral particles, called aggregate, combined with a paste of water and a cementing material, most commonly portland cement, which changes or cures from a soft and plastic into a hard and rigid mass.

Cements

The cementing matrix may be one or a combination of several natural or industrially-produced materials. Among them are:

Asphalt

Asphalt occurs naturally or as the end product of the distillation of crude oil. It may be heat-softened or emulsified, mixed with aggregates, deposited, and formed, hardening in place.

Gypsum

Naturally-occurring gypsum is heated or "burned" to drive off part of the water or crystallization. When mixed with water, it hardens by replacing the water driven off. Retarders control the hardening action, allowing aggregates to be mixed in and the mass to be deposited, and to harden in suitable forms. Gypsum is also widely used as plaster, as masonry mortar, and in gypsum building board.

Lime

When calcium carbonate in the form of limestone, shells, and other raw materials is heated or "burned" to calcium oxide or "quicklime" and "slaked" with water, the resulting calcium hydroxide or lime putty can be mixed with aggregate to form a kind of concrete. Hardening is generally too slow to make useful concrete but lime may be mixed with other ingredients to provide useful cements for concrete (see hydraulic cements, below).

Hydraulic-Setting Binders[1]

These natural or artificial cementing materials react with water and harden either by themselves or when combined with other materials.

Pozzolan. These volcanically-produced materials, including volcanic glass, pumices, tuffs, and diatomite, are used alone, combined with other materials, or as extenders for portland and other cements. Pozzolan combined with lime has been used for centuries.

Fly ash. This by-product of the burning of coal is similar to pozzolans, and forms a useful hydraulic cement when combined with lime. It is also a good extender for other pozzolans and for portland cement.

Blast-furnace slag. This by-product of steel manufacture provides a hydraulic cement when "activated" by combination with lime, sodium and potassium hydroxide, and portland cement. Thus it is similar to other pozzolanic materials. It is a basic ingredient in the manufacture of some portland cements.

Bauxite. Hydraulic-setting materials are made by fusing bauxite and limestone at high temperatures and grinding the resulting clinker.

Limestone. When impure limestone, especially if it has clayey ("argillaceous") constituents, is "burned" the resulting impure lime may have hydraulic characteristics.

Magnesia. Sorel cement, a combination of magnesium oxide and magnesium sulfate, hardens by crystallization when mixed with water, similar to the hardening of gypsum. Sorel cements and magnesium oxychloride are widely used as binders for wood and other fibers, and as flooring material. They are sensitive to moisture.

Portland cement[2]. By all standards, the most important cement for concrete is portland cement. Five types of portland cement are recognized by the American Society for Testing and Materials and other materials authorities throughout the world.

-Type I. General purpose or ordinary cement, useful for the great majority of applications and supplied unless some other type is specified.

-Type II. Modified. Similar to Type I but less heat given off during curing and at a slower rate than Type I.

Type III. High early strength. Hardens more rapidly and attains higher strengths in a few days than other cements. Used where high early strength and hardness are needed. Ultimate strength the same as other cements, but heat of hydration and shrinkage upon curing are greater.

-Type IV. Low heat. Generally opposite to Type III, lower rate of curing and development of strength, less heat given off during curing, less shrinkage, same ultimate strength as other cements. For use where minimum heat and rate of heat evolution are important, as in mass concrete, e.g., dams.

-Type V. Sulfate resistant. Used where alkaline soils, sulfate-containing groundwater, and seawater might cause rapid deterioration of other types of portland cement.

Aggregate[3]

Aggregate is the second and, by volume, the predominant constituent of concrete. Aggregate imparts strength and hardness, reduces shrinkage, contributes to workability and plasticity, and reduces cost. The combination of cement paste and aggregate gives concrete its properties.

Ideally the sizes are graded to fill virtually all the voids among the particles, allowing for a thin cement paste to bind the particles firmly together. In practice this ideal is approximated by mixing various proportions of large and small particles.

The terms sand and gravel are frequently used. Because they usually refer to naturally disintegrated material, and much aggregate is derived by mechanically crushing rock or stones, the terms fine and coarse are better. The dividing line between the two is arbitrary. Fine aggregate passes through a standard 3/8-inch (0.94 cm) mesh and almost all passes through a standard No. 4 (4.75 mm.) mesh sieve. The lower limit of fine aggregate is the No. 200 sieve (75 μm). The upper limit of size for coarse aggregate depends upon the use of the concrete. In mass concrete boulders and cobbles may be permissible, whereas in reinforced concrete the aggregate must easily pass between reinforcing rods, and in topping for floors nothing coarser than fine aggregate may be permitted.

Sources

Conventional mineral aggregates are derived from natural sources and from industrial processes. "Natural" aggregates are sand and gravel, and crushed stone and crushed gravel. Any of the igneous, sedimentary, and metamorphic rocks can be employed, but those rocks that tend to form flat, elongated, or flaky particles (slate, gneiss) are less desirable than those that provide chunky, more nearly spherical particles. The most common industrially-produced aggregate is blast-furnace slag. Crushed brick is used in some places. Crushed concrete rubble has been employed with variable success.

Lightweight[4]

The foregoing conventional aggregates result in heavy concrete. Lower weight is frequently desirable, especially if the strength and hardness of standard concrete are not needed. Lightweight natural or industrially derived aggregates are therefore employed. Among them are: expanded shale, clay, and slag; sintered fly ash, scoria, and pumice; and perlite, vermiculite, diatomite, and cinders. Whereas standard high-density "stone" concretes weigh in the vicinity of 2240 to 2400 kg/m^3 (140-150 pcf), lightweight concretes usually range from approximately 1360 to 2080 kg/m^3 (85 to 130 pcf). With better grades of aggregates such as expanded shale, clay, and sintered fly ash, and the higher weights, strength may be equivalent to some heavy-weight concretes. Aggregates such as perlite, vermiculite, and similar materials tend to result in lower strengths.

Foams. For still lower weight, pastes of cement and fine aggregates may be blown into foams. This is achieved mechanically, as by beating in air, or chemically, as by adding aluminum powder which reacts with the cement to give off hydrogen gas which causes the mass to rise like bread dough. Foamed plastic or glass pellets may be mixed with cement paste, with or without very fine aggregate.

Agricultural wastes

In developing regions without volcanic rock or other suitable lightweight aggregate, experiments have been conducted with the use of various agricultural wastes as the aggregate. The successful manufacture of concrete blocks from a lightweight concrete made with rice husks has been reported.[5] A report has been published in which a rice husk concrete weighing 103 pcf developed a compressive strength of 1500 psi while a 73 pcf concrete developed a compressive strength of 335 psi.[6] The stronger sample was judged usable for the precasting of reinforced elements, and the weaker concrete suitable for concrete blocks for interior partitions.

Other agricultural waste materials found in developing regions which may be suitable for such lightweight concretes include coffee bean husks, coconut shells, and bagasse.

Water

The third essential ingredient in concrete is water. Enough water must be added to hydrate the cement and provide a workable concrete, i.e., one that is plastic enough to flow in a form or mold and fill it completely without voids. Excess water is subsequently removed as the concrete dries after curing.

Water that is drinkable is usually satisfactory for concrete. Deleterious substances to be avoided in water include, for example, excessive salt (seawater is generally not satisfactory), acids such as tannic acid, sugar (an inhibitor), and excessive clay. If the water is potable, these substances are usually not in excess.

Admixtures

The straight cement-aggregate-water mixtures may not result in concrete that has all the necessary properties. Admixtures are incorporated to supply missing features or to enhance others. Among the important ones are air entrainers, accelerators, finely divided minerals, pozzolans, and colorants.

Air-entraining agents

The disintegrating effect of repeated freezing and thawing can be greatly reduced by entraining great numbers of small discrete air bubbles of proper size and distribution in the concrete. Optimum quantities are 3-8% by volume. Examples of air-entrainers are neutralized Vinsol resin (lignin residue) and sulfonated hydrocarbon salt of triethanolamine.

Accelerators

Concrete may harden too slowly for some purposes. Setting can be accelerated by a variety of soluble chlorides, carbonates, silicates, fluosilicates, hydroxides, and organic materials. Calcium chloride is the most commonly used. It also lowers the freezing temperature. However, not more than 2% by weight should be used because it may cause corrosion of steel reinforcing, it increases drying shrinkage, it lowers the resistance to freezing and thawing and to attack by sulfates, and can promote electrolytic corrosion when steel and aluminum (such as conduit) are both present in concrete.[7]

Fine mineral admixtures

When a concrete mix is poorly graded in aggregates and may therefore not be as workable or plastic as is desirable, finely divided materials can help to increase the volume of paste and the workability of the concrete. Such additives may be inert (ground quartz, limestone, hydrated lime, talc), cementitious (natural cement, hydraulic lime, blast-furnace slag), or pozzolans.

Pozzolans

These by themselves often have little or no cementitious action, but when combined with lime a chemical reaction occurs that provides a hydraulic cement. To the extent that this occurs, the portland cement content of a concrete can be reduced. If high strength is not needed, the portland cement may be eliminated. The plasticizing action of pozzolans may lead to a reduction in water content, thus improving the water to cement ratio.

Proportioning[8]

Since the usual objective in proportioning concrete is to achieve maximum density and minimum voids, an effort is made to grade the coarse and fine aggregates so as to fill the crevices as nearly as possible, utilizing the cement paste to fill the last remaining openings as well as to bond the particles together.

For small simple applications, fairly rough proportions are often adequate. Thus, taking one bag of cement as one cubic foot and as the basis of measurement, a typical fairly rich mix might call for one bag (one cubic foot) of cement, two cubic feet of sand (fine aggregate) and 4 cubic feet of gravel (coarse aggregate), or the familiar 1-2-4 mix. For less rich concrete the proportions might be 1-2 1/2-5 or 1-3-6. In each case, sufficient water must be added to provide a workable mix. Since approximately the same amount of water is needed for equal volumes of mix, the ratio of water to cement is higher for the leaner mixes, with a corresponding reduction in strength.

For applications in which strength, density, or both are important, the mix is more closely controlled. The desired strength, for example, is specified. The water-cement ratio is determined. Coarse and fine aggregates are screened to determine the distribution of sizes, from which the fineness modulus is determined. The proper combination of fine and coarse aggregate is developed together with the amount of cement and water to provide the desired property. If needed, the types and amounts of admixtures are also determined on the basis of the mix and the desired end result.

Mixing, Placing, Curing[9]

No matter how carefully a concrete mix may be designed to meet a certain specification such as strength, it can fail if it is not properly mixed, placed, and cured.

Mixing

Concrete may be mixed at the job, either by hand or machine, or it may be purchased already mixed. Hand mixing is becoming increasingly rare, but must be used frequently enough to make an understanding of its technique necessary.

Hand mixing, representative technique:

- -Lay five or six planks on the ground to form a mixing platform.
- -Station two men, one at each side at the upper end of the platform.
- -Deposit a wheelbarrow of sand at upper end. Three cubic feet are comfortably carried in a wheelbarrow.
- -Spread on it the proper amount of cement.
- -Deposit two wheelbarrows of aggregate.
- -Turn the dry mass over twice, down the platform.
- -Add proper amount of water to give a workable mixture. This usually runs about 5 to 8 gallons per sack of cement.
- -Turn mass over once more and shovel into wheelbarrow for depositing in forms.

Machine mixing. Exactly the same proportions are used, but the actual labor of turning over the mass is done by a machine consisting of a rotating drum containing oblique blades. Machine mixing speeds the whole operation and provides a pacer for the job but does not necessarily result in better concrete.

Workability

"Workability" is one of the most important attributes of good concrete. Plasticity is another term. It means the property of flowing easily (not running like water) into and filling all the parts of a form completely without segregation of the ingredients. Good, workable concrete does not show free water on the surface; when poured out of a bucket it forms a pancake curling under at the edges, and it quakes when shaken or prodded. Small variations in water content make large differences in workability, strength, and density.

Depositing

Concrete must not be dropped from a great height, best not over 3 to 4 feet. For greater height a trough or chute should be used. Dropping, causes sand, water-cement paste, and aggregate to separate. Individual loads should be spread out, not merely dumped. Spreading remixes the ingredients which may have started to separate and mixes the entire mass in the forms uniformly.

Working

Once in the forms, the concrete must be worked into all corners and any entrapped air must be removed. This is done by spading and rodding the whole mass or by using a mechanical vibrator. If working is neglected, "honeycombs" or pockets form in which there is no concrete. Too much working tends to separate the ingredients, the lighter material rising to the top or to the face of the form as a milky scum, consisting of over-watered or "drowned" worthless cement which soon disintegrates.

Curing

Concrete should not be permitted to become too hot, too cold, or too dry. The best curing temperature is between 10 and 21°C (50 and 70°F). Above that range the speed of set is accelerated but strength and hardness are impaired, largely because of too rapid evaporation of water. High temperatures are therefore best offset by constant sprinkling.

Low temperatures in themselves do not affect concrete, they merely retard the set and stop it completely below freezing temperature. Once frozen, concrete is merely dormant, and after it has thawed, continues to set with little harmful effect. Repeated freezing and thawing, however, may ruin the concrete. Therefore, concrete which has once frozen should be kept heated until it has passed well beyond the final set. Hard concrete rings when struck with a hammer; frozen concrete gives a dull thud and shows a wet spot where the hammer strikes.

Concrete permitted to dry before it has completely set soon disintegrates because hydration of the cement is incomplete. After the initial set, therefore, concrete should be kept as wet as possible (submerging is excellent treatment) until well after the final set and after forms are removed. Several days to a week of wetting and covering with wet burlap or plastic film, or other coating, assures good concrete.

Concrete Blocks

Standard concrete has long been made into blocks laid up in much the same fashion as brick to provide walls, partitions, piers, and other parts of the building. Many different shapes and sizes are available.

Soil-cement and other types of stabilized earth are also commonly made into blocks, in addition to the ancient widely-used adobe blocks. To expedite manufacture and improve quality of such blocks, a number of hand or power-operated molds have been devised. One of the best known and widely used is the CINVA-ram developed by the Centro Inter-Americano de la Vivienda operated by a combination of American countries in Caracas. In this simple device, a measured amount of material is placed in the mold and quickly compacted by a hand-power, lever-operated ram. The compressed block is ejected and set aside to cure.

Because the laying of such blocks requires a considerable amount of skill, there have been numerous attempts to develop self-aligning blocks, often interlocking, so that even unskilled labor can lay up an acceptable wall. This is especially important for self-help construction. Another common objective with interlocking blocks is to eliminate the mortar joint usually required. Evidently, such blocks must be made with a considerable degree of precision.

Concrete in Special Climatic Conditions

Many developing countries have tropical climates. As a result, the behavior of concrete building materials in hot climates has been an area of considerable research activity. Among the problems which current research is trying to solve are the following:

- Concrete cured at the elevated temperatures common in tropical regions (30°C and more) shows significant loss in ultimate strength.

-Concrete cured in dry, hot regions faces the possibility of dessication of the concrete before the concrete is reasonably cured, thus resulting in drastic losses of strength.
-Concrete exposed to ocean water or breezes in tropical climates has a tendency to deteriorate rapidly.
-Laboratory tests for the quality control of concrete do not duplicate satisfactorily the field conditions in tropical regions, and consequently, new more direct testing methods are needed.

The RILEM Conference on Materials and Structures (November 1963) dealt extensively with problems.[6,10,11,12] The prevention of the deterioration of concrete under extreme climatic conditions by the admixture of various synthetic resins, was the subject of numerous papers at the 1967 RILEM Conference on Synthetic Resins in Building Construction.[13]

Steel Reinforcement of Concrete in Tropical Climates

The deterioration of steel reinforcing rods in concrete is known to occur at accelerated rates in tropical climates. Reinforced concrete structures exposed to ocean breezes in tropical regions may deteriorate at dangerous rates. In one case study, for example, a 180 m concrete bridge in northern Morocco was judged unsafe for use after only 35 years, owing to the extreme deterioration of the steel reinforcement. The deterioration was shown to be the result of anode-cathode reaction between the steel reinforcement and porous areas of the concrete rich in salts carried to the concrete by ocean fogs.[5] Problems of this type might indicate possible research into inert coatings for the steel reinforcing rods. Indeed, epoxy resin coatings have been shown to be effective in protecting steel from deterioration under tropical conditions.[14]

References

1. Bushnell, C. E., "Hydraulic-Setting Cement Binders", Appendix E in *Roofing in Developing Countries*, National Academy of Sciences, National Research Council, Washington, D.C., 1974.

2. Waddell, J. J. (Ed.), *Concrete Construction Handbook* (2nd Ed.), McGraw-Hill Book Company, 1974, Chapter 1, Cement (J. J. Waddell).

3. Waddell, J. J. (Ed.), Chapter 2, Aggregates (F. E. Legg, Jr.).

4. Waddell, J. J. (Ed.), Chapter 31, Lightweight Concrete (Perry H. Petersen).

5. Cockcroft, John, *Technology for Developing Countries*, Overseas Development Institute, Ltd., London, 1966.

6. Arenas, P. G. and Orjuela, D. C., "Lightweight Concrete Made with Portland Cement and Rice Hulls" in *An Inter-American Approach for the Seventies: Materials Technology*, American Society of Mechanical Engineers, New York, 1970.

7. Waddell, J. J. (Ed.), Chapter 4, Admixtures, Curing Mediums, and Other Materials (Raymond J. Schutz).

8. Waddell, J. J. (Ed.), Chapter 11, Proportioning Mixes (J. F. Artuso).

9. Dietz, A. G. H., *Dwelling House Construction* (3rd Ed.), M.I.T. Press, Cambridge, Mass., 1971.

10. Chen, W. F., Menta, H. C. and Slutter, R. G., "Sulfur and Polymer-Impregnated Brick and Block Prisms", *Journal of Testing and Evaluation,* American Society for Testing and Materials, Philadelphia, Pennsylvania, pp. 283-292.

11. Cox, F. B. and Geymayer, H. G., *Expedient Reinforcement for Concrete for Use in Southeast Asia,* U. S. Army Corps of Engineers Technical Report C-69-3, February, 1969.

12. RILEM Bulletin, *Behavior and Durability of Building Materials in Hot Countries,* September, 1964.

13. RILEM Bulletin, *Synthetic Resins in Building Construction,* 1967.

14. George, J. and Hajela, R. B., *Epoxy Resin Protection for Mild Steel Reinforcing Rods in Brickwork,* Synthetic Resins in Building Construction, RILEM, Paris, 1967.

7.8 CEMENT-ASBESTOS

Cement-asbestos products for building consist of a mixture of portland cement and asbestos fiber, wetted, and pressed into a board or sheet, or formed with other shapes such as pipe. Boards may be flat, corrugated, or have other configurations. [1,2]

Organic fiber is added in some cases to promote resiliency and ease of machining. Curing agents, water-repellent admixtures, pigments, mineral granules, and mineral fillers may be added, depending upon the end use of the product and the desired appearance.

Depending upon the degree of durability required for outdoor or indoor exposures, sheets may be of general utility grade, easily cut and worked and nailable (below specified thicknesses) without drilling, or may be harder, denser, smoother, less flexible, stronger, and require more drilling and machining than the utility grader. Other grades are made specifically for roofing and side walls. Allowable water absorption and strength and stiffness when wet and dry depend upon the expected exposure.

For housing, especially in warm areas, many large deeply-corrugated sheets have been developed that are capable of spanning the full distance from wall to wall, or wall to ridge, of small houses, under moderate loads, as described in other chapters of this book.

Although asbestos is the principal, and frequently the only, fiber employed, organic fibers are incorporated, as indicated above. Glass fiber is also found, and mixtures of fibers are becoming increasingly common.

References

1. Dietz, A. G. H., "Building Materials", Section 2 of *Building Construction Handbook,* (3rd Ed.), Frederick S. Merritt, McGraw-Hill Book Company, New York, 1975.

2. *Specifications for Cement-Asbestos Products*, ASTMC220, 1, 2, 3, American Society for Testing and Materials, Philadelphia, Pa.

7.9 CLAY PRODUCTS

Materials

The very fine constituent of soil known as clay (see section on Soils) has long been employed in making building products by forming the wet plastic raw material into desired shapes and then heating it to high temperatures ("firing" or "burning") to convert it into a hard, infusible, insoluble mass, resistant to weathering and fire.

Six major classifications of clay are:

1. China clay or kaolin, for fine grades of pottery;
2. Ball clay, for pottery, ceramic tile, terra cotta;
3. Fire clay, for refractory materials, fire brick, ceramic tile, and structural tile;
4. Bentonite, for foundries and petroleum industries, and for slurry methods of excavation;
5. Fuller's earth, for absorbent and filtering agents, and in oil refining;
6. Miscellaneous clays, for brick, structural clay tile, terra cotta, ceramic tile, and cement.

Shale, or consolidated clay, is used in addition to clay.

As may be inferred, the clays used for brick and other building units are quite variable in composition, and properties of the resulting units vary correspondingly, not only because of composition but because of differences in manufacturing.

Manufacture

Clay and shale brick are made by three different processes, known as soft mud, stiff mud, and dry pressed. In all three processes, the raw material is molded to desired shape, dried, and burned in kilns.

In the soft-mud process clay is mixed with water to a soft consistency and pressed into molds. If the mold walls are wet with water to prevent sticking the bricks are called water struck; if the mold walls are sanded, the bricks are sand struck.

In the stiff-mud process, high-consistency clay is forced through a die as a continuous ribbon whose cross section is equal to either the end or the flat side (bed) of the unit. It is cut by taut wires into individual end-cut or side-cut units.

In the dry-pressed process "dry" consistency clay is forced under heavy pressure into multiple or gang molds. This process produces the greatest accuracy.

Extruded bricks frequently have holes formed in the ribbon during extrusion. Such holes not only lessen the weight of the brick, they promote uniformity in burning and reduce both the burning time and tendency to crack. Pressed or molded bricks often have indentations or "frogs" in the beds. These depressions help to act as keys in the mortar beds.

After forming and drying, brick and other clay units are burned in kilns. The durability of the bricks depends upon the degree of burning achieved. Bricks nearest the fire are burned the hardest and may be overburned, dark, and misshapen although extremely durable. Bricks farthest from the fire may be too lightly burned, soft, light in color, and known as "salmon" bricks, lacking in great durability and strength but good enough for mild conditions. Hard or well-burned bricks have had the correct degree of burning and are suitable for general use outdoors and indoors.

Although the great majority of bricks are various shades of red, cream, and buff because of the oxides naturally present in the clay, other colors are achievable by introducing different oxides and mixtures of clays.

Surfaces, in addition to the water-struck and sand-struck surfaces usual with common brick, include combing, scoring, stippling, hammering, and various degrees of roughness to provide textures on face bricks or to enhance the bond to mortar. Ceramic glazes are glossy surfaces fused to the brick, clear or colored, glossy or matte. Salt glaze is a lustrous finish obtained by introducing the vapors of various salts during burning.

The American Society for Testing and Materials divides common brick into three grades:

- SW: for rigorous conditions including frequent exposure to heavy rainfall followed by freezing and thawing, extremes of temperature;
- MW: for average exposures to moisture and to minor freezing conditions;
- NW: for exposure to minimum moisture and freezing conditions.

Other Brick

Fire Brick, a refractory brick made of fire clay containing approximately 50% alumina, and used for fireplaces, etc.

Sewer Brick, special clay or shale brick for construction of sewerage for household and industrial wastes as well as storm water.

Concrete Brick, made by pressing a dry mixture of portland cement and sand into molds, removing the molds, and curing in aqueous fog or steam.

Sand-Lime Brick, made of a mixture of sand and hydrated lime, pressed into molds, and cured by steam which promotes bonding of lime and sand by a chemical reaction between them.

7.10 MORTAR

Materials

Mortar is a plastic mass consisting of mineral particles and a binder or cementitious material. It is used to fill the interstices between masonry units. As it hardens, it bonds to the units, binding them together and at the same time sealing the interstices.

Common mineral particles or aggregate (see section on Concrete) are sand, crushed stone, slag, and burned shale. Other aggregates are used for special purposes.

Cementitious materials are usually hydrated or hydraulic lime and portland cement, or a variety of other cements, such as pozzolonic, natural, masonry, and slag. White portland cement is used for light colors.

The third essential constituent, water, must be clean and free of deleterious substances such as alkalis, acids, organic matter and excessive amounts of clay and silt. A convenient stipulation is that it be drinkable.

Color is generally obtained by adding inorganic pigments not to exceed 10-15% by weight of cement (2-3% if carbon black).

Admixtures may be employed for various purposes. Among them are air-entrainers, freezing-point depressants, water-repellants, and water reducers.

Composition

Composition of mortar varies greatly with the intended use, and the constituents are chosen accordingly.

Lime putty has great sand-carrying and water-retaining capacities in addition to providing plastic workable mortars that bond well to many masonry units. The putties by themselves tend to shrink markedly but the sand helps to offset this. Compressive strength is only moderate, and the material hardens slowly. High-magnesia lime putties may swell considerably upon hardening and have a disruptive effect.

Portland cement provides mortars that develop high compressive strength and harden quite rapidly by hydration of the cement. Such mortars tend to be harsh and less workable than lime mortars, shrink and swell with changes in moisture content, and frequently tend to break the bond with the masonry units.

Many of the other cementitious materials listed above, of natural or industrial origin, harden more slowly than portland cement, and the mortars develop less strength, but are less harsh and more plastic than straight portland cement mortars. Because these cements are often more variable in composition than portland cement, the mortars also vary more.

Mortars are commonly combinations of lime, various other cements (mainly portland) and aggregate. By combining lime and portland cement, an attempt is made to overcome the low strength and slow hardening of lime but to take advantage of its workability and bonding power, while utilizing the strength and quick hardening of portland cement (or other cements) but minimizing harshness and reduced workability. Just what combinations to employ depends upon the ultimate use.

ASTM classifies mortars by types A, B, and C with corresponding property and composition specifications. Typical applications of these mortars are:

Type A. High strength, for general use, especially reinforced masonry and masonry below grade.

Type B. Medium strength, for general exposed masonry above grade.

Type C. Low strength, for non-bearing walls of solid masonry units, partitions, and load-bearing walls where compressive stresses are below 100 psi.

Bonding

The bonding or adhesion between masonry units and mortar is evidently of crucial importance. It is affected by the water retentivity of the mortar, the absorption or rate of "suction" of the masonry unit at the time it is laid, the workability or consistency or flow of the mortar, and the manner in which the point is made. Water retentivity should be high; and the water content as high as consistent with workability. Correspondingly, the rate of absorption of brick should be moderate, but the brick should not be non-absorbent. Too much suction results in a dry interface, too little suction leaves it too wet, and in either case, poor adhesion results. When the mortar joint is made, it should be full and pressure should be applied ("shoved" joint) instead of just buttering the ends of the brick.

7.11 FUTURE TRENDS

For the immediate future, materials now used will undoubtedly continue to be employed. These will include soil, wood, bamboo, clay products, gypsum, concrete, and similar familiar materials. As time passes, adaptations and combinations can be expected to emerge as new industries come into production, and industrialization of the building process increases in scope.

Soil, for example, will undoubtedly continue to be used widely, but modifications can be expected to increase in importance. Soil stabilized for enhanced durability should become more important, with increased use of block. Burned clay products such as brick and tile can be expected to become more common, especially in those areas in which fuel is increasingly available. High-strength mortars should make masonry construction more efficient and economical than it now is.

Concrete, already firmly established as the pre-eminent material in many areas, especially for urban housing, can be expected to continue in that position. Modifications, as by impregnation of concrete blocks, may become important for demanding applications if economy can be demonstrated. Lightweight concrete to reduce weight of tall structures, should increase in importance, provided the necessary aggregates can be had economically. More or less unconventional aggregates will probably be used increasingly as local supplies of conventional aggregates are exhausted. Precasting and prestressing are expected to increase, particularly in urban areas.

Conservation of energy, especially in energy-short areas, should become increasingly important. This means energy needed to produce and install materials, and energy for the operation of buildings. Materials requiring relatively little energy to produce should be favored. Wood demands about the least energy per unit weight of the common materials. Gypsum requires less energy to produce than portland cement. Byproduct and natural materials having cementitious properties, such as fly ash and pozzolans, should find increasing use where available.

For energy conservation, insulating materials are needed. These include lightweight foams, fibrous masses, and other low-conductivity materials. On the other hand, mass helps to absorb excess energy, such as solar energy, and to release it when the source of excess, such as the sun, is removed. Mass should be obtained with low-cost materials. These need not necessarily have great strength.

It is to be expected that composite materials will become increasingly prominent as one means of extending the availability and usefulness of materials. Fibers, both natural and synthetic (e.g., sisal, jute, bamboo fiber, glass, asbestos) embedded in a variety of matrices such as mortar and resins, can provide strong, tough, lightweight structural materials. By-product and waste materials; such as agricultural and industrial wastes, can be combined with binders to provide blocks, boards, and molded shapes. The principle of the structural sandwich, utilizing thin facings of hard strong materials with thick cases of lightweight materials, can provide lightweight efficient structures. Indigenous and low-cost materials can be employed.

To summarize, conventional materials can be expected to be used in the future, together with modifications, combinations, and now little-used or unfamiliar materials. A revolutionary "new" material that will sweep away all others is not in sight, but evolution and adaptation will occur.

One development that can bring about intensive adaptation is solar energy for heating, cooling, and refrigeration. This can be expected to become highly important in favorable areas. Where it does, materials for the capture, storage, and release of that energy will be of utmost importance, and now-unfamiliar materials will play an important role.

8

CRITERIA FOR SEISMIC DESIGN OF UNREINFORCED MASONRY AND ADOBE LOW-COST HOUSING

Reza Razani

8.1 INTRODUCTION

Motivations

During the past fifteen years more than 30,000 people have lost their lives, and in excess of 50,000 adobe and masonry building units have been destroyed in earthquake stricken towns and villages in rural Iran. During this century more than one million people have died due to earthquakes throughout the world. In Table 8.1, the number of earthquake fatalities during this century (from 1900 to 1973) in 41 developing and developed countries is given. More than 90% of these earthquake deaths have been caused by the collapse of unreinforced masonry and adobe buildings. A major percentage of the deaths has been inflicted on the low-income people in the seismic regions of the less developed countries.

In view of all these human losses in recent times, it is unfortunate that no concrete effort has been made to study the problems relating to minimization of damage or destruction of low-cost masonry and adobe buildings. Research and development for improving the seismic resistance of this type of buildings has not seriously attracted the attention of research workers and of research sponsors. Extensive amount of earthquake engineering related research and development has been carried out in the research centers and universities of the developed and developing countries. These research and development activities are needed to gain a better understanding of the earthquake phenomena, and the mechanism of resistance against the destructive effects of the earthquakes. However, most of these research activities deal with modern buildings, especially with multi-story reinforced concrete and steel structures or other major constructed facilities, where a high level of investments are concerned. So far only a very small amount of research has been carried out on improving the seismic behavior of low-cost masonry and adobe buildings in few developing countries, notably India. It is surprising that even in some seismically active less developed countries, where on the average each three or four years, thousands of people perish under the debris of falling masonry and adobe buildings, most research and development on earthquake engineering deals with modern engineered buildings.

Review of the technical papers published in earthquake engineering journals and in the proceedings of the past regional or world conferences on earthquake engineering points out that while the collapse of unreinforced masonry and adobe low-cost housing (UMALCH) has caused more than 90% of the earthquake fatalities during this century throughout the world, only less than 5% of the published technical papers deal with this type of buildings. The percentage of the total research funds allocated worldwide to the study of seismic behavior and design of these type of buildings is much less than 5%.

TABLE 8.1. Total number of fatalities due to earthquake in various countries (1900-73)[1]

	Country	Total Number of People Killed
1.	China	269,279
2.	Japan	163,245
3.	Italy	92,465
4.	U.S.S.R.	67,089
5.	Iran	58,176
6.	Peru	57,308
7.	Turkey	56,799
8.	Pakistan	54,448
9.	Chile	37,827
10.	Nicaragua	13,502
11.	Morocco	11,511
12.	India	8,523
13.	Taiwan	6,969
14.	Argentina	6,228
15.	Ecuador	5,864
16.	El Salvador	2,989
17.	Costa Rica	1,726
18.	Colombia	1,587
19.	Greece	1,563
20.	Algeria	1,499
21.	Yugoslavia	1,326
22.	Mexico	1,302
23.	Burma	831
24.	U.S.A.	791
25.	Venezuela	577
26.	Albania	551
27.	Afganistan	534
28.	Romania	469
29.	Libya	450
30.	New Zealand	276
31.	Israel	264
32.	Bulgaria	252
33.	Lebanon	136
34.	Ethiopia	85
35.	Cyprus	40
36.	Guatemala	43
37.	Portugal	22
38.	South Africa	17
39.	Tunisia	14
40.	Spain	8
41.	Iraq	6
	Total	926,591

[1]Data of this table is given by Ambraseys[12]

Many research workers in the developed and developing countries do not seem to be motivated to do research on improving the seismic resistant of UMALCH. Some of them believe that there is no feasible solution for the problem of seismic strengthening of UMALCH, and the present problem will gradually wither away in the process of development, industrialization, and urban or rural renewal and reconstruction. Therefore, in their opinion, this "passing phase" does not merit intensive research and involvement. Some research workers handle only those problems that can be easily modeled and can lend themselves nicely to the rigor of high level mathematics, dynamic computer analysis, and the use of high technology, high cost, experimental investigation. Only those structures made of steel, reinforced concrete or reinforced masonry with known engineering properties and familiar, well-behaved, stress-strain diagram or hysterisis loop, attract their attention.

Seismic design of UMALCH is constrained by social, economical, financial, and technological factors. The modeling of this type of structures is complex, the engineering property of the material is undependable and unknown, and the seismic behavior of the structure is complex and erratic. The author believes that finding feasible and implementable solutions for the problem of earthquake resistant design of UMALCH for less developed countries is a great scientific and technological challenge for the research workers and earthquake engineers in the developed and developing countries. Without doubt, it is an important service to the less developed countries and to humanity in general.

The Magnitude of the Problem

Due to their low-cost and traditional use, and their dependence upon low level technology and indigenous materials, masonry and adobe buildings are very popular in most rural and poor urban regions of Iran and in many other less developed countries (LDCs). In these regions due to the existing technological, economical, financial, and sociocultural constraints these types of construction cannot be replaced quickly and easily with modern earthquake resistant construction. Therefore, it is certain that in most LDCs their use will be continued for many more decades to come, at least until the future industrialization and development of these countries removes these constraints.

A large percentage of buildings in the LDCs are of UMALCH type. The magnitude of the problem can be better understood if, as an example, the present condition of housing in Iran is presented. The construction materials, design methods, and building technology presently used in the construction of UMALCH in Iran and few other LDCs are discussed in more details in References[1] and [2]. In Table 8.2 the type of construction in four large cities in Iran (Teheran, Tabriz, Isphahan, and Shiraz) on November 1966 is shown. It is seen that almost 99% of the buildings are of masonry and adobe types. Independent study by the author has shown that less than 3% of these buildings have some kind of seismic reinforcement. Therefore, it can be concluded that about 97% of buildings are unreinforced. In this table an estimate of the extent of potential damage and destruction due to probable future earthquakes of various intensities is shown. The assumed percentage of destruction of various types of construction due to earthquakes with intensities VII, VIII, and IX are shown in the first three rows of this table. These assumed percentages are based on the results of various damage surveys of the past earthquakes in Iran and in other LDCs such as Chile and Turkey. It is seen that an earthquake with intensity VII, VIII, or IX in a large city of Iran such as Tabriz may destroy about 25, 45, or 65% of all the existing buildings respectively. The potential danger in small towns and rural areas is greater because the percentage of adobe buildings and unreinforced brick masonry buildings with flat timbered roofs or with roofs made of dome-shaped or cylindrical vault roofs is much larger in these areas.

TABLE 8.2. The Type of Construction in Four Large Iranian Cities and an Estimate of the Extent and Percent of Destruction Due to Probable Future Earthquakes of Various Intensities

Type of construction		Reinforced concrete	Brick masonry & steel	Brick masonry & wood	Adobe & wood	Total	% ruined
Assumed percent of destruction	VII	0	10	20	30	___	___
in earthquakes w/intensity of	VIII	10	20	40	50	___	___
	IX	20	40	60	70	___	___
Tehran							
No. Buildings		2,835	262,446	73,855	15,210	354,346	___
Percentage		0.8	74	20.9	4.3	100	___
Destruction	VII	___	26,240	14,780	4,700	45,720	13
in earthquake	VIII	280	52,480	29,560	7,600	89,920	25
intensity:	IX	560	104,960	44,340	11,300	161,160	45
Tabriz							
No. Buildings		45	3,448	21,710	36,576	61,779	___
Percentage		0.1	5.6	35.2	59.1	100	___
Destruction	VII	___	340	4,340	11,000	15,680	26
in earthquake	VIII	5	680	8,680	18,300	27,665	44
intensity:	IX	10	1,360	13,020	25,600	39,990	64
Isfahan							
No. Buildings		290	6,623	9,212	38,790	54,915	___
Percentage		0.6	12.1	16.8	70.5	100	___
Destruction	VII	___	660	1,840	11,650	14,150	26
in earthquake	VIII	30	1,320	3,680	19,400	24,430	44
intensity:	IX	60	2,640	5,520	27,100	35,320	64
Shiraz							
No. Buildings		424	11,486	13,976	3,194	29,080	___
Percentage		1.5	39.5	48	11	100	___
Destruction	VII	___	1,150	2,800	960	4,900	17
in earthquake	VIII	40	2,300	5,600	1,600	9,540	33
intensity:	IX	80	4,600	8,400	2,230	15,310	52

Data on Construction Type is based on Nov. 1966's National Census of Iran

In Table 8.3, the distribution of buildings constructed during 1972 and 1973 in urban areas of Iran on the basis of construction material used is shown. This table shows that in the smaller urban areas the percentage of unreinforced masonry and adobe buildings with timbered or masonry roofs is very large. The table also shows a trend in the construction from adobe toward brick masonry with roofs made of steel I-beam and shallow brick arches. This trend is strong in large cities notably in Teheran, where the percentage of buildings with reinforced concrete or steel frame is also increasing. This table shows that in 1973 about 95% of all the buildings in rural areas of Iran were of reinforced masonry and adobe type.

In Table 8.4 the distribution of buildings constructed during 1972 and 1973 in urban areas of Iran on the basis of the number of stories is shown. It is seen that in Teheran and in large cities there is a movement toward high-rise buildings. However, about 74% of the buildings constructed in all urban areas of Iran during 1973 had one story and about 94% were of one or two stories.

From the study of these tables it can be concluded that low-rise UMALCH will be used in Iran for many more future decades. Therefore, there is a real and urgent need for development of a method for seismic strengthening of low-rise (one to three story) UMALCH buildings for Iran and for other seismically active LDCs.

The Main Causes of Death and Destruction

Unreinforced masonry and adobe structures generally have a brittle lateral-load-resisting system and a destructible vertical-load-supporting system of the shear-bearing type in the form of unreinforced bearing and partition walls. These type of buildings have a very poor seismic performance. Total or partial collapse of these structures, especially the collapse of the roofs has been the main cause of the loss of life and property during the past earthquakes.

The seismic behavior and the failure modes of unreinforced masonry and adobe buildings during the 1972 Qir earthquake in southern Iran which represent the typical behavior of this type of structures is discussed in detail in Reference 8.1. Some of the important causes of the failure of the walls and the subsequent collapse of the roof of these type of structures are as follows:

1. The brittleness and the poor strength of the earth or clay materials used in the construction of adobe buildings.

2. The brittleness and the inferior quality of the mortar used in the construction of brick, concrete block or stone masonry buildings, and the poor quality of the bricks and concrete blocks.

3. Lack of sufficiently strong connection between the various elements of the structure, such as between walls, partitions and roofs. In addition, the brittleness and low strength of the individual elements especially the walls. Often, these structures do not behave as a rigid box. Under a moderate shaking, various elements become disconnected from each other and fail, gradually causing the failure of the whole system.

4. The brittle destruction of the unreinforced load-bearing walls, partitions, and spandrels over the openings due to the effects of shear-compression and shear-flexure. The failure modes caused by these effects are very brittle and highly strength degrading. Therefore, under a few cycles of shaking the wall elements crack and partially or totally disintegrate. Brittle cracking and subsequent fast disintegration of the load-bearing

TABLE 8.3. The Distribution of Buildings Constructed During 1972 and 1973 in Urban Areas of Iran on the Basis of the Construction Materials Used

Type of Building	Teheran		Large cities		Other urban areas		All urban areas	
	1972	1973	1972	1973	1972	1973	1972	1973
Unreinforced adobe and earth blocks	___	___	2.3	1.7	22.9	21.5	13.2	12.7
Unreinforced brick masonry with wooden timbers	___	___	17.3	15.9	21.9	25.3	17.2	18.7
Brick Masonry(1) w/roofs made of steel I-beam & shallow brick arch	97.0	95.8	72.1	75.5	41.8	42.0	59.3	59.7
Unreinforced masonry & adobe walls w/roofs made of wooden truss & tin sheet covering, masonry dome, masonry cylindrical vaults, etc., others	0.4	0.7	7.6	5.3	10.8	9.7	8.3	7.1
Sub-total	97.4	96.5	99.3	98.4	97.4	98.5	98.0	98.2
Buildings w/steel or reinforced concrete frame	2.6	3.5	0.7	1.6	2.6	1.4	2.0	1.7
TOTAL	100.0	100.0	100.0	100.0	100.0	99.9*	100.0	99.9*

* - Data given in the reference does not add up to 100.

(1) A sample study by the author carried out in Shiraz showed that less than 3% of this type of buildings have some degree of seismic reinforcement, the rest can be classified as unreinforced.

Information of this table is obtained from the following report:
Ali-Asghar Pashmini entitled, "Twenty-Year Program for Urban Settlement in Iran", Revision No. 1, Ministry of Housing and Urban Development of Iran, Mehr 1354 (Sept. 1975).

TABLE 8.4. The Distribution of Buildings Constructed During 1972 and 1973 in Urban Areas of Iran on the Basis of the Number of Stories

Number of stories	Teheran		Large cities		Other urban areas		All urban areas	
	1971	1972	1971	1972	1971	1972	1971	1972
One-story	36.5	25.5	68.7	68.3	94.0	89.7	76.8	73.7
Two-stories	44.9	45.4	29.4	28.7	5.8	9.3	19.5	20.4
Three-stories and over	18.6	29.1	1.9	3.0	0.2	1.0	3.7	5.9
Total	100.0	100.0	100.0	100.0	100.0	100.0	100.0	100.0

Information of this table is obtained from the following report: Ali-Asghar Pashmini entitled, "Twenty-Year Program for Urban Settlement in Iran," Revision No. 1, Ministry of Housing and Urban Development of Iran, Mehr 1354 (Sept. 1975).

shear-walls and partitions not only destroy the lateral strength and rigidity of the structure but also destroys the vertical load-carrying capacity of the bearing walls and results in the collapse of roofs supported on such walls.

5. Disintegration of the heavy, brittle, non-monolithic and non-rigid timbered roofs or roofs made of unbraced steel I-beams and shallow brick arches (the popular roofing system in urban areas of Iran) or roofs made of joists and hollow blocks due to falling and deformations caused by their total, partial, or uneven collapse.

6. Separation of the roofs from the walls due to the insufficient length of bearing of the beams on the supporting walls and the subsequent failure of the bearing areas.

Costly Advanced Methods of Seismic Strengthening are not Feasible for the LDCs

Various recommendations for designing earthquake resistant low-cost buildings in rural and urban regions of Iran and in other LDCs have been proposed in the past by local and foreign experts. Some experts have recommended that the use of traditional type of adobe and masonry buildings should be totally abandoned and these buildings should be replaced as soon as possible by engineered buildings with wooden, steel-framed, reinforced concrete, or reinforced masonry earthquake resisting structures. The low level of income, shortage of skilled manpower, lack of proper supervision, scarcity of high quality construction materials, and the high cost of non-indigenous imported construction materials in Iran and in the poor regions of many less developed countries makes the speedy implementation of these recommendations infeasible. Other recommendations such as the construction of earthquake resistant cylindrical, conical, or dome-shaped buildings with non-traditional forms and plans are also impracticable and unpopular. Masonry structures having monolithically braced roofs with reinforced concrete tie beams on top of all the load-bearing walls and reinforced concrete tie columns at the intersections of the walls, and proper reinforcements within the walls and around the openings have been proposed by many experts for the use in urban masonry buildings. This method and other design methods based on seismic strengthening using reinforced masonry design concepts,[3] is not easily applicable in poor rural or urban regions due to the relatively high cost and the low level of local technology. Even in large cities where these types of seismic strengthening has been implemented, the effectiveness of these methods of strengthening of the buildings against earthquakes is questionable due to inferior material and poor workmanship and supervision. In many cases in urban areas of Iran the resources have been used with no assurance of obtaining reliable results.

Seismic Protection Policy and Seismic Strengthening of UMALCH

In Section 8.2 of this chapter, the earthquake protection criteria for low-cost housing in seismically active LDCs are discussed in detail. It is found that due to shortage of construction resources such as capital, skilled manpower, advanced equipment, and material and the existence of high priority key development projects, the LDCs cannot afford to spend or tie down a large portion of their construction resources on extravagant earthquake resistant design of residential buildings. In most LDCs due to low level of income, inferiority of construction material and technology, low level of public education, lack of good workmanship and professionalism, poor supervision and safety consciousness, etc. the upgrading of the present UMALCH to a level that the resulting low-cost buildings can resist damage due to high intensity earthquakes is technically very difficult to achieve and economically and socially impracticable to implement. Constructing earthquake resistant buildings which are designed according to a rigorous and advanced specification, and which contains numerous complicated details in a less developed environment especially when shortage of skilled technicians prevails is very costly and almost an impossible task to accomplish.

Preventing the collapse of the roof of UMALCH, which is the main cause of death during earthquakes is found to be feasible to implement in most LDCs. Under high intensity earthquakes, roof collapse can be prevented by incorporating a simple braced skeleton system within the UMALCH. This system is discussed in Section 8.3. The function of this skeleton is to tie together various parts of the buildings so that under earthquake motion roof collapse does not occur. The main components of this skeleton are:

1. A horizontal in-plane diagonal bracing system for tieing together the roof beams so that a monolithic roof with sufficient in-plane rigidity is obtained;

2. An auxillary system of ductile wooden, steel or reinforced concrete columns embedded within the masonry walls so that in case of destruction of these walls these ductile columns can support the vertical load of the roof;

3. These columns are braced together laterally by means of a vertical system of diagonal bracing in such a way that the lateral instability, excessive drift, and subsequent collapse of the structure during or after the earthquake is prevented.

It is concluded that a suitable policy for earthquake protection in the LDCs is to minimize the number of deaths due to roof collapse during earthquakes by strengthening the UMALCH against collapse by the aforementioned braced skeleton. It is not feasible to protect the UMALCH against earthquake damage, however, it is proposed that in LDCs, national earthquake damage insurance be instituted to cover the cost of repair and rebuilding of damaged or destroyed UMALCH in case of earthquake.

Research and Development Needs

At present, development of a locally feasible, low-cost earthquake resistant building system is urgently needed in Iran and in many seismically active less developed countries. Research and development on improving the seismic resistance of low-cost adobe and masonry buildings can be carried out in the following directions:

A. Improving the mechanical properties of low-cost construction material, in particular in developing suitable, low-cost materials to be used for the construction of building blocks, and panels for walls, floors and roofs. A few examples of recent research and development in these areas have been reported in Refs. 2, 4, 5 and 6. More research on strengthening and stabilization of soils for construction of low-cost housing is needed. The objective of stabilization is to increase the strength, insulation, impermeability, erosion resistance, vermin and termite resistant, and durability of the soils used for construction of adobe blocks, wall elements, mortar, and plaster used in masonry, adobe and earth buildings.

Cement, lime, gypsum, certain types of asphalts, oil, polymers and other admixtures have been used for stabilization purposes. There has been some recent breakthrough in stabilization of lateritic soils for construction of strong and durable adobe blocks. Development of low-cost material for construction of foundation, building blocks, wall panels, elements for roof and floor systems, mortars, plaster for interior and exterior use, floor covering, etc. from available, indigenous earth materials is urgently needed.

B. Developing low-cost, light-weight roofing systems for various environments and weather conditions. The presently used timbered roofs or roofs made of unbraced panels, joists, steel I-beams and shallow-brick arches or hollow blocks are very heavy and have insufficient strength, resilience and rigidity. Their disintegration during the collapse of roof is responsible for a large number of deaths. Use of low-cost, light-weight materials in conjunction with innovative and feasible designs can provide a suitable roofing system. This system should reduce the danger of death during roof collapse, and at the same time should provide sufficient strength, durability and insulation against heat and rain which are necessary for safety, maintainability and the comfort of the occupants.

C. Improving the methods of design of low-cost masonry and adobe buildings to better resist the effects of earthquakes. Research in this area should be carried out in the three following directions:

I. Determination of suitable methods for design of UMALCH against roof collapse under the action of earthquake. In this direction the study of the causes of the roof collapse and the identification, design, feasibility analysis, evaluation and assessment of suitable low-cost methods for increasing:

(1) the monoliticity of roof and floors, (2) the earthquakeworthiness and strength of the auxiliary vertical load-resisting system, and (3) the resistance of the auxiliary lateral load-resisting or bracing system of the structure, in case of the destruction of load-bearing and partition walls under the action of earthquakes of various magnitudes and intensities must be carried out.

II. Determination of suitable architectural forms and plans for UMALCH, which inherently are more resistant against lateral earthquake loads. During many past earthquakes it has been observed that some UMALCH buildings with conical or dome-shaped roofs, round floor plans, round-shaped openings or spandrels, symmetrically planed floor and resisting walls of uniform stiffness in each direction had a better relative resistance against earthquake damage. In this direction, it is necessary to carry out experimental and theoretical study on the optimum seismic-resistant forms and plans of this type of buildings. Due consideration should be given to the feasibility and local social acceptability of the results which are to be recommended for use in any region.

III. Determination of suitable low-cost methods for reinforcing and strengthening UMALCH buildings and their various structural elements against earthquake damage and disintegration. Theoretical and experimental studies and field observations of some past earthquakes has shown that the use of wooden, steel, or reinforced concrete grade-beams between foundations and horizontal tie-beams on top of the walls beneath the ceiling and the use of vertical tie-beams at corners of the buildings and at the intersection of walls have some beneficial effects in reducing the earthquake damage and collapse of buildings. Horizontal, vertical and diagonal reinforcing of walls and partitions which are made of bricks or block elements with metal strips, reinforcing bars, bamboo and wood elements is found to be very effective in reducing the wall cracking and in increasing the seismic resistance of the building. The beneficial effect of reinforcing the walls around the doors, windows and openings also have been very significant. The results of some past studies carried out in India and elsewhere in this area are reported in References 7 and 8. More comprehensive experimental investigation should be carried out in this area so that suitable low-cost methods for seismic strengthening of masonry and adobe wall elements are obtained.

8.2 EARTHQUAKE PROTECTION CRITERIA FOR LOW-COST HOUSING IN SEISMICALLY ACTIVE LESS DEVELOPED COUNTRIES

Level of Seismic Protection should Depend on the Level of Development

The earthquake protection criteria in a less developed country logically should be different from that used in a techno-economically developed country where capital and skilled manpower are abundant, and where the use of technologically advanced construction material and methods are customary and accceptable. The main differences between the industrially developed and less developed countries in the availability of various resources used in construction and in the dimension and importance of their construction problems are shown in Table 8.5. In this table the less developed countries are classified into two groups; those with a high foreign exchange earning such as oil producing countries and those with a low level of foreign exchange and income earnings.

In order to accelerate their rate of development, LDCs generally assign a high priority to investment in the development of basic infrastructure, and in those projects responsible for producing basic essential goods and services, especially the industries for high-demand necessary consumer-goods, and the industries which produce construction and other capital products.

Due to the shortage and unavailability of various resources such as capital, manpower, machinery, and material in LDCs (as reflected in Table 8.5), a sharp competition will arise concerning the use of these scarce resources between the aforementioned high priority and key development projects and the consumer products including

TABLE 8.5. The Main Differences in Construction Environment Between the Developed and Less Developed Countries Which Affect the Choice of Seismic Protection Policy for Low-Cost Housing.

Resources and needs	Industrially developed countries	Less developed countries: Those with high foreign exchange earning such as oil-producing countries	Less developed countries: Those with low foreign exchange earning
(1) MONEY:			
1.1 Capital available for construction	Abundant	Abundant	Scarce
1.2 Average family income	High	Moderate	Low
1.3 Effective interest rate	Low	Moderate	High
1.4 Public construction expenditure v.s. private sector	Low	High	Very High
(2) MACHINERY:			
2.1 Construction technology	Capital intensive	Labor intensive moving toward mechanization	Labor intensive
2.2 Modern construction equipment and machinery	Plentiful	Available, but shortage of services and operators	Scarce
2.3 Ratio of average cost of renting construction equipment to average labor wage	Low	Moderate-high	High
(3) MATERIAL:			
3.1 Modern construction material	Plentiful	Imported types available but shortage of skilled manpower constrains their use	Scarce
3.2 Quality of available construction material	Good	Poor	Poor
3.3 Ratio of average cost of material per average labor wage	Moderate	High	High
(4) MANPOWER, EMPLOYMENT, WAGES:			
4.1 Indigenous skilled manpower	Plentiful	Scarce	Scarce
4.2 Foreign skilled manpower	___	Available	Unavailable
4.3 Indigenous unskilled manpower	Scarce	Plentiful, but due to high demand, they are in short supply	Plentiful
4.4 Unemployment	Small among skilled Moderate among unskilled	Negligible among skilled Moderate among unskilled	Small among skilled High among unskilled

TABLE 8.5 cont.

4.5 Wages	High	Very high for skilled Moderate for unskilled	Moderate for skilled Low for unskilled
(5) PROBLEMS AND NEEDS:			
5.1 Population growth	Low	High	High
5.2 Rate of urbanization	Slow and steady with local shifts	Very high	Moderate
5.3 Urban housing shortage	Moderate	Very high	High
5.4 Rural housing shortage	None	Moderate	High
5.5 Level of competition of residential housing with key development projects in using available resources	Not competitive	Not competitive for money but highly competitive for manpower, machinery and material	Highly competitive for all resources except manpower

the residential housing industry. Under this condition when a less developed country tries to accelerate the rate of its development, it has no other alternative except to restrict the level of scarce resources which are allocated for consumption or use in the construction of residential buildings, especially the government sponsored low-cost public housing projects. The national objective at this stage of development is to provide a technologically safe, economically feasible, and socially acceptable level of housing for the low-income people while utilizing the minimum amount of available resources.

During the present race for the rapid industrialization and development of infrastructure, Iran and most other less developed countries cannot afford to spend or tie down a large portion of their needed construction and development resources in extravagant earthquake resistant design requirements for residential buildings. For these countries, the best national policy may require the assumption of certain amounts of calculated risk during a short prescribed period of time. This period encompasses the early stages of industrial, economical, and social development when the scarce resources such as capital, skilled manpower; and construction materials are needed for other high-priority and high-yield key development projects. The important national policy question at this stage of development is: What degree of seismic protection should be required, and what should be the national earthquake protection criteria for a less developed country such as Iran?

The present rate of population increase and the rapid urbanization of LDCs have created a tremendous shortage of housing especially for low-income families in the urban area. At present the construction technology in LDCs is in a transition period from a traditional labor intensive to a semi-industrialized form. The present level of production of construction materials and technical manpower is absolutely insufficient to cope with the present housing demand.

For industrialized and developed countries the earthquake protection criteria has been more or less similar to the criteria adopted by the Earthquake Engineers Association of California[9] which is as follows:

1. The structures should resist minor earthquakes without damage.

2. Structures should resist moderate earthquakes without structural damage but with some non-structural damage.

3. Structures should resist major earthquakes, of the intensity of severity of the strongest experienced in the region without collapse, but with some structural as well as non-structural damage. In most structures it is expected that structural damage, even in a major earthquake, could be limited to repairable damage.

The imposition of the requirements of items two and three of the above protection criteria in the construction of low-cost housing in a less developed country will increase by a large percentage the consumption of money, manpower, and construction materials per unit building produced. This will substantially aggravate the shortage of new housing in the country. At the present period of scarcity of resources, it is definitely wasteful to take away the precious brick, cement, concrete, steel, and construction technicians from many needed and high-yield key development projects and employ them uselessly in the construction of foundations, wall, or roofs of low-cost residential housing in anticipation of a potential risk which may never materialize.

The protection criteria of developed countries should be adopted in Iran and in other similar LDCs when;

1. The industrialized construction technology is developed to such an extent that an excess supply of productive capacity and skilled manpower becomes available;

2. Consumption of these resources in construction of low-cost residential housing will not take them away from needed national development projects, but will create needed jobs and employment opportunities.

Proposed Earthquake Protection Policy and Criteria for Low-cost Housing in Less Developed Countries

Based on the discussion of the previous section the author believes that under the present socio-techno-economic conditions prevailing in Iran and in most less developed countries, their earthquake protection policy should have the following basic objectives:

1. It should stress a very high level of protection for key development projects which are either parts of basic infrastructure, or are responsible for the production of basic essential goods and services, especially those industries which produce construction and other capital products. A good level of protection should be provided also for emergency and critical type buildings and other expensive public projects.

2. It should stress an acceptable and optimal level of protection for residential housing, especially the low-cost public housing, and other commercial and public buildings which are of moderate costs and importance. The main goal of this policy should be the protection of the life of the people. However, it should not result in an unnecessary expenditure of scarce resources in the upgrading of the level of the protection of these buildings against damage.

The main cause of the loss of life and property during earthquakes is the collapse of buildings. Therefore, in less developed countries the primary objective of the earthquake protection policy should be collapse prevention. At present an earthquake with an intensity of VI (MM Scale) seems to be an upper limit of safety for the traditional rural, and low-cost urban masonry and adobe buildings in Iran.

Prevention of the collapse of the roofs of these buildings from earthquakes of an intensity of VII, or VIII is an objective which is technically and economically feasible. Prevention of the structural damage in these buildings from earthquakes of an intensity of VII and VIII is an objective which is technically very difficult to achieve and is prohibitively costly and economically infeasible. Satisfying the latter objective in a country of the size of Iran needs a large expenditure of much needed capital and manpower resources which will slow the tempo of the industrialization and basic development of the country.

The construction materials and technology used in low-cost housing in Iran and in many LDCs are of such inferior quality that their upgrading to such a level that the resulting low-cost buildings can resist damage due to moderate earthquakes and satisfying items two and three of the protection criteria for industrially developed countries described earlier is technically very difficult to achieve and will increase the cost of the structure substantially.

Most LDCs are confronted with certain aspects of socio-techno-economical backwardness, such as; a low level of public education, lack of good workmanship and technical professionalism, poor supervision and safety consciousness, etc. These shortcomings in conjunction with a low level of construction technology, inferior quality of construction materials, and high intensity of the use of unskilled or low skilled labor, lead to the production of poor quality housing, a low level of structural safety, and a minimum of attention to details. Under such conditions, especially when skilled laborers and technical supervisors are scarce, the implementation of construction of a building which is designed according to a rigorous and technologically advanced specification, and which contains numerous complicated structural and non-structural details is either an impossible task, or it needs constant supervision and many work rejections which increase the costs and difficulties to an unbearable level.

Prevention of seismic roof collapse can be achieved by means of the development of a simple braced skeleton system (this system will be discussed in Section 8.3 of this paper) which is attainable in the construction environment of LDCs. However, prevention of structural and non-structural damage and cracking to low-cost buildings, not only is costly but it needs a high level of technical skill, total attention to numerous and relatively complicated details, and constant supervision, which may not be attainable in the construction environment of LDCs.

Based on the above reasoning, the objective of earthquake protection policy for low-cost residential housing in the LDCs can be summarized in the following slogan:

> *"Save the lives of the residents by preventing seismic collapse of the roofs, do not worry about property damage; the houses can be repaired or rebuilt later."*

Destructive earthquakes do not happen in all regions of the country at the same time. The frequency of occurrence of earthquakes, their severity, and their epicentral location generally have certain stochastic characters. Therefore, the property damage due to a probable severe earthquake in any location is a probable risk. The expected cost of this risk can be calculated for each region and each type of housing using available methods.[10,11] The optimum policy regarding the protection of low-cost housing against property damage due to an earthquake can be obtained from a cost-benefit analysis. In this analysis, the study of the frequency of the occurrence of destructive earthquakes in each region during the past decades, the available historical data on destructive earthquakes during the past centuries in that region, and the expected cost of loss of buildings versus the estimated cost of upgrading the strength of the buildings to resist earthquake damage should be considered. Based on limited studies the author has concluded that for present conditions in Iran the most economical way to face the expected cost due to probable earthquake damage and to prevent the wastage and misuse of precious resources is by means of a national earthquake-damage insurance. In case of an earthquake the structural and non-structural damage to buildings should be compensated for by the insurance, however no roof collapse should occur during the earthquake.

To satisfy the objectives of the aforementioned protection policy, the earthquake protection criteria described in items 1 and 2 below are proposed for all classes of buildings in Iran and in other seismically active LDCs, during their present rapid rate for development and industrialization:

1. All buildings should be designed according to the protection criteria shown in Table 8.6. In this table the buildings are classified into four major categories. The criteria for earthquake protection against damage and roof collapse is given in terms of the maximum safe earthquake intensity. The intensity of the earthquake (expressed in MM scale) is chosen as a measuring criteria for earthquake protection. This choice is due to the fact that the damage and collapse of buildings (especially low-rise masonry and adobe buildings) are closely related to the local intensity of the earthquake. In Section 8.4 of this paper, a suitable ground motion spectrum for earthquakes of various intensities is proposed so that a direct and comprehendable numerical relationship between earthquake loading and intensity can be obtained. The proposed ground motion spectrum and structural response spectrum for earthquakes of various intensities are shown in Figs. 8.2 and 8.4 respectively. On the basis of these spectra and the seismic protection criteria for LDCs shown in Table 8.6 the proposed design spectrum for seismic designs of UMALCH (class D) and other classes of buildings in LDCs is obtained and is shown in Fig. 8.5. The effects of the magnitude of the earthquake which influence its duration are included in the criteria by specifying moderate and major earthquake classifications for each case.

2. All new or old residential and temporary buildings in rural and urban regions (buildings of class C and D and private buildings of class B) should be insured against earthquake damage. The premium for this insurance can be collected from the homeowners yearly or on various occasions such as when they apply for building permits, mortgage their homes, transfer or sell their homes to a new owner, apply for rental permits, apply for utilities, etc. For each building the premium should be proportional to the cost of the building. Discounts should be given to buildings with good quality earthquake resistant structures, and for buildings located in the regions which historically have been less seismically active and are subject to less seismic risks. The funds collected from homeowners by this national insurance should be deposited in a publicly chartered Disaster Chest. The interest obtained from this fund may be spent for the promotion of the industrialization of building, research and development in the construction of low-cost housing, improvement of building materials and technologies, development of plants and factories for the production of emergency or disaster-type low-cost housing, research and development in lowering the cost of the earthquake resistant design of buildings, the training of technicians, inspectors, and needed manpower for supervision and control of the construction of residential buildings, etc. The funds collected for earthquake insurance in each country and deposited in the Disaster Chest should be protected by a suitable geographical distribution of the investments in the national and international regions so that the chances of the loss of a major portion of the collected funds in any local calamity is minimized.

TABLE 8.6. Proposed Seismic Protection Criteria for Iran and Other Similar Seismically Active Less Developed Countries

Class	Type of buildings Description and examples	Damage Criteria: No structural damage should occur under any moderate or major earthquake[1] having a local intensity[2] less than or equal to the following:[4]	Collapse Criteria: No roof collapse should occur under any moderate or major earthquake[1] having a local intensity less than or equal to the following:
A	*Emergency Buildings;* police and fire stations, hospitals and clinics, transportation centers, electric generating stations, water storage and utility facilities, very important buildings such as archives, museums, treasuries, etc., *expensive key projects and buildings.*	IX[3]	X[3]
B	*Community and Public Buildings;* educational, recreational, industrial, commercial, religious, and assembly buildings, government offices, private and public buildings and projects of moderate cost and importance.	VIII	IX
C	*Residential* rural or urban *Low-Cost* or public *Housing.*	VII	IX
D	*Temporary Buildings;* warehouses, animal buildings or miscellaneous other buildings with small frequency of occupation by people.	VI	VIII

[1] Having a magnitude of less than or equal to 7.5 on the Richter Scale.

[2] Local intensities are described on the Modified Mercali Scale.

[3] For class A buildings, major earthquakes of Magnitude 8 should be considered.

[4] Buildings of classes C and D should be nationally insured against earthquake damage.

8.3 DESIGN OF UNREINFORCED MASONRY AND ADOBE LOW-COST BUILDINGS AGAINST ROOF COLLAPSE DUE TO EARTHQUAKE

According to the proposed earthquake protection criteria for the LDCs the designers' primary goal will be to prevent structural collapse during high intensity moderate or major earthquakes.

Conditions for Prevention of Roof Collapse

A structure becomes less susceptible to collapse under high intensity earthquakes if:

(a) The roof or floors of the building remain monolithic with sufficient in-plane rigidity during and after earthquakes;

(b) The vertical load-carrying system of the structure can survive the earthquake without its function being impaired;

(c) The lateral load-resisting system of that structure during the main earthquake and succeeding after shocks and until structure is repaired can retain enough residual capacity to resist safely the lateral forces due to lateral instability (P-Δ effect), wind loads, and future earthquake lateral loads.

The above three conditions can be materialized by proper design as described in the following sections respectively.

The Need for a Monolithic and Non-Disintegrable Roof and Floor System

The roof or floors of the building should remain monolithic and connected during the main earthquake and subsequent aftershocks. The components of the roof should not separate from each other and should not disintegrate under earthquake action. Monolithic and rigid reinforced concrete floors satisfy this requirement. Non-monolithic floors made of components such as reinforced concrete joists and blocks, steel joists or I-beams and shallow brick arches, wooden timbers and flattened bamboo mats plus earth covering, etc. must be made monolithic by properly tying and connecting all the components together.

In general it is necessary that the ends of the roof beams, which are supported on bearing walls, to be connected together by providing a suitable *horizontal tie-beam system* (HTBS). The function of this system is to connect the ends of all the beams together and to support individual beams in case of local failure of the bearing walls. The tie-beams of various portions of the roof or floors should be connected together properly so that all portions of the roof become attached to each other. In addition to this horizontal tie-beam system, it is necessary to provide an *in-plane bracing system* (IPBS) for the roof or floors, so that they will behave as a monolithic slab with some degree of rigidity.

For roofs made of prefabricated panels or of reinforced concrete joists and concrete or tile blocks, it is appropriate to provide a system of reinforced concrete tie-beams for connecting together the ends of the roof joists or panels. These tie-beams which are placed on top of the bearing walls may be prefabricated or cast in place. A four to six centimeter thick layer of reinforced concrete covering can be used as the in-plane bracing system of the roof or floors. This layer of concrete is cast on top of the joists and blocks of the roof connecting them rigidly with the horizontal tie-beam system. This reinforced concrete system provides overall connection, monolithicity and a good degree of in-plane rigidity.

For roofs made of steel joists or I-beams, with blocks or shallow brick arches the system of reinforced concrete tie-beams and in-plane bracing described above can be used. However, techno-economically it may be more feasible that in some cases the tie-beams are constructed of steel I-beams and the in-plane bracing system are constructed of x-shaped cross-bracing made of round or flat steel bars. Welding or bolts can be used to connect the tie-beams, the cross bracing systems and the floor beams to each other.

For low-cost roofs which are made of wooden timber and flattened bamboo mats plus earth covering, wooden timbers can be used for the construction of the tie-beam system. The in-plane bracing system can be provided with x-shaped cross bracing made of long flat boards or steel bars. Connection between the tie-beams, roof timbers, and cross bracings, can be provided by a sufficient number of spikes, nails, bolts or other connections. Design of Low-Cost HTBS and IPBS systems for various types of roofs and various applications are presently under investigation by the author.

The Need for an Auxiliary Earthquake-worthy Vertical-load-carrying (EWVLC) Column System

In addition to masonry or adobe shear-bearing walls, the building should possess an additional independent *earthquake-worthy vertical load-carrying (EWVLC) system*, capable of supporting the vertical loads when the bearing walls have failed. This system should work as the last line of defense against collapse of the roof or floors. The EWVLC system can be materialized if appropriate number of ductile and earthquake-worthy wooden, steel, or reinforced concrete columns with sufficient axial compressive strength exist within the structure to support the vertical load of the monolithic roof when the load-bearing masonry or adobe walls fail. Since these columns are the second line of defense against collapse, therefore, they can be designed to support the dead and live loads by using a reduced design load-factor of about 1.10. These columns should be suitably distributed within the building plan so that the load of the monolithic roof or floors is uniformly supported by these columns without formation of long spans.

The columns can be located within the bearing walls or partitions, at intersections of the walls, around the openings, or as individual free-standing columns. The frames of doors and windows, if properly designed, can be used as columns for this purpose. The lower end of these columns should be supported by the foundation in such a way that after the destruction of surrounding walls due to lateral earthquake motion the columns do not lose their attachment with their foundations. Individual foundations under these columns should be designed to support the column loads without appreciable settlement or failure. The upper part of these columns should be connected to the tie-beams at the roof and floor levels. In case of the failure of load-bearing walls these connections should have sufficient strength to transfer the loads of the roof and floors to the columns without failure. In two or three story buildings it is better to use columns which extend to all the stories without a splice. Column splicing should be avoided if possible. Proper connections between the horizontal tie-beam system at floor levels and the proposed columns provides a vertical load-carrying skeleton which should be able to safely carry the weight of the monolithic roof and floors and to survive the effect of earthquake without substantially losing its safe vertical load-carrying capacity. Proper care should be exercised so that the adjacent walls or non-structural elements do not shorten the effective length of the columns thus changing their behavior from ductile to brittle. This possibility which may cause the premature failure of the columns can happen especially in structures with reinforced concrete columns or wall panels. Due to the interaction of adjacent walls the effective lengths of these columns are shortened, as a result, the ductile flexural failure mode of these long elements is changed to brittle shear failure mode. Designs of Low-Cost EWVLC Column System for various applications are presently under investigation by the author.

The Need for an Auxiliary Ductile Counter-collapse Lateral Load Resisting (ADCLR) System

The existence of monolithic roof and floors supported by a system of earthquake-worthy vertical load carrying columns does not insure the lateral stability of the buildings during or after a destructive earthquake. If during the main earthquake shock, most shear-bearing walls and partitions fall down, then, due to aftershocks, wind, lateral instability, or continuation of the mainshock, the whole building may extensively sway in an arbitrarily direction and collapses. In order to insure the lateral stability of the buildings an independent earthquake-worthy lateral load resisting system should be provided within the structure. This system may consist of ductile cross-bracing panels made of wooden members, steel bars, or cables embedded within the walls or partitions, or of rigid reinforced concrete shear walls. This auxilliary lateral load resisting system should have sufficient capacity to resist safely the lateral forces due to the lateral instability (P-Δ effect), wind loads, and future aftershocks.

When architectural considerations permit, the braced panels should be placed within the building in such a manner so that the center of lateral stiffness and the center of mass of each floor coincides, or be very close to each other. In such a case the undesirable effects of torsion during the earthquake is minimized.

Design of the ADCLR system in the form of diagonal bracing for earthquakes of various intensity and duration is presently under investigation by the author. The objective of this study is to relate the design yield level of the bracing system and the collapse time of the structure to the earthquake magnitude and intensity. Thus, for various levels of earthquake protection suitable level of yield requirement for the design of the bracing system can be obtained.

8.4 PROPOSED DESIGN SPECTRUM FOR UNREINFORCED MASONRY AND ADOBE BUILDINGS

During the past decade in various parts of the world many reinforced concrete or masonry low-rise buildings of shear-bearing type or with brittle shear-resisting elements have failed under the action of earthquakes. These buildings have been designed for loads recommended by the current seismic codes. At the same localities tall buildings with ductile lateral load resisting elements designed on the basis of the same codes have survived these earthquakes undamaged. A study of these failures shows that the design base-shear traditionally recommended in most codes for seismic design is suitable for tall, long-period structures with ductile lateral-load resisting elements. In order to obtain the same probability of damage or failure for brittle and ductile structures under the same earthquake, it seems that a much higher load factor should be used in the design of brittle-type structures. In a limited degree this necessity has been reflected in the current seismic codes by requiring a higher base-shear for structures with shear-bearing walls and a higher load-factor for design of shear walls. However, the experience from the past earthquakes have shown that these provisions are not sufficient. In the past few years there has been a tendency by many practicing earthquake engineers to use a much higher design base shear for low-rise rigid type structures in which the lateral-load carrying system has a brittle shear failure mode. Ideally, the design load factors should depend upon the load deformation and ductility characteristics of the resisting elements. Establishing a quantitative relationship between the ductility of a structural system and its design load-factor needs further research.

In technologically advanced countries large portions of theoretical and experimental earthquake engineering research have been devoted to the study of tall multi-storied structures with ductile lateral-load resisting system of framed or shear-walled types. Only in the past few years there has been some progress in the art of seismic design of structures with brittle shear-resisting elements made of reinforced concrete or reinforced masonry. So far, with all the human and property losses caused by the destruction of brittle unreinforced low-cost masonry and adobe buildings throughout the world no significant progress in the art of seismic design of these structures have been reported. As a result at present in Iran and in most seismically active less developed countries (SALDCS) the design base-shear and the load factors used for tall, ductile buildings are used for design of low-cost masonry and adobe buildings.

In this section of this chapter a design spectrum is proposed which can be used for the seismic design of low-cost masonry and adobe buildings. From this design spectrum one can obtain the design loads which are necessary to satisfy the protection criteria discussed in Section 8.2. The concepts and the assumptions used in developing this design spectrum is not limited only to low-cost masonry and adobe buildings, they can be extended to any low-rise, rigid-type buildings with brittle shear-bearing walls or shear-resisting elements. From the proposed design spectrum the following information can be obtained:

(a) If a designer wishes a structure to resist collapse under a major earthquake or under any earthquake of local intensity of VIII or IX (MM scale), what level of the maximum structural drift and lateral load should he use for the design of the counter-collapse bracing system of that structure?

(b) If a designer wishes to protect a structure against damage and cracking under the action of earthquakes with local intensities of VI, VII, VIII, or IX (MM scale), what level of design lateral load should he use for the design of shear walls of the buildings for each level of earthquake intensity.

Assumed Ground Motion Spectrum

In Table 8.7 the assumed maximum ground acceleration, velocity, and displacement for earthquakes with various intensities are shown. For determination of the entries of this table it is assumed that a standard earthquake having a maximum ground acceleration of about 0.5 g produces an earthquake of intensity IX in MM scale. It is further assumed that the relationship between the earthquake intensity and ground acceleration can be obtained from the following simple formula.

$$\text{Log}\ (a) = 0.30I \qquad (8.1)$$

Where a is the ground acceleration in cm/sec^2, I is the intensity of the earthquake in MM scale.

TABLE 8.7. Assumed Values for Maximum Ground Acceleration, Velocity, and Displacement for Earthquakes of Various Intensities

Earthquake Intensity I	Maximum Value of Ground Motion						
	Acceleration a			Velocity $v^{(3)}$		Displacement $d^{(3)}$	
MM Scale	$^{Cm}/_{sec^2}$ (1)		$^o/_o g$ (2)	$^{IN}/_{sec}$	$^{Cm}/_{sec}$	IN	Cm
VI	63	≃	0.063 g	3"	7.5	2.25"	5.7
VII	126	≃	0.126 g	6"	15.2	4.5"	11.4
VIII	251	≃	0.25 g	12"	30.5	9"	22.9
IX	501	≃	0.5 g	24"	61.0	18"	45.7
X	1,000	≃	g	48"	121.9	36"	91.4

[1]Relationship between maximum ground acceleration and intensity is obtained from equation Log(a) = 0.30I, where a is in $Cm/_{sec}{}^2$ and I is in MM Scale.

[2]For simplicity it is assumed g = $981^{Cm}/sec^2 \approx 1{,}000^{Cm}/sec^2$.

[3]The values for maximum velocity and displacement is assumed to be proportional to the corresponding values for the standard earthquake (a = 0.50 g, V = 24 $IN/_{Cm}$, d = 18").

Relationship between maximum ground acceleration and earthquake intensity has been studied by many investigators. Ambraseys[12] has collected many of these formulas and has plotted the recorded maximum acceleration versus reported intensities for many earthquakes which have occurred between 1933-73. These information and various proposed relationships between ground acceleration and earthquake intensities are given in Fig. 8.1. Equation 8.1 is also plotted on this figure. It is apparent that this simple equation provides a relationship between maximum ground acceleration and earthquake intensity with a degree of simplicity and conservatism sufficient for design purposes.

For a standard earthquake having a maximum ground acceleration of 0.5 g the maximum ground velocity and displacement are assumed to be 24 in/sec (61 cm/sec) and 18 in, 45.7 cm respectively. These values are given in Reference 13. The ground acceleration for earthquakes with intensities of VI, VII, VIII, and X are obtained from equation 8.1. The relative values for the ground velocity and ground displacement for these earthquakes are assumed to be proportional to that of the standard earthquake.

For example, from equation 8.1 the maximum ground acceleration for an earthquake with intensity of VII is 126 cm/sec^2 ≅ 0.126 g. The maximum velocity for this earthquake will be: (0.126 g/0.50 g) x 24" = 6"(15.2 cm). The maximum displacement for this earthquake will be: (0.126 g/0.50 g) x 18' = 4.5" (11.4 cm). The result of these calculations are shown in Table 8.7. The assumed ground motion spectrum for earthquakes with nominal intensities VI, VII, VIII, IX and X are shown on Fig. 8.2.

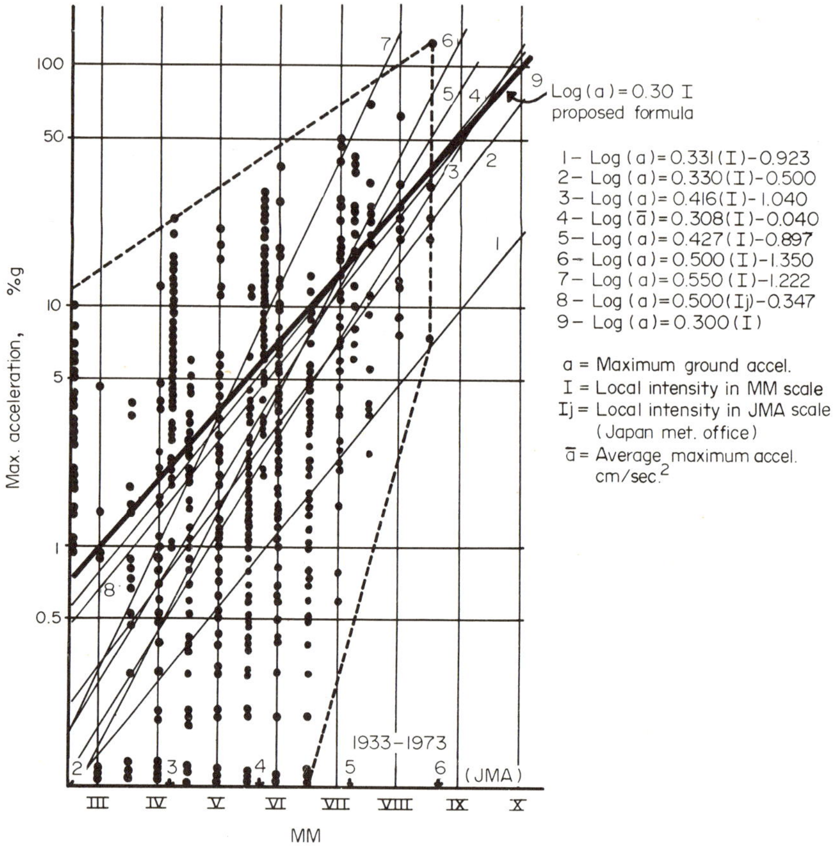

Fig. 8.1. Relationship between Maximum Recorded Ground Accelerations and Reported Intensities for Earthquakes Occurring during Period 1933-1973 (Figure and information is given by Ambraseys[12])

Structural Response Spectrum for Earthquakes of Various Intensities

In Table 8.8 the basic data needed for plotting the response spectrum of structures is given. These data are obtained by multiplying the quantities given in Table 8.7 for each type of earthquake intensity in an appropriate amplification factor. This factor depends upon the level of damping within the structure which is represented by the percentage of critical damping. The assumed value of damping and the corresponding amplification factor is given at the footnotes of Table 8.8. The amplification factors are obtained from Reference 13. For low-cost masonry and adobe buildings the damping ratio during earthquake action varies and depends upon the state of cracking and disintegration of load resisting walls. For various stages of cracking and destruction of this type of buildings, an estimate of the average value of damping and the range of natural frequency of structure at each stage is given below.

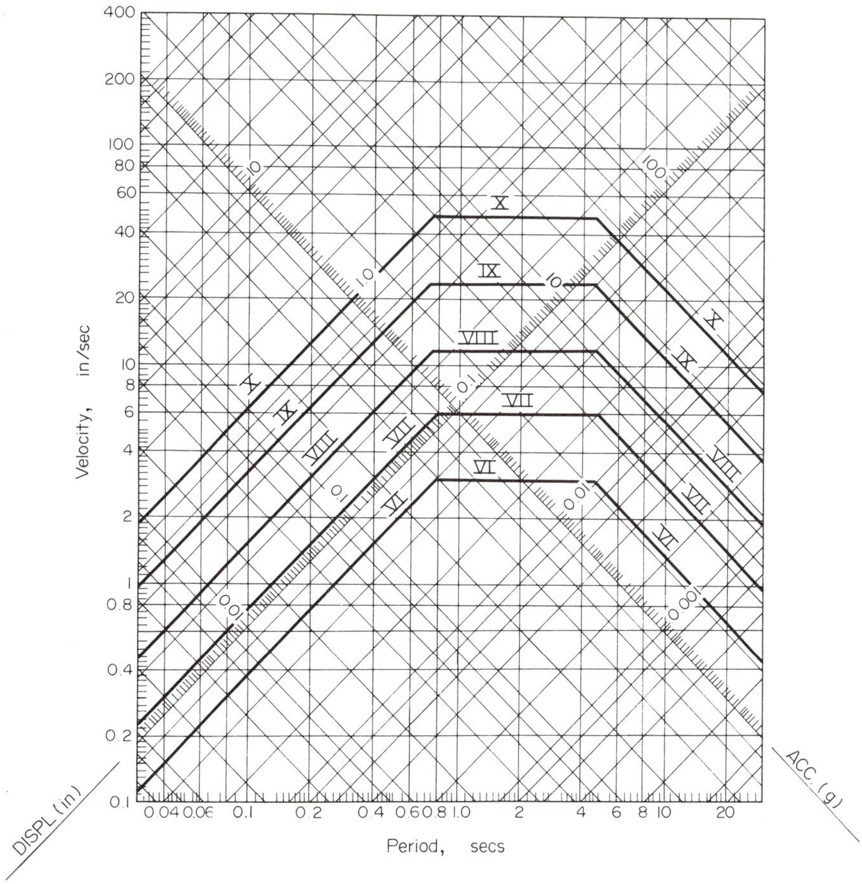

Fig. 8.2. Assumed Ground Motion Spectrum for Earthquakes with Nominal Intensities VI VII VIII IX and X in Modified Mercali Scale

Behavior of structure during successive stages of cracking and destruction

The behavior of a low-rise masonry and adobe structure under earthquake action generally passes through the following stages:

Stage I - Before cracking:

At the start of the earthquake the resisting walls are uncracked, the building is very rigid, the natural frequency of the building especially those of 1-3 stories is very high (6 cycles/sec or more), the structural damping is very low (less than 2%). The vertical loads of the roofs are carried by the shear-bearing walls. As earthquake shaking progresses with increasing intensity the resisting walls will reach the yield and cracking limits.

Stage II - Initiation and intensification of cracking:

At the initiation of cracks the structure becomes slightly more flexible and damping increases. This gradual softening of structure continues until all resisting walls have cracked. As cracks spread within the walls and become wider and larger the stiffness of the walls is reduced but their energy absorbing capacity increases. As a result the structure becomes more flexible, the natural frequency of the building is gradually reduced to about 2 cycles/sec and the damping increases many folds. At some point along this stage the structure passes through its maximum lateral load carrying capacity. For reinforced masonry, some authors[13,14] have reported a very high damping ratio at this stage. No experimental data is available to the author at present. Therefore, for non-reinforced masonry and adobe structures the damping ratio at this stage is assumed to be 10%. The vertical load of the roof is still supported by load-bearing walls.

Stage III - Disintegration of walls:

As shaking continues the cracked non-reinforced masonry and adobe walls start to disintegrate and to fall down either as a whole or in parts. The structure becomes very flexible (natural frequency between 0.4-2 cycles/sec) and it starts to sway in large amplitudes. The damping increases at the beginning of this stage but reduces considerably as all the resisting walls crack. During the disintegration and falling the walls do not absorb as much energy as during cracking the average damping ratio at this stage is assumed to be 10%. At this stage the load-bearing walls cannot carry the vertical load of the roofs. This load gradually will be supported by the auxilliary system of earthquake-worthy vertical load carrying (EWVLC) columns which has been provided within the structure. Thus, as the structure passes through this stage the vertical loads will be gradually transferred to these columns. The counter-collapse bracing system also will become gradually effective and will resist the further increase in the amplitude of the drift of the structure.

Stage IV - Excessive drift of the structure:

At this stage all the resisting walls have fallen down or have ceased to resist the lateral movement of the roof. Only the auxilliary counter-collapse bracing system is left to resist the increasing amplitude of the sway and the drift of the building. At this time the system of EWVLC columns carries the vertical load of the roof. The natural frequency of the structural system is very low (less than 0.4 cycles/sec). The damping is also very low, because the only energy absorbing is provided by counter-collapse bracing system which does not absorb much energy due to the gradual plastic elongation of its diagonal members. The average damping ratio at this stage before collapse is assumed to be about 7%. If the ductile diagonal members of the counter-collapse bracing system do not fail due to excessive elongation before the earthquake stops, then, the building roof and columns will stand in a drifted form. In this case the counter-collapse bracing system should resist further increase in the structural drift and should prevent the collapse of the building roof due to instability (P-Δ effect) and aftershocks. These four stages of structural failure are shown in Fig. 8.3.

Structural response spectrum

The structural response spectra for earthquakes of intensities VI, VII, VIII, IX and X is shown in Fig. 8.4. This figure is obtained from the data shown in Table 8.8, in which for each type of earthquake the corresponding ground motion acceleration, velocity and displacement is multiplied by the amplification factor of 1.5, 1.5 and 1.2 respectively. On this figure also, the various stages of the response and destruction of the low-rise unreinforced masonry and adobe buildings are shown.

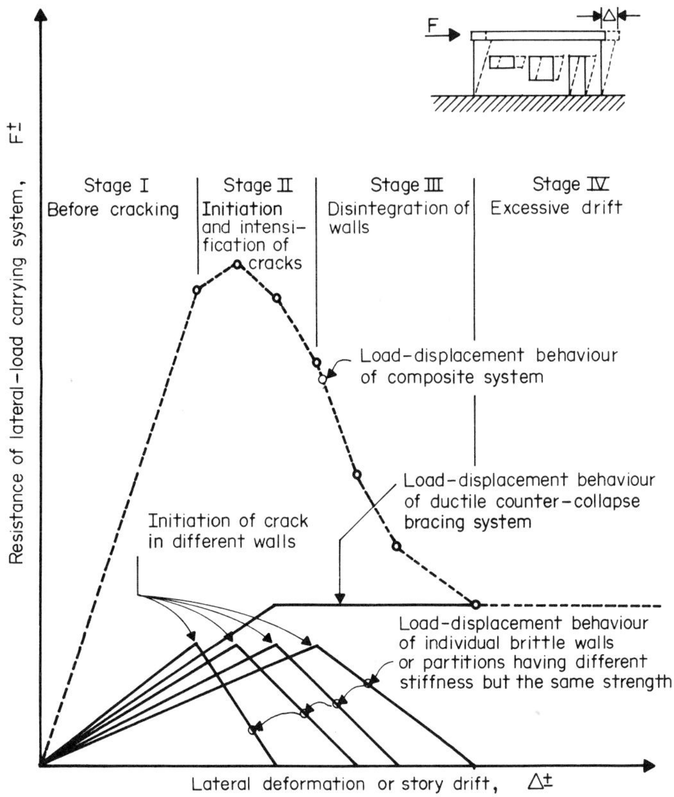

Fig. 8.3. Various Stages in the Load-displacement Behavior and Failure of Structures with Brittle Lateral-load Carrying Elements

Proposed Design Spectrum

The proposed design spectrum for seismic design of all classes of low-rise unreinforced masonry and adobe buildings in the LDCS are shown in Fig. 8.5. The classification of the buildings was discussed in Table 8.6 and is based on the earthquake protection criteria for buildings in less developed countries. The design spectrum shown for buildings of classes C and D should be used for design of UMALCH.

8.5 REDUCING EARTHQUAKE DAMAGE TO UNREINFORCED MASONRY AND ADOBE BUILDINGS

In the proposed earthquake protection criteria for Iran and LDCS it was pointed out that under high intensity moderate and major earthquakes, it is not economical to design the UMALCH against damage. It was recommended that instead of this approach a national earthquake damage insurance should be instituted. Increasing the strength of low-cost masonry and adobe buildings against earthquake damage may become advantageous when:

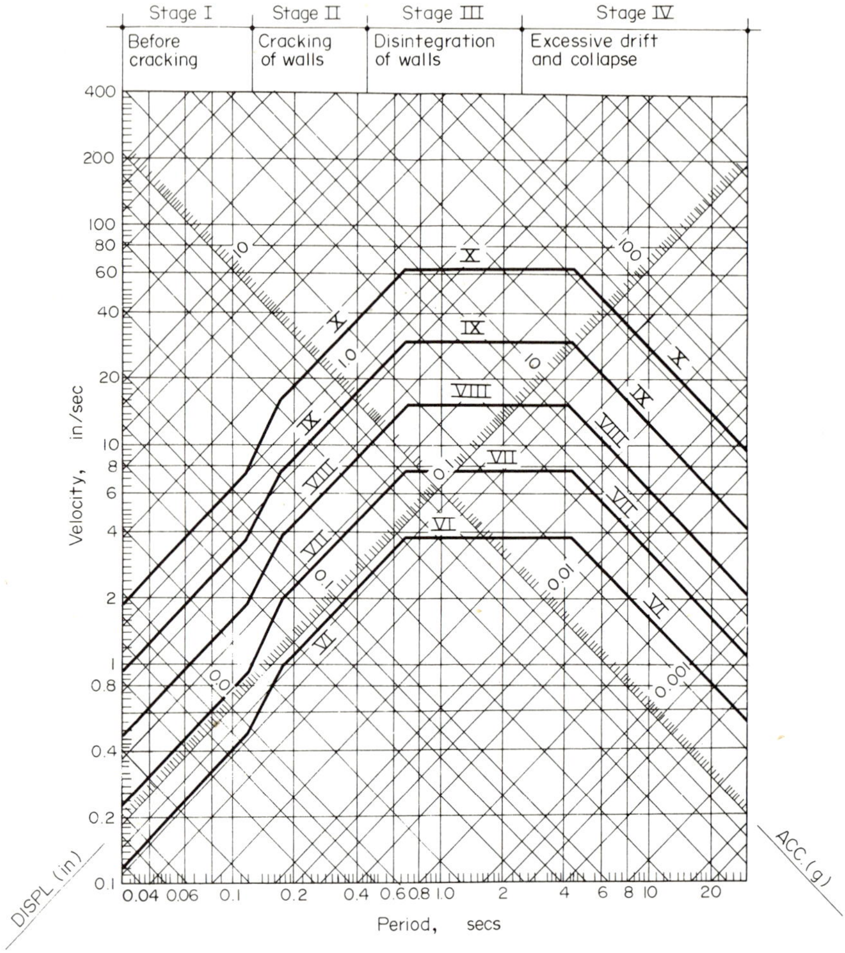

Fig. 8.4. Structural Response Spectrum for Earthquakes with Nominal Intensities VI VII VIII IX and X in Modified Mercali Scale

1. National insurance against earthquake damage is non-existent;

2. Some degree of strengthening of the building may result in lower premium for the damage insurance;

3. For some particular reason, a building owner wants to reduce the possibility of earthquake damage to his building, although it may be more economical for him to insure his building against the expected cost of a probable future earthquake damage.

Various methods of strengthening are available that their use will increase the chances of a building to survive a moderate and even a major earthquake without damage. Some of these strengthening methods require the use of high quality

TABLE 8.8. Assumed Values for the Maximum Acceleration, Velocity, and Displacement Response of Brittle-type Structures for Earthquakes of Various Intensities

Earthquake Intensity I MM Scale	Maximum Values of Structural Response (1)						
	Acceleration a (2)			Velocity v (3)		Displacement d (4)	
	$^{Cm}/_{sec^2}$		$^{o}/_{o}g$	$^{IN}/_{Sec}$	$^{Cm}/_{Sec}$	IN	Cm
VI	95	≃	0.095g	3.9"	9.91	2.7"	6.9
VII	189	≃	0.189g	7.8"	19.81	5.4"	13.7
VIII	376	≃	0.376g	15.6"	39.62	10.8"	27.4
IX	751	≃	0.75g	31.2"	79.25	21.6"	54.9
X	1,500	≃	1.50g	62.4"	158.50	43.2"	109.7

[1]Entries of this Table are obtained from corresponding values of Table 8.7 multiplied by many amplification factors given below.

[2]Acceleration governs the design when the structure is rigid, i.e. at the stages I (before cracking) and II (at the initiation and intensification of cracking), average structural damping at stage II is assumed to be 10% of critical damping. The corresponding acceleration amplification factor is 1.5.

[3]Velocity governs the design when the structure is at stage III during disintegration of walls. Average structural damping is assumed to be 10% of critical damping. The corresponding velocity amplification factor is 1.3.

[4]Displacement governs the design of stage IV during excessive drift when the structure is totally flexible. Average structural damping at this stage is assumed to be about 7% of critical damping. The corresponding displacement amplification factor is assumed 1.2.

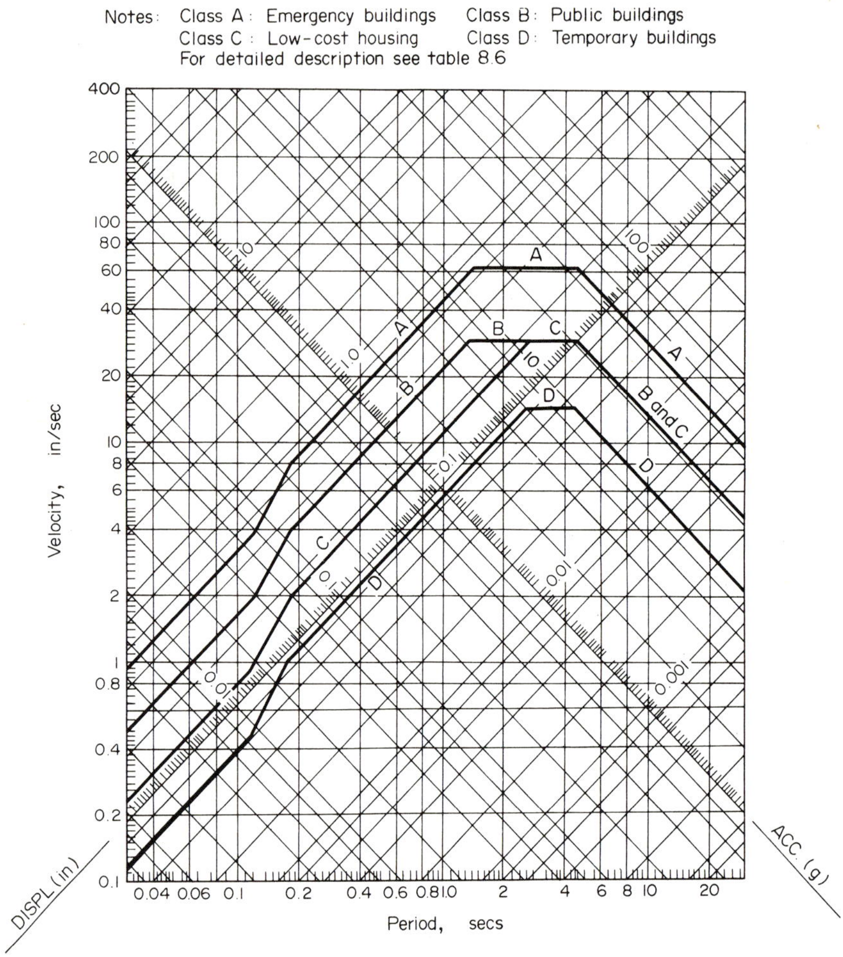

Fig. 8.5. Earthquake Design Spectra for Classes A, B, C, and D Buildings in Less-developed Countries

construction materials, technology, and supervision, which in turn increases the costs of the building substantially. Some of these methods are as follows:

1. Using higher quality masonry blocks or adobe;

2. Using high quality sand-cement mortars;

3. Reinforcing the walls, partitions, and spandrels according to the requirements of Reinforced Masonry Codes;[3]

4. Using thicker shear-walls and a higher ratio of the length of resisting shear walls to floor area;

5. Using reinforced concrete horizontal tie-beams, and vertical tie-columns with infilled high quality masonry or reinforced concrete shear walls.

Other strengthening methods also exist which are based on superior design techniques and stress a more effective and optimal deployment of the existing construction materials without the necessity of using extra resources or high quality technology. In this paper these latter methods which their use in the structure do not increase the costs will be discussed. The basic concepts behind these design methods are obtained from the theoretical study of the optimum architectural form of structures with brittle lateral load resisting elements, and from the field observations of the failure mode of masonry and adobe buildings during the past earthquakes.

The underlaying design philosophy and the optimum architectural form of structures with brittle lateral load resisting elements can be recognized better if the differences between the structural behavior of buildings with ductile or brittle load resisting components is understood. In technologically advanced countries, a large portion of theoretical and experimental earthquake engineering research has been devoted to the study of tall, multi-story structures with ductile lateral load resisting system of frame or shear-wall type. So far, no significant progress in the art of the seismic design of a structure with brittle elements has been reported.

In seismic design of brittle structures many traditional assumptions and approximations have been carried over from the design of conventional ductile structures against static or dynamic loadings. The most important one is the assumption of redistribution of forces between the resisting elements due to successive yielding of these elements. This is a fundamental assumption in plastic design and in seismic design of ductile structures. This assumption is not valid for brittle structures. Therefore, for this type of structure, many accepted methods of seismic analysis and design and many analytical and design simplifications and rules of thumb based on ductile behavior must be revised. In ductile structures the maximum load that a parallel row of resisting elements can carry usually is equal to the sum of the yield load or ultimate resistance of all these elements. The maximum load carrying capacity of a system consisting of rows of brittle elements is usually less than the sum of the ultimate resistance of all the elements, because, each overloaded member fails in a sudden and premature brittle mode. Therefore, the members would rarely reach their ultimate strength all at the same time. Sometimes, especially under alternating loads, the brittle elements of the lateral load resisting system may fail one after the other in a successive mode, similar to the so-called unbuttoning phenomenon which occurs in long rows of rivets and bolts.

In analysis and design of structures with brittle lateral load resisting elements the following recommendations should be observed:

(a) The loads and stresses in brittle resisting elements should be obtained by an accurate elastic method of analysis;

(b) Any obstruction or stiffening of elements during or after construction which would change the distribution of the lateral load within the brittle elements of the structure from their estimated design values obtained by an elastic analysis must be avoided or their action must be included in the elastic analysis.

(c) The effect of torsion due to an earthquake must be minimized or its effect must be considered in an elastic analysis.

(d) The structural design of a building with brittle components is optimum and has the highest value of damage resistance if the building fails by simultaneous failure of the brittle components under the critical load conditions (the so-called One-Hoss-Shay design philosophy), but remains elastic under less than critical load conditions. For structures with brittle shear resisting elements, this condition can be achieved if the lateral stiffness of all the resisting elements are equal.

From the above theoretical studies and the survey of the damages during the past earthquakes, it can be concluded that structures with a relatively low and uniform level of shearing or tensile stresses in all members under lateral earthquake loads have the best resistance against damage. Structures having dome-shaped, cylindrical, conical, or other axi-symmetrical forms, especially those with an initial state of compressive stress, lead to a more uniform and low tensile stress distribution pattern, and thus, are often more damage resistant than similar structures of conventional forms.

In buildings with conventional architectural forms a state of low and uniform level of shearing and tensile stress under earthquake loading can be achieved if the lateral load resisting elements are so proportioned that their lateral stiffness are equal in each direction. Such proportioning results in equal level of stress in all members under seismic loadings. A building in which under earthquake loading in each direction all the lateral resisting elements such as walls, partitions, wide columns, spandrels, etc. reach their cracking and ultimate load carrying capacity at the same time has an optimum design. Such a building structurally has the highest amount of resistance against cracking and damage, however if it fails all of its resisting members also fail at the same time.

The structures that their brittle lateral load-resisting elements have such geometries that will not give rise to points of a stress concentration have a good resistance against earthquake damage. The form of the openings in a masonry wall has a significant influence in the pattern of stress concentration and in the seismic behavior and damage strength of that wall. Small openings with round corners seems to be better than those with sharp corners. Sharp changes in the stiffness of the walls or shear-resisting elements give rise to points of stress concentration and should be avoided as much as possible.

REFERENCES

1. Razani, R. and Lee, K. L. *The Engineering Aspects of the Qir Earthquake of 10 April 1972 in Southern Iran.* A report to the National Science Foundation, published by the National Academy of Sciences for the National Academy of Engineering, Washington, D. C. 1973.

2. Reps, W. F. and Simiu, E. (Eds.) *Design, Siting, and Construction of Low-Cost Housing and Community Buildings to better withstand Earthquakes and Windstorms,* Report No. NBS BSS-48 prepared for the U.S. Agency for International Development, Washington, D.C. 20523, Jan. 1974.

3. Amrhein, J. E. *Reinforced Masonry Engineering Handbook - Brick and Other Structural Clay Units,* Published by Masonry Institute of America, Los Angeles, California, 1st edition 1972.

4. Razani, R. and Behpoor, L. *Some Studies on Improving the Properties of Earth Materials Used for the Construction of Rural Earth Houses in Seismic Regions of Iran,* Proc. 4th Symposium on Earthquake Engineering, Rooskee, India, Nov. 14-16, 1970; pages 81-89.

5. *Manual of Asphalt Emulsion Stabilized Soil Bricks,* International Institute of Housing Technology, California State University, Fresno, California, 1972.

6. Dietz, A. G. H. "Materials for Low-Cost Housing Construction", Chapter 7 of this book.

7. Arya, A. S. "Construction of Small Buildings in Seismic Areas", *Indian Journal of Power and River Valley Development Special Number, 1971*, Proceedings of the Symposium on Konya Earthquake of Dec. 11, 1967 and related problems; pages 69-76.

8. Krishna J. and Chandra B. *Strengthening of Brick Buildings in Seismic Zones*, Proceeding 4WCEE, Santiago, Chile, January 1969, Vol. 3; pages B-6-11 to B-6-20.

9. *Recommended Lateral Force Requirements and Commentary*, Seismology Committee, Structural Engineers Association of California, 1968 edition.

10. Grandori, G. and Benedetti, D. "On the Choice of the Acceptable Seismic Risk," *Journal of Earthquake Engineering and Structural Dynamics*, Vol. 2, 3-9 (1973).

11. Esteva, L. *Seismic Risk and Seismic Design Decisions*, MIT Symposium on the Earthquake Resistant Design of Nuclear Reactors, Cambridge, Massachusetts (1969).

12. Ambraseys, N. N. *Notes on Engineering Seismology*, Proceedings of the NATO Advanced Study Institute Congress, Izmir, 1973, on "Engineering Seismology and Earthquake Engineering" edited by J. Solnes, Published by International Publishing Company, 1974; pages 33-40.

13. Newmark, N. M. and Rosenlilueth, E. *Fundamentals of Earthquake Engineering*, Englewood Cliffs, N. J., Prentice-Hall, 1971.

14. Mallic, D. V. *Structural Damping of Brick Masonry in Different Mortars*, Proc. Second Earthquake Symposium, University of Rooskee, India, Nov. 10-12, 1962.

9

HIGHER EDUCATION IN LOW-COST HOUSING

Floyd O. Slate

9.1 INTRODUCTION

During the two Roving Workshops and related conferences, it was observed that low-cost housing technology has only small, and surely insufficient or even negligible, input, guidance, and analysis by the non-technological fields, such as sociology, anthropology, economics (other than financing), law, and human ecology. The same situation that was observed in the four Asian countries and the U.S.A. is almost universal. The result is that low-cost housing programs are being organized, planned, supervised and carried out, in most cases by architects or engineers (or by planners), without proper consideration of related social and "human" aspects, and that serious errors, deficiencies, and failures are occurring, world-wide.

It was further observed that the programs of higher education and research in the five Network nations, and in the world in general, do not provide for integrated, multi-disciplinary education and research in complex, applied fields such as low-cost housing. Thus, higher education is not leading or pointing out the way in this new direction, or providing the important, needed multi-disciplinary aspect of training for these traditionally technological fields.

The purpose of this chapter is to discuss the place of higher education (including university research) in the field of low-cost housing: to discuss its current status, influences, failures, deficiencies, successes, needed changes, newly introduced programs, trends, and possible future directions. To this educator, with his inevitable prejudices, it seems obvious that universities, colleges, research and training institutes, and other such organizations must provide advanced education and training, as well as research, in this difficult, complex field, which involves inter-relationships that break across traditional academic and practicing disciplines, and which, again in the opinion of this writer, demands a multi-disciplinary approach to the solution of many of its problems. In addition to opinion, some speculation will necessarily be involved in this chapter.

It is assumed that content of courses of study and directions of research in higher education will influence strongly what will actually happen in the future in the "real" practical world, when the advanced students later assume positions of influence and decision.

9.2 WHERE WE ARE NOW IN ACTUAL HOUSING PRACTICE

The situation is grim, considering the present seriously inadequate housing of the world's poor, with the added specter of massive future growth of their numbers, combined with rapid urbanization. Total additional housing needs by the end of this century may well exceed 200 or even 300 million units, more than the total world supply at this time, and even more than the total built throughout human history.

Surprisingly little has been done toward providing low-cost housing for low-income people, particularly for the great numbers of poor families who do not qualify for any kind of loan because they are not landowners or have no regular income. Much of what has been built has been subsidized housing for civil servants, who are largely people of middle or low-middle income, and who include even some in high-income levels. Most of the remainder of low-cost housing, both governmental and private, has gone to middle-income people who can afford to make appreciable purchase or rental payments. In general, the poorer people have remained outside the mainstream of help in housing.

The World Bank's 1972 working paper, "Urbanization", states that "Perhaps the most salient feature of the housing situation is the stark fact that typically well over half of the urban population cannot afford minimal 'permanent construction' housing, even if financing arrangements are made available or limited subsidies given."

Squatters now make up about one-third of the population of a large part of the world's medium-sized and larger cities. Refugee squatters pose a particularly large and difficult problem, and one that shows no sign of being ended, for new refugee groups fleeing from new conflicts feed the total faster than earlier groups can be taken care of.

Of the too-few attempts to house the poor, most have failed. Those projects that are built for the very poor, such as squatters (not *all* of whom, in fact, are poor), are traditionally segregated from other housing, and often placed on undesirable, waste land — in a swamp or desert, for example, apparently to "get rid" of those people. In such a segregated area, perhaps a few miles out in the desert, people have little chance to work or even to beg, and rather than starve in their new, better, subsidized housing, they often abandon it and return to the city as squatters again. Abandoned, and therefore wasted, housing projects of this nature are not uncommon.

Among the important problems in addition to population pressure of the poor, is that of land tenure. Various economic problems include financing of construction, from initial cost through — usually — at least partial repayment by the tenant or new owner; unfortunately, seldom is any thought given to creating an environment in which the occupant can earn a living. Frequently, the housing units are not suitable for the climate, or are poorly sited, and thus fail to meet the needs of the intended occupants. All too often there is a lack of understanding of, or even of concern about, social aspects such as customs of the people or their own preferences and desires.

Frequently, perhaps usually, little concern is given to use of indigenous rather than imported materials, resulting in problems of foreign exchange. This is illustrated by the ever-present corrugated steel roof found on poorer housing in all parts of the world.

A related problem is the frequent use of capital-intensive rather than labor-intensive schemes for housing construction. This is usually a misguided approach, since unemployment and poor housing usually go hand-in-hand. An unfortunately popular approach in some developing countries is to import unmodified or slightly modified foreign standards and technologies, including complete large capital-intensive industrialized housing factories, rather than to develop use of indigenous materials with designs and construction techniques chosen to be compatible with local materials, labor, and conditions, including at times self-help by the future occupants.

It is probable that one of the greatest social problems is unhappiness or resentment resulting from segregation, with its resultant identification of people as "poor," or "destitute," or "wards of the state."

Finally, in many or most cases, the poorer occupants are given no equity and have no chance to develop any. Thus, they have less incentive to care for and upgrade their living quarters, as compared to the case where they might foresee the possibility of some future profit for themselves or their children through some sort of progressive ownership (designed to prevent early sale or sub-rental).

This short over-view of typical problems and failures will now be related to how problems of low-cost housing are typically approached at the present time. The approach, which has failed and is still failing badly, is primarily a unidisciplinary approach. Typically, an architect, or an engineer, or more recently a planner, or even occasionally an economist will conceive, plan, and execute a project, based on his own training and experience, with little input (or little input that is actually used) from any discipline other than his own. He usually cannot even communicate effectively with people of other disciplines and has little understanding of their approaches and problems. Then, to top off the situation, an organized literature does not exist to define and analyze this complex field, with coordinated multi-field information on the current stage of knowledge and experience; thus, a person of a given discipline is not able to depend effectively on the published literature to enable him to digest and consider views of other disciplines as related to his own. He is alone, because of his own training, because of lack of communication among disciplines, and because of lack of an organized, coordinated literature.

It is the opinion of the author that much of the problems, failures, and lack of progress in the field of low-cost housing is related to the limited vision of a uni-disciplinary approach. This typical approach is a direct result of our traditional systems of higher education.

9.3 WHERE WE ARE NOW IN EDUCATION AND RESEARCH

For many years there has been a trend toward more and more highly specialized educational programs to produce highly trained specialists in narrow fields of knowledge. The theory apparently has been, with considerable truth, that with limited time to study, the professional can delve deeper into his specialized subject if his study is limited to that subject and others closely related, without dilution or distraction from other fields. This approach is now almost universal, especially for graduate study. It has worked very well for highly developed fields that are closely defined and well organized, such as some specialities of medicine, physical sciences, and engineering, in those particular specialities where broad perspective is not needed or sometimes even of value.

These highly trained specialists have often become the leaders of our educational programs, with great influence on other educationalists and researchers, resulting in a self-feeding system that becomes more and more specialized, even in cases where breadth might be vital. Research publications have paralleled the trend, becoming increasingly specialized. As a result, each field has developed a literature and even a jargon of its own, that outsiders find difficult to understand. And thus, specialists from different fields often or usually have difficulty working together or even communicating with each other on broad problems encompassing all their specialities.

In the past several years, there has been an attempt, in engineering for example, to broaden and "humanize" undergraduate students by requiring them to take a certain minimum number of courses in the humanities or liberal arts. This approach has been worthwhile, in terms of greater breadth at a very shallow level and in opening up new perspectives, but it has not helped much in terms of cooperation among highly specialized workers.

The highly- and narrowly-trained specialists tend to operate in practice as they were trained in the universities. Advanced university training labels the specialist, and effectively defines his areas of competence and therefore of endeavor. When a complex field such as low-cost housing is encountered, with its many interrelated problems involving several of the traditional disciplines, trouble arises. The problems are deep enough that a high level of training is required, and the problems are complexly interrelated, yet a group of specialists cannot work as a team in a coordinated fashion — they usually cannot even understand each other. Generalists are of little value, either as members of, or leaders of, teams.

Traditionally, a "house" belongs to the architect (or to the "builder"), and a "public works" belongs to the civil engineer. As a result, the great majority of low-cost housing projects are organized, controlled, and carried out by either an architect or an engineer. An economist often has a strong voice in financing schemes, but neither he nor anyone else considers the general economic well-being of future occupants. Sociologists and anthropologists are little heard in circles where decisions are made. The legal aspects of land tenure are not viewed in terms of social and economic long-term impact. Even if the architect or engineer has a multi-disciplinary team or committee, the main thrusts of the project are usually based on his own traditional discipline and academic training.

At the present time, it is very difficult for a graduate student to take any course (except sometimes a required modern language) outside his major and closely related minor subjects. His advisors and his peers will encourage him to specialize as much as possible. The examination systems pressure him to do so. Finally, there are no, or almost no, multi-disciplinary courses at a fairly high level, aimed at examining problems and solutions of a real, complex problem such as low-cost housing. The only ones known to the author, that integrate teaching and research from various disciplines rather than merely have separate "guest" lectures from different fields, are the two current courses at Cornell University in low-cost housing for developing nations.

Thus, our higher education and research programs are sadly deficient in meeting the needs of this important area. They are failing to integrate thinking and understanding among specialized disciplines.

9.4 PROBLEMS AND TROUBLES WITH PRESENT SYSTEMS

The complex problems of low-cost housing require the coordinated thinking and solutions of various fields of knowledge, but our current systems of practice, which are in turn based on our current systems of higher education, do not provide for, or even easily allow, such coordinated approaches. The problems include traditional academic and practicing responsibilities and authorities for the low-cost housing area, lack of and inability for communication among the various highly specialized disciplines that should be involved, and even a certain antagonism between some fields concerning methods of thinking about and approaching problems.

9.5 WHAT IS NEEDED

Some innovations are needed in higher education to provide backgrounds for teams of specialists to work together effectively in a coordinated manner on the complex problems of low-cost housing. It can be expected that applied practice will follow educational precedents. The approach must be inter-disciplinary, to blend and integrate the fields. It must involve courses applied to real problems, such as low-cost housing, and not merely a course in sociology for an architect or engineer. Research, as well as teaching, must be multi-disciplinary. Specialists must be able to understand each other and each other's literature. Finally, we must develop new leaders who are specialists in a specific field and who are simultaneously sufficiently knowledgeable about the approaches and problems of other pertinent disciplines to be able to work effectively with them. In the opinion of the author, the new breed of "planners", an off-shoot from architecture, is generally not meeting the needs.

These needs will require some new approaches in teaching and research at the graduate level in universities and research institutes. Teachers from various disciplines must integrate their efforts to focus on real problems and students from various academic and personal backgrounds must integrate their study efforts and viewpoints — all this must be done at a level of serious study, not just as "stimulating seminars". A series of guest lectures in an architecture or engineering course will not do.

In summary, what is needed are trained-in-depth specialists in each of various disciplines, who also receive training in multi-disciplinary courses and research to learn each other's approaches and problems and to learn to communicate with each other in the field of low-cost housing, and who then can work as a coordinated team under the direction of a specialist who is also trained to use a multi-disciplinary approach. If necessary, to avoid longer schooling, it might be acceptable to reduce slightly the amount of specialized study in order to include the multi-disciplinary study. Specialists can continue to advance themselves in their own field by self-study, but breaking across into a new field is extremely difficult and usually requires help, as from a course of study. Finally, a multi-disciplinary team, just as any other team, must have a leader with authority. A headless committee will be useless. In the author's opinion, the leader should usually be an engineer or architect.

The needs for a multi-disciplinary approach to education involving low-cost housing are probably also present, perhaps in lesser degree, for existing educational programs in housing in general, including the industrialized countries and higher-cost housing.

9.6 THE CORNELL PROGRAM IN LOW-COST HOUSING

As an example of what is being done in higher education in low-cost housing at one university, a detailed description will be given of the program at Cornell University. This is given as an example, and not necessarily as a recommendation.

History

Before 1972, some individual courses in housing involving a single discipline, usually in architecture or human ecology, had been given at Cornell University and elsewhere, but apparently no attempt had been made to combine and integrate disciplines for such a course. The author planned in 1971 and early 1972 for such a course, to be given in Civil and Environmental Engineering with help by a multi-disciplinary faculty group, starting in Spring 1973, and the course was entered into the catalog in early Spring 1972. Meanwhile, Cornell's Program on Policies for Science and Technology in Developing Nations, based on a 211(d) grant from the U.S. Agency for International Development, encouraged the author to broaden the course to involve an even greater amount of input from the disciplines outside of civil engineering.

The author agreed and a group of faculty members from various disciplines at Cornell University was assembled and met together in late Spring, 1972 to discuss the possibility of establishing a truly multi-disciplinary course in low-cost housing for developing nations. Only people with actual working experience in low-cost housing (and closely related areas) in developing countries, were invited to participate, in an attempt to form a down-to-earth group with practical knowledge. There was immediate agreement that such a course was needed, and general interest in participating. Interest was also expressed in cooperating as multi-field teams, as appropriate and feasible, in projects in low-cost housing, from study to design and construction. A series of meetings was held through the Summer of 1972, and general agreement was reached for an outline for the course and for a method of operation of the group. A detailed description of the course, including an outline of topics, was released to the students before their pre-registration in Fall 1972, for their Spring 1973 semester, when the course was first given. The course, an integrated blend of several disciplines, was apparently the first of its kind given anywhere. It was not simply a course in architecture or engineering with a few guest lectures or seminars by sociologists and economists — a kind of course that does not get to the heart of the matter.

Courses of Study

Development of course

The major objectives of the first course were to present and analyze various aspects of low-cost housing, including architecture, engineering, physical planning, physical sciences, economics, and sociology, and to attempt to integrate these aspects into a realistic, practical picture. The faculty group felt that many projects have been bungled because key aspects of the over-all problem of low-cost housing have been ignored — such as social problems or problems of general economic well-being of the future occupants. All too often an architect or an engineer alone, without adequate background in or input from other vital fields of knowledge, has developed and carried out a project, with serious consequences. It was felt that people, both students and faculty, from the several key disciplines should become familiar with each other's problems, as well as work at depth in their own field, so they could then not only communicate readily, but work together jointly as teams to attack problems in low-cost housing.

The course was listed for credit under three different colleges of the University — Architecture, Civil and Environmental Engineering, and Human Ecology. Thus, the same course was given under three different departmental numbers, all with the same title. This was done to encourage students from various fields to take the course, as well as to satisfy administrators who wish to claim student class hours for their own courses in their own departments.

Students

The class for the first offering consisted of about ninety students ranging from seniors (with a few juniors) to Ph.D. candidates. About one-fourth were graduate students and about one-fourth were from developing countries (more than one-half of the latter were graduate students). Countries represented were: Afghanistan, Bangladesh, Canada, Republic of China, Ethiopia, Ghana, Greece, India, Iran, Italy, Mexico, Nigeria, Pakistan, Peru, Rhodesia, El Salvador, Sierra Leone, Thailand, and the U.S.A. Fields of major study represented were: Agricultural Engineering, Agriculture, Architecture, Arts and Sciences, Business and Public Administration, Chemical Engineering, Civil and Environmental Engineering, Hotel Administration, Human Ecology, Industrial Engineering, Industrial and Labor Relations, Mechanical Engineering, and miscellaneous, with by far the largest group from Civil and Environmental Engineering, perhaps because the author (chairman of the faculty group) was from that area.

Faculty

The faculty for the first offering consisted of six members of the Cornell faculty, as follows: an engineer, an architect, a sociologist, an economist, and architect-physical planner, and a chemist-civil engineer.

In addition, there were lecturers and discussions by guest speakers from outside Cornell: an architect-anthropologist lectured on his experiences in Ceylon; an architect in government service spoke on U.S.A. foreign help and on his work in Latin America; an Afghani architect described his work in Afghanistan; and an architect-sociologist spoke about general principles and about his work in Peru.

In addition to the experience in developing countries in low-cost housing of the guest speakers, the central faculty group had such experience as follows: Afghanistan, Central America, Colombia, Costa Rica, Ghana, Nigeria, Pakistan, and Puerto Rico.

Conduct of course

The students from the various fields met together for common lectures and discussion periods, led by the multi-disciplinary faculty group. These meetings were held one hour every week plus one and one-half hours every other week, with the former meetings mostly for lectures and the latter meetings dedicated mostly to panel and class discussions. The class was divided into small teams (usually four to six people) for in-depth study of specific projects. In the alternate weeks (between the discussion periods) each team met with the appropriate faculty member. The course-wide lectures and discussions, and associated reading and study, provided general background, and acquainted people of one discipline with methods and problems of other disciplines, as well as of their own. The segregated, team meetings made actual use of such cross-disciplinary study, on a specific project, and also provided for in-depth work by each student in his own discipline.

Subject matter

The major topics covered in the lectures and discussion periods were: Historical Aspects, Indigenous Housing, Housing Problems in the Developing Nations, Current Practice and Malpractice, Site Selection and Planning, Use of Indigenous Materials, Design of Housing, Construction Technology, Housing Production, Case Studies, and Policy Proposals. Aspects of economics and sociology were brought into the picture as an integral part of each topic. Faculty members from the different disciplines coordinated and interrelated their presentations and approaches as much as feasible (or tried to), by teaming together for common lectures and panel discussions.

The textbooks used were: Abrams, *Man's Struggle for Housing in an Urbanizing World;* and Rapaport, *House Form and Culture*. Lengthy reading lists were distributed. A large selection of appropriate reading and reference material was placed on reserve in the library.

Student teams and term projects

The assignment of a specific project to a small team as the major term assignment was used in an attempt to encourage (or even force) people of different disciplines actually to work, and work closely, together (rather than merely to listen to lectures together and to engage in joint discussions). Each team, usually composed of four to six people, was assigned a topic or project for a term report to be prepared jointly by the team members. As feasible, the team chose or defined its own assignment, or more commonly, the team was more or less voluntarily assembled around a generalized problem assignment which the team defined in detail. As much as possible, the teams were deliberately formed in such a manner as to mix disciplines; for example, an attempt was made to have on a team one or two engineers (we had too many engineers), an architect and/or planner, an economist, a sociologist, and perhaps another specialist. An attempt was also made to mix undergraduate with graduate students and especially to include a student from a developing country on each team (if the latter was a graduate student, he would usually be the team leader and the team project would involve his home country).

Each team was assigned to the faculty member best qualified to advise and guide it (by actual experience, by study, and by interest), although all students and teams were free to, were encouraged to, and often did confer with other members, even outside the low-cost housing group. A team normally met at least once every two weeks with its professor, and often met at other additional times without him.

To aid in the formation of teams, a tentative list of projects was presented and students asked to choose first, second, and third choices. Tentative teams were then assembled, with strong consideration being given to mixing disciplines, mixing graduate and undergraduate students, and mixing foreign and U.S.A. students. The tentative teams met to determine if they were compatible in interests and could work together; some shuffling and re-shuffling occurred.

In order to provide the teams with a focus for their detailed study and reporting, it was suggested that a team could assume that it was a team of consultants with the task of preparing a coordinated report. Each team was required to define the nature and scope of its project, with faculty help. One approach was to prepare a report assumed to be for a Ministry of Public Works, perhaps for use by the ministry in helping to prepare a five-year plan in low-cost housing for a specific location. Other approaches might include preparing a report for a large industry interested in building or up-grading company or plantation housing, or for a developer or builder, or for a financing institution, or for a building research center.

Division of efforts and responsibilities among team members varied, but generally the author's teams were advised to prepare jointly an extensive and carefully written introduction which defined the problem, states its scope, reported on pertinent literature or known similar problems or projects, and stated the approaches to be used. Each team member would, usually, then write a chapter in depth in his own discipline, sometimes working with others. Finally, the team would prepare a joint concluding statement or joint recommendations.

Thus, for the term projects, the intent was that each team would work jointly as a multi-disciplinary team of consultants or experts to prepare a substantial report on a specific problem or project.

Some of the titles of projects submitted by the student teams were:

An Analysis of Uncontrolled Settlements
Contracting Systems

Environmental Systems
Housing for Reconstruction of Managua
Impact of Urbanization on Traditional Living Patterns
Indigenous Materials for Low-Cost Housing in Arid and Semi-Arid Areas
Indigenous Materials for Low-Cost Housing in the Wet Tropics
Low-Cost Housing for the Region Around Delhi, India
Low-Cost Housing Project for Lagos, Nigeria
Migrants
Planned Communities
Site Influence
Standardization and Modular Components
Structural Systems for Low-Cost Housing
Suitability of Modern Technology
Unit Design

Problems and deficiencies

Planning and organization by the faculty group extended over part of the previous spring semester, intermittently through the summer, and through the fall semester preceding the course. Four to six people were involved, and serious and vexing problems arose in getting all, or even most, of the people together for meetings. It is vital to have a person in charge who will move strongly to organize, to hold together interest, and to prepare in advance suggested details of plans and procedures. Since the multi-disciplinary faculty group was widely scattered, both in space and in interests, it was difficult to hold such a group together, and for the group to mold and give a meaningful, useful course. Careful organization and careful coordination are imperative.

It is far easier to have one teacher and one group of students from a single field, with guest lectures that are not integrated directly into the course, or to hold a series of only loosely related seminars or lectures or discussion periods for a varied group of students, but neither approach is adequate for low-cost housing. The former is not broad enough for the compound-complex problems involved, and the latter while perhaps stimulating or fun, does not provide a coordinated picture in depth, as is needed. The author is of the opinion that many students and many faculty members have been "turned off" by some multi-disciplinary discussion courses of a dilettante nature (usually called seminars), and thus, unfortunately, tend to avoid all multi-field approaches.

There is a possible, or even probable, danger of some "proprietary interest" in a given subject (such as low-cost housing) by a department or by each of different departments, or by individual faculty members. At Cornell, the former problem was partially solved by listing the course for credit in each of these different departments (in Engineering, in Architecture, and in Human Ecology) so students could sign up where they wished, and so each department involved could be credited with student hours taught.

There is a particularly serious problem involving graduate students. Cornell, as most schools, has no specifically-named major subject for graduate work in Low-Cost Housing. These students find it difficult to identify such courses with a major field of study, and often have to take them as overloads or extras in addition to a full schedule or courses for a more traditional major. Perhaps introduction of new areas of major study in low-cost housing (multi-disciplinary) would be appropriate and helpful — this needs careful consideration.

Although the course and its sequence of topics were carefully organized *on paper* before the course started, it was found that, because of the large number of faculty members and guest speakers involved, sequences were changed to meet schedules of availability of the many people speaking, and the resulting presentations of material out of sequence led to loose, or at times poor, organization of the course as actually given. This, combined with the fact that the students were given no small, sequential assignments as the semester progressed (only the one term project), led to confusion among many students about the order of topics and organization of the course. There was in the middle of the course a temporary but serious drop in interest and effort by many of the students. The faculty made a strong effort to tie the course together in a logical manner by carefully planned summation statements and discussions during the final class meetings. Thus, a course given jointly by several people may easily become disorganized unless very carefully planned and controlled; the author admits to some deficiency in preventing such disorganization.

A major problem was the lack of an organized, coordinated literature in the field of low-cost housing, to help students in their study and their research. An annotated bibliography, to delimit and cover the field, was being prepared at this time, and will be described later.

The student enrollment was too great for sufficient individual contact between student and professor (although some active students, as usual, managed to get much time). The faculty group had hoped for about 30 students, was surprised by the enrollment, and decided to try to deal with the large class instead of limiting the number of students. The largeness of the class interfered with class-wide discussions during discussion periods.

Special problems arose among the student teams. These included the difficulty of communication between and among students from different disciplines; widely varying points of view on the approaches to some of the problems; conflicting social, political, or national points of view; difficulty in scheduling the numerous joint meetings required; and even severe personality conflicts. As a result, several teams reorganized, a few teams disbanded or individual members split off, and two or three students worked entirely alone. Of course, there was the inevitable problem of some members of a team working hard while others worked much less, although, perhaps because the course was elective and not required, no cases of attempts of a "free ride" were noted.

Successes

The most apparent success was the superb quality of some of the term reports, indicating that great interest was aroused and that great effort was applied. Evaluation statements from students at the end of the course were generally favorable to highly favorable, while containing criticisms and suggestions, and most urged that the course be continued in future years. Also, the faculty members certainly learned much from each other (as well as from some of the advanced students).

Conclusions

The faculty group, at the end of the course, recognized various problems or failures, as well as some highly successful results. Overall, the course was judged to be much more of a success than a failure, and it was decided to give the course again, continuing the cooperative, multi-disciplinary approach, and modifying and hopefully improving the course from the lessons learned.

It is certain that much time and effort are required to bind and hold together such a multi-disciplinary approach to make a coordinated, useful course. If such time and effort are not available, it is probably best to avoid trying to give a course such as this one.

Course given again

This course was given again in the Spring Term, 1975, two years later. This represents the approximate turnover time for graduate students (total time for Masters and course-work time for post-Master Ph.D.s). This time the enrollment for credit (some additional people audited for no credit) was about 60, with about the same percentage as before for foreign students and somewhat higher for graduate students. Most of the reduction was in undergraduate civil and environmental engineers — the interested students from this group were screened carefully and those lacking a high level of motivation were discouraged from taking the course, since civil and environmental engineering students were present in too great a proportion during the first course offering. This size of class was more manageable than 90, but was still too large for good discussion periods.

The faculty were the same, except that since the economist was no longer at Cornell, a planner-economist became an active member. More students from Planning were in the course the second time. The course was not listed as a course for credit in Human Ecology, but a listing for planning was added, Architecture and Civil and Environmental Engineering being retained as listings for credit.

This time, the announced list of topics was: Housing Problems in Developing Countries, Social Problems and Economic Aspects, Indigenous Housing, Current Housing Practices, Site Selection and Planning, Design and Construction, Construction Technology, Housing Production, Case Studies, Policy Proposals for Regions or Environments, Possibilities for Technology Transfer, and Possibilities for Future Research.

The same general approach as before was used, including guest speakers.

Again, some term reports were outstanding. It was becoming obvious that some faculty members were losing student interest through poor presentations (if not poor material); they need to improve their delivery and other aspects of presentation. This is a touchy and difficult problem. Also, it was clear that many of the students were enthusiastic about open discussion periods. Both times the course was given there were some advanced graduate students who had actual working experience in low-cost housing in their own countries.

At the end of the second offering of the course, the faculty group indicated that it intended to continue to offer the course in the future, probably at two-year intervals. Overall, the response and interest of the students had been good, and in some cases highly enthusiastic. The course is being given again in 1977.

Courses of Study

In the Spring Term, 1974, in the year between offerings of the first course, a second, more specialized course was given. It was called Workshop: Site Selection, Physical Planning, Materials, and Design, and was designed primarily for joint, integrated study by mixed teams of architects and civil engineers.

This second course was given because, in addition to the need for a multi-disciplinary approach to the overall problem, specialists must also work at depth in their own specialities, within the framework and guidelines set up by the multi-disciplinary approach. It was planned that the specialists taking the second course would have already taken the first, broader course. Further, the second course was in fact a "bi-disciplinary" approach, since it attempted to integrate the efforts and approaches of engineers and architects, who traditionally, all over the world in modern times, do not often study together, coordinate their efforts well, or even communicate well with each other, even though their duties and goals are closely related. A major purpose was to encourage, even force, architects and engineers to work together throughout a project, and to work in detail and at depth in their fields.

Enrollment was limited to 16 students, architects and engineers plus one city and regional planner and one economist. The faculty consisted of two architects and a chemist-civil engineer. Four teams were formed, each to work on one of the following projects: Re-housing of Squatters in Limon, Costa Rica; Low-Cost Housing for Industrial labor at Lyari, Karachi; Low-Cost Housing in Lagos; and Low-Cost Housing (general). Detailed study was to be made, including detailed design of structures.

No formal lectures were given. Once each week, during the first half of the semester, the class met together for informal discussion sessions. Otherwise, each team worked on its own project, in consultation with an assigned faculty member, plus other faculty members as appropriate.

Overall, the course was successful. Architects and engineers did learn from each other and learned something of each other's thinking and approaches. Most of the people had previously had the general course in low-cost housing. The more specific, detailed work they did in the second course, along with their additional study of the subject, brought many of them to a fairly high level of knowledge and ability, to the point where they should be able to work immediately and effectively in low-cost housing. The course was not repeated in the Spring Term, 1976, during the author's Sabbatic leave, but he hopes it will be given again later, probably at two-year intervals, in the years alternating with the large general course.

Annotated Bibliography

A major problem in giving the first course, and of doing either broad-scale study or intensive research in low-cost housing, was the lack of an organized literature of the field. The vast field, with its diffuse boundaries, had not even been defined or delimited.

In a first effort toward gathering and organizing the widely-scattered literature, involving several different disciplines, the author and two assistants prepared an annotated bibliography* of about 1,000 entries, covering such aspects of low-cost housing as architecture, materials, design, services, health, sociology, economics, planning, construction, urbanization, urban renewal, rural housing, co-operatives, self help, management, political aspects, slums, squatters, migrants, re-location, environment, surveys and analysis, pilot projects, country and regional studies, case studies, and policies and programs. The bibliography is far from exhaustive and has missed some important publications, but it gives for the first time a reasonably representative picture of the kinds of literature available, lists many of the more important writings on the subject, and in effect defines and delimits the field. It has been most helpful, both to faculty and students, for study and research, despite its limitations and shortcomings.

---

*See Reference at the end of this chapter.

Other Activities

Among other activities have been some field projects. A demographic survey was planned and conducted in a squatter colony in Limon, Costa Rica. The colony is to be re-located to make way for an industrial park. The survey makes available some alternative approaches for the relocation. The fairly extensive questionnaire program, conducted by especially trained local people after mass meetings called to explain and discuss the survey, covers among many other things such items as skills available for self-help, preferences of the future occupants for design, problems they foresee, and their own suggestions for solving the over-all problem.

Studies for housing delivery systems and related items have been under way in Ghana for about two years. Reports are now forthcoming. The results should be highly useful to Ghana for deciding which of possible strategies to adopt.

The author has had a most interesting, enjoyable, and useful association with Technology and Development Institute of the East-West Center, in connection with its Low-Cost Housing Technology program and the two Roving Workshops it has conducted into the Orient and Hawaii.

Laboratory research at Cornell has included studies on coconut fibers in concrete, design of earth houses to reduce earthquake damage, and modification of existing earth houses to reduce earthquake damage.

All these other activities have contributed to the teaching program, sometimes in a major way. Some of them developed from the teaching program, then provided material useful in later courses.

The Future at Cornell

Enough students, faculty members, departments, and administrators have shown active interest to indicate that Cornell will probably continue and increase its activities in this area, using a multi-or inter-disciplinary approach, as well as the more traditional uni-disciplinary approaches.

9.7 POSSIBLE EDUCATIONAL FUTURES IN LOW-COST HOUSING

Trends Now Apparent or Indicated

Unquestionably, there is in educational circles already a strong and growing interest, worldwide, in the use of multi- or inter-disciplinary approaches in attacking large problems of broad scope. The recent emphasis on problems of the environment and of energy, as well as of low-cost housing, has pointed up the vital necessity of integrated approaches from various disciplines. A few schools have already started to introduce courses that mix and hopefully integrate disciplines, such as the course just described at Cornell. Even though the attempts have yet been very few (according to the author's criteria for such courses), there is every indication that there is developing a trend for such educational programs.

In Asia, educational programs of this nature have not yet been developed (to the author's knowledge). However, he has been told by people from the Asian Institute of Technology and the Institute of Technology, Bandung that these organizations are interested in introduction of multi-disciplinary teaching in low-cost housing.

Trends in Practice, Related to Education

Some people in the field, from all sectors — private, public, and popular — are already aware of the interest in and need for multi-disciplinary approaches. This awareness may be greater among the younger people, who have recently finished their schooling. Already there is much talk about integrating the inputs from various fields of knowledge for solving low-cost housing problems. In some cases, partially integrated efforts are already being used, although with some inputs missing (for example, analysis and recommendations by anthropologists, joint study of implications of land tenure and equity by lawyers and sociologists, social implications of forced segregation, and considerations of general future economic well-being of occupants of new projects).

The interest and potential "demand" already exists among the practicing professionals, for multi-disciplinary approaches. The pressure is on the educational institutions to meet those interests and "demands" by creating a "supply" of appropriately educated (or trained) people.

Dreams of Future Educational Programs

The following thoughts are offered, involving opinion and speculation. As a first step, it would seem appropriate for universities to plan teaching programs to bring engineers and architects closer together, so they can better understand each other's concepts, approaches, and problems (after all, both are in reality builders) and so they can work together cooperatively — instead of separately — with effectiveness for the benefit of their structures and the users of those structures. Most of our educational programs now segregate architects and engineers, even those programs that give lip service to integration of the two. Perhaps some of the approaches of Frank Lloyd Wright at Taliesin deserve a new look.

Perhaps we place too much emphasis on the traditional definitions and scopes of the disciplines as specialities, by insisting that all advanced graduate students adhere to a relatively fixed, rigid program of study and research. Some recent trends away from this rigidity can be noted. For example, environmental engineering is in some places now being treated as a field somewhat separate from civil engineering, with study programs being oriented more toward actual problems of the environment, partially at some expense to what have traditionally been considered to be important or even vital aspects of an education in civil engineering. Other examples of educational programs being oriented more directly toward real and major problems could be cited.

Consideration could be given to introduction of a new graduate program with major study in low-cost housing, or introduction of this major in some of the traditional fields. Short of this, it would be very helpful to have listed and implemented graduate programs of minor study (MS and Ph.D.) in low-cost housing, coordinated among fields of major study such as civil engineering, architecture, planning, sociology, economics, etc. In either case, students from the various traditional disciplines would presumably meet, study, and work together on what in actuality is their common problem — low-cost housing.

The author has toyed with and been intrigued by the idea of "extension" services in low-cost housing by educational (and possibly research) institutions. The idea and method has been well developed in agriculture by American universities. Trained professionals, sometimes on the staff of a university, go into residence in local communities where needed (as "county agents" in agriculture). They become part of the community, become familiar to and friends with the local people, and thus no longer being viewed as outsiders and academic do-gooders or meddlers, are able to work effectively with the local people to pass along new information and to help in general. This approach in agriculture has had a tremendous impact in the U.S.A.

It should be possible for educational institutions to train people in low-cost housing, particularly concerning materials, construction, maintenance, up-grading, and health, then for these trained people to be on the payroll of either the educational institution (if supported by the State) or a governmental agency and be assigned to live for extended periods in areas where their knowledge and skills are needed. Perhaps much or most of the training should be of the technical school nature. As residents of a community, they could win the confidence of the local people (a vital point always, and especially so in slum areas or areas of low educational level). They could help greatly in programs of upgrading existing housing through self-help. Whenever outside help of any kind became available, they would be in a position of great influence among the local people and could well tip the balance toward success rather than failure of a project — since we have (or some have learned that the opinions, attitudes, and actions of the occupants themselves usually ultimately determine success or failure of a housing project.

The U.S.A.s International Cooperation Administration (ICA) foreign aid program used an approach a little like this in its "village aid" program of the middle 1950s, which usually worked well. Local, trained professionals (not Ph.D.s!) travelled through the villages, tried to, and often or usually did, win the confidence of local people of influence, then were able to make suggestions for improvements and help initiate them.

In the opinion of the author, the county-agent idea, or some appropriate modification of it, could have a major impact on the problem of low-cost housing for low-income people.

Committees Need Heads

In all the talk about teams composed of people from various disciplines, all working together in a coordinated manner, it must not be forgotten that a team or committee must have someone in charge. Presumably, this should be a professional trained in depth in some aspect of low-cost housing, but it is not at all clear which discipline this should be. Educational programs will probably not resolve this question and perhaps should not try to do so. On the basis of our current traditions of education and application, it would seem that an architect or civil engineer would usually be best fitted for the job. Political considerations may control, for governmental projects. Ultimately, it is likely that personal characteristics and abilities will be found to be more important for a team leader than which discipline he was trained in, assuming that he is at least trained in low-cost housing with some background in the multi-disciplinary approach.

9.8 HOW TO SET UP AN EDUCATIONAL PROGRAM

Perhaps an appropriate first statement is that there is no one "best" or "correct" way to establish a multi-disciplinary teaching and research program in low-cost housing, or if there is, the author surely does not know it and in any case feels somewhat inadequate in trying to write this section. Surely, the approach will vary widely, depending on people and facilities available, faculty interests and capabilities, local, regional or other needs and interests, support, and other factors. Any attempt to list a series of steps to be carried out would be futile.

However, a few notes will be given, based on the Cornell experience, in case some of them might be of interest or help to others.

First, it would seem vital to have one or two people strongly interested in establishing such a program, who have extensive backgrounds in the subject, and who are willing to work hard and long beyond their other normally assigned duties. There must be, in at least one or two other disciplines at first, capable and experienced people who can be invited, who would like to join in a team effort and who would be willing to take on an over-load beyond regular duties.

There must be some interest, encouragement, and hopefully support, from administration at the level of a deanship and encouragement, or at least lack of opposition, from chairmen of the departments involved.

The choice of faculty to be invited to participate is crucial. In addition to interest, background, and ability, each person must be willing to work in a team and not to insist on being a prima donna or in carving out a piece of the program as a personal possession. Still, freedom absolutely must be left for each member to do individual work on his own, as he deems appropriate. The key is that he not steal pieces of the group effort, but that his individual efforts support and reinforce his group efforts. This is a touchy matter, and illustrates that not all otherwise qualified people are appropriate for a group effort.

The dangers of a "proprietary" outlook are not limited to individual faculty members, but extend to departments and colleges, involving not only location of traditional subjects for teaching, but also involving credit for student course-hours and therefore budget credits for such student course-hours.

Since the approach is new, a very large number of meetings may be required to determine a structure and a method of operation for the group, and more importantly, to determine first a teaching approach, then organization of a course, and finally details of who does what and when.

Not much money is needed for such a course, particularly if laboratory experimentation is not extensive and most particularly if no travel is involved. Of course, availability of funds can allow a great variety of improvements, expansions, and extensions. Most libraries do not have sufficient literature on the subject — this may be the area of greatest need for money, and will involve time delays.

Since the field is vast, and since the assembled faculty will be small and limited in background and competence, it is almost certain that important and even major portions of the field must be omitted. The author is convinced that the course coverage must correspond with what the faculty can do well, rather than with an idealized coverage of topics which may include items outside the interest and abilities of the faculty. Thus, the Cornell courses have, for example, omitted coverage of the all-important legal aspects of land tenure, as well as other items. The students should be told of such omissions and advised that they should in the future consider them.

Actual case studies, well illustrated, are often of great value in teaching. If the faculty group is fortunate enough to have among their students some mature people at graduate level who have had practical experience in low-cost housing, the faculty should surely seek out and consider their suggestions and evaluations, but not be ruled by them.

In summary, the need is great, the task is difficult, and each case will be different and require its own unique approaches and solutions. Hopefully, as new courses get under way in new places, the pioneering faculty groups from different schools can meet together, exchange ideas and information, and learn from each other's experiences.

9.9 CONCLUDING STATEMENT

The central theme of this chapter is a plea that a coordinated, multi-disciplinary approach be used for working on large-scale, broadscope problems of low-cost housing, and that educational and research institutions take the initiative by setting up appropriate programs of study leading to the multi-disciplinary approach. The author hopes, and believes, that the East-West Center's Technology and Development Institute, through its Low-Cost Housing Program, has helped in a significant way to encourage these ideas.

ACKNOWLEDGEMENTS

The principal faculty members who have developed and given the courses are: Dr. Franklin J. Ahimaz, Engineer, Assistant Dean of Engineering and Assistant Director of Cornell's Program on Policies for Science and Technology in Developing Nations; Mr. Peter Cohen, Architect, Associate Professor of Architecture; Dr. James W. Converse, Sociologist, Assistant Professor of Rural Sociology; Mr. Charles B. Daniels, Economist, Assistant Professor of Consumer Economics and Public Policy; Mr. Henry W. Richardson, Architect and Physical Planner, Assistant Professor of Architecture; Mr. Darrel Williams, Planner and Economist, Assistant Professor of Policy Planning and Regional Analysis; and Dr. Floyd O. Slate, Chemist and Civil Engineer, Professor of Engineering Materials.

Dr. Ahimaz and the author worked closely together in developing the Cornell program; the contributions of Dr. Ahimaz were great.

REFERENCE

Low-Cost Housing for Developing Countries — An Annotated Bibliography 1950-1972, by Floyd O. Slate assisted by Mary Acton and Thandiwe Chinamora, Cornell Univ., June 1974, 214 pp.

10

LOW-COST HOUSING INFORMATION EXCHANGE

Fredrich J. Burian and Eduardo Q. Canela

10.1 INTRODUCTION

During the past decade much attention has been devoted to the search for viable systems for transferring housing technology information on a global scale. The search, however, has concentrated on strategies for developing large-scale information exchange networks, initiated and maintained by information resource rich institutions/agencies/nations. Substantial and growing capital and manpower resources have been invested in these large-scale supplier-initiated networks as more economically developed (information rich) nations add information transfer programs to their development assistance portfolios. Yet, policymakers and development architects in developing countries are raising serious questions regarding the viability and effectiveness of such globalized supplier-dominated systems/networks as a necessary ingredient of efforts designed to alleviate the human shelter and settlement problems of low-income people which have perenially plagued the Third World

The concern in developing countries seems to revolve around questions of usability of the information inventories available through developed country suppliers. Information on high technology, capital/energy intensive housing and settlement solutions employed in industrialized countries cannot be easily used in most developing countries where the need is for intermediate technologies and capital/energy-saving approaches. Even so, the supplier dominated systems/networks appear to be proliferating.

It usually is the practice in supplier-initiated networking to begin with information inventories and/or services already being used in developed countries and seek out potential clientele institutions in the Third World that fit the supplier's perception of appropriate new users.

The supplier characteristically selects initial information users/clients in developing countries with great care and nurtures their participation through grants and subsidies. The selection of linkage partners is often influenced by broad development assistance policies of supplier countries. Since the supplier is comparatively resource-rich, the initial linkage units need not exhibit a very high level of demand for the information offered. This networking strategy usually shows strong initial promise since considerable resource input required for the network start-up is subsidized by the supplier while the developing country user's contribution is comparatively small.

Unfortunately, supplier-initiated networks have a difficult time sustaining themselves; they rarely reach a point where the supplier can reduce its capital and material input without collapse of the system. Failure of supplier-initiated information networks to sustain themselves can be in large measure traced to strategies used to initially link information users to networks. If information services are subsidized and the participating institutions can maintain a passive or dependent role in the exchange process, there is no way to validate real need on the part of the network participants for the information short of reducing the information flow and measure the inclination of the users to demonstrate need by "buying" back the reduced services.

Characteristically, however, when the supply of "free" information service in a supplier-dominated network is reduced, or when charges are levied against the users, demand is substantially reduced. In such a situation, rather than finding the network participants willing to now pay for information, the supply-network managers find the participants turn away from the network. Such networks require the constant infusion of resource/effort on the part of the supplier to maintain momentum. If the network is built to return initial investment, it will go bankrupt. If it is funded by a non-profit corporation/foundation/governmental agency, it will continually operate at a loss. Unfortunately, this is the sorry state of affairs of many information networks sponsored by resource-rich suppliers.

A perhaps more favorable but untried approach is to encourage small-scale information users in developing countries to establish linkages with information generators/sources of their own choice. Under this approach, users, given access to a large selection of potential information suppliers, initiate linkages with suppliers they themselves judge as potential sources of information compatible with their own fields of interests and levels of development. Armed with appropriate initiation and linkage strategies, these units can probe a large number of potential information suppliers in both developed and developing countries, inducing them to reveal comparability in interests and development.

In time, these units could filter through a substantial number of potential sources identifying a relative few where more formalized information exchange arrangements could evolve. Each of the original information seeking units would become the hub of a mini-network screened for specific fields of interest and level of development. By mutual agreement, several of these mini-networks could then link themselves into larger network configurations, covering entire countries, even regions.

It is only at this point that supplier-initiated resources should be invested — only when the information users have validated their own need for information by seeking-out their own networks of suppliers and engaging them on a *quid pro quo* basis. Input from resource rich suppliers would thus stimulate saturation coverage of an information domain validated in the Third World.

Success of this bottom-up approach to information exchange networking rests with the adoption of uniform procedures for building exchange arrangements by a number of information seeking units spread widely over the globe. The objective of this chapter is to present such a uniform set of procedures for consideration by users of housing/human settlements technology information. The chapter describes an overall strategy for inducing mini-networks built around the specific information needs and exchange capabilities of small-scale housing/human settlements information units. This technique will provide small-scale units with information and data screened for relevance to their own needs.

10.2 USER-INITIATED LINKAGE ESTABLISHMENT

User-initiated linkage formation is the first stage of viable information network development. It is a stage that must grow out of the information user's own initiative. It is a stage that cannot be properly induced by resource-rich information suppliers seeking markets for their inventories. User-initiation is a dynamic process of building information-exchange arrangements where small-scale information users link themselves with relevant sources of information they judge compatible with their own areas of interest and level of development. The key to the process rests with the small-scale information user's ability to clearly profile its own areas and level of research/development interests and its information, inventory and exchange capabilities. It should be emphasized that this information exchange strategy will work only if the small-scale, linkage-seeking institution itself has information to offer others no matter how humble the offering. Anything less takes the control of information exchange out of the hands of the user and returns it to the supplier.

The linkage establishment strategy described here is a three-level process: profiling, expansion of information sources base, and information exchange. The process of linkage establishment begins when an institution, possessing limited information exchange capabilities and resources, commits itself to seeking out potential information suppliers in the specified field of human interest, e.g. housing and settlement technology. Activated by the commitment, the institution generates a list of potential information exchange partners and initiates direct correspondence with them revealing the nature of its information offerings, its information acquisition and dissemination capabilities, and the identities of its current external information sources. The initiating institution thus profiles its information use and exchange capacities for potential information suppliers and requests these sources in turn reveal the level, nature, and scope of their own information capacities. If the exchange of correspondence is properly controlled, the initiating institution will receive profiles primarily from information suppliers possessing information holdings and services which are compatible with the information needs and processing capabilities of the institution initiating the correspondence. Ideally, the end result of this *quid pro quo* profiling process is equitable information exchange arrangements where the initiating institution becomes a source of information for a growing array of relevant institutions and in like measure, becomes the recipient of information already substantially screened for its own sociotechnical and economic development level.

Expansion of the information source bank is accomplished by asking institutions being profiled to send lists of their own most valuable external information sources to the initiating institution. The initiating institution then selects new institutions from these lists and implements profiling steps with these newly acquired sources. In this way, the initiating institution induces an expanding network of information linkages tailored to its specific areas of interest and level of development. Based on profiles of an expanding set of potential information sources, the initiating institution can negotiate the most appropriate information exchange arrangements with each participating institution revising the exchange pattern as less relevant sources are replaced by more appropriate sources.

A suggested hierarchy of information exchange arrangements has been developed to identify several exchange thresholds that institutions pass through as exchange partnerships mature.

These three levels correspond to stages of the user-initiated information exchange strategies described in this chapter. Level I exchanges are the result of mutually satisfactory profiling activities; Level II exchanges mark the flow of published information; and Level III exchanges represent reciprocal arrangements for the exchange of services.

TABLE 10.1 Levels of Institutionalization of Information-Exchange Arrangements

LEVEL I :		Profiling
	a.	Correspondence Exchange
	b.	Profile/Questionnaire Exchange
LEVEL II :		Expansion
	c.	Publications Exchange
	d.	Current Awareness List Exchange
	e.	Bibliography/Abstract Exchange
	f.	Inter-Library Loan/Exchange
LEVEL III:		Exchange
	g.	Literature Search Service
	h.	Technical Inquiry Service
	i.	Data Sharing Service
	j.	Extension/Training
	k.	Consultancy

Schematically, the overall strategy for developing linkages with sources of information can be represented as a three-level interaction between an initiating institution and a group of institutional sources of information individually contacted by the former, as shown below:

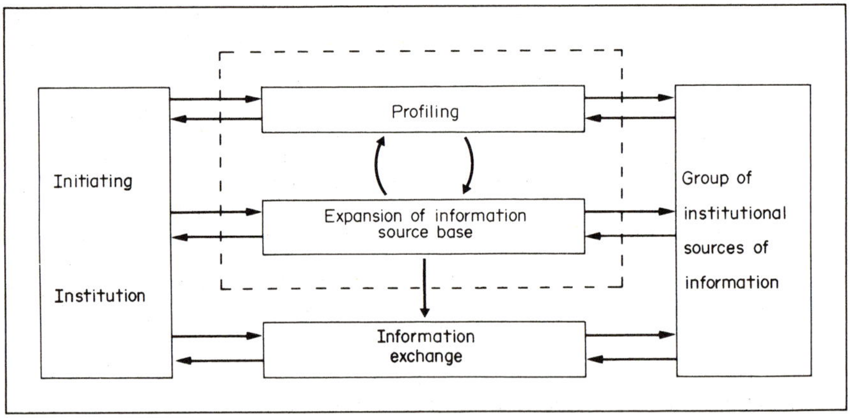

Fig. 10.1. Schematic Model of Information Linkage Establishment Process

The three linkage establishment steps outlined in Fig. 10.1 (Profiling, Expansion of information source base, and Information exchange) are sequentially implemented. However, the profiling and expansion steps proceed reiteratively, generating an ever-increasing number of potential information sources, while the information exchange step narrows the scope of actual information exchange arrangements to a highly relevant subset of total number of information sources/contacts.

10.3 IMPLEMENTING A LINKAGE ESTABLISHMENT STRATEGY

The establishment and expansion of a user-initiated strategy for linking sources of housing/settlements technology information can be described in operational terms. The three step strategy introduced earlier can be decomposed into five major task clusters or subsystems outlined below:

PROFILING

Subsystem 1: Setting-up an initial information sources bank. The initiating institution (a small information unit) creates a preliminary file or bank of names and addresses of potential information sources. The initial identification of contacts is made by the staff members and researchers working within the initiating institution.

Subsystem 2: Initial contacts and profiling of the sources of information. The basic profiling activities are developed for establishing linkages with various information sources. Activities here include profiling the information exchange capabilities of the initiating institution, establishing contact with potential external information sources, soliciting source profiles, and tabulating results.

EXPANSION

Subsystem 3: Expanding the initial information sources bank. The tabulations of information resource profiles from responding institutions are evaluated and new sources are identified. The initial information sources bank is adjusted to incorporate an expanded method of classifying/indexing the various sources of information.

Subsystem 4: Generating additional sources of information. The initiating institution installs the capability of keeping itself abreast of new research, development and applications compatible with its special interests. This subsystem reiterates on the initial information sources transforming the system into a dynamic information identification-utilization cycle.

EXCHANGE

Subsystem 5: Initiating information exchange. The analysis of profiles enables the initiating institution to select high value information sources with which to implement information exchange. This subsystem provides guidelines for developing information exchange arrangements. A schematic diagram of the sequence of performing the task cluster is shown in Fig. 10.2 below.

10.4 SUBSYSTEM 1: SETTING UP AN INITIAL INFORMATION SOURCES BANK

Subsystem Objectives:

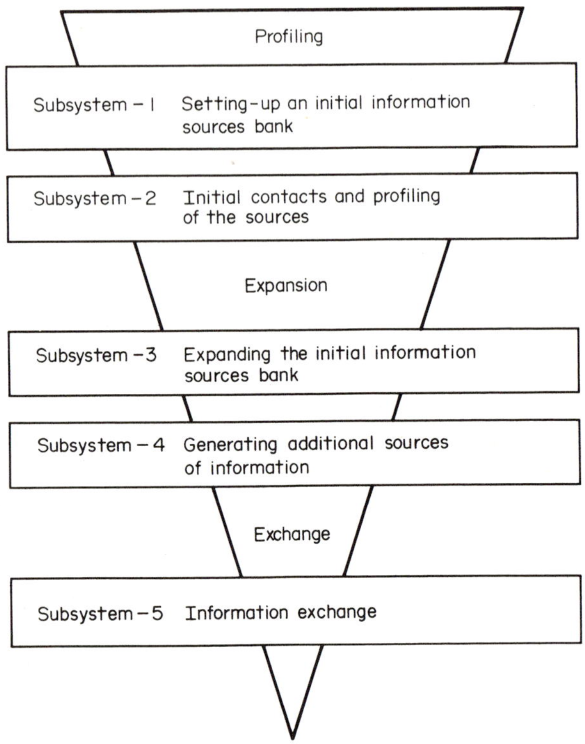

Fig. 10.2. Sequence of Performance of the Task Clusters

After completion of the operations involved in this subsystem, the information manager will be able to:

(a) Generate a provisional information sources list in the area of housing and human settlements;

(b) Establish a provisional updatable card file of information sources in the area of housing and human settlements;

(c) Acquire experience in setting-up a manually operated information sources bank;

(d) Revise or adapt the subsystem's structure to accommodate unique local information needs and resources.

Subsystem Description:

Subsystem 1 is made of four major activities; viz.,

1. *Generation of an Initial List of Potential Sources of Information in the Area of Housing and Human Settlements.* To operationalize this subsystem, the

information manager should initially identify contacts and sources of information frequently used by practitioners and researchers in his/her own institution and local setting. The output of this activity is a list enumerating sources of information and their business addresses.

2. *Transfer of the Entries in the List of Initial Information Sources into a Carding Deck.* The output of this activity is a deck of entry-cards consisting of a single card for each information source identified. Initially, the information source entry-cards are used to track the progress of linkage formation activities, thus control points are included on the cards, permitting the user to monitor the status of linkage negotiations with the sources. A sample entry card format is shown in Fig. 10.3.

Sample initial information sources bank Entry card

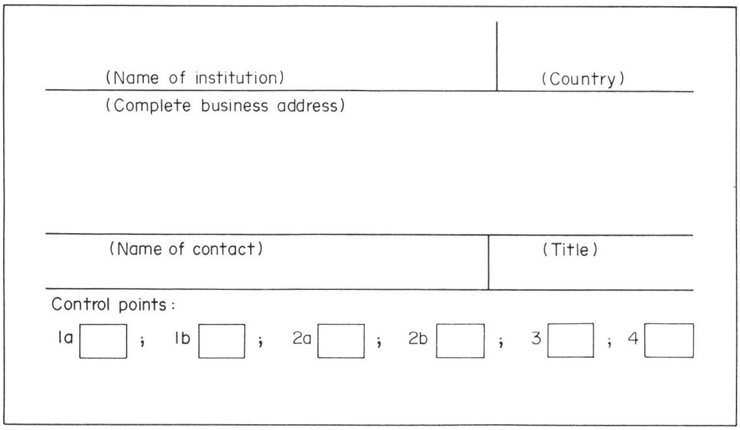

Fig. 10.3. Sample Initial Information Sources Bank Entry Card

NOTES ON THE CONTROL POINTS:

The numbers (la, lb, 2a, 2b, 3 & 4) correspond to the subsystem used in establishing linkages with various sources of information. At the completion of each of the subsystem task cycle, the person responsible for the completion should date and initial the appropriate box provided.

3. *Selection of a Reasonable and Practical Card Filing System.* The following strategies can be used to organize the initial information sources deck of entry cards:

   (a) Alphabetic System — Use the name of the institution as a basis for arranging the entries into alphabetic order for easy retrieval;

   (b) Country-Alphabetic System — Arrange the different countries involved alphabetically, then alphabetically arrange all the institutions falling under each country;

   (c) Other applicable card filing system (subject areas if known, etc.).

Although choice of filing system is left to the discretion of the information manager, the following criteria should be used in deciding what type of filing system to use;

   (a) Time of retrieval must be relatively short;

(b) Costs of maintaining the initial information sources file should be low;

(c) The file should complement existing information facilities and systems.

4. *Utilization of the Information Sources File.* Even in its initial form, the information sources file can be used to access specific information sources, in any or a combination of the following situations:

(a) When a user knows the name of an institution supplying the information, but not the full address;

(b) When the user requests a list of information sources in housing and human settlements for a specific country or group of countries;

(c) When a user requests specific sources of information, based on broad areas of concern (subjects), which are hinted in institution names on the entry cards.

Examples:

SUBJECT	*INSTITUTIONAL NAMES OF POTENTIAL INFORMATION SOURCES*
Ecology	Agence de Presse Rehabilitation *Ecologique*
	Adirondack *Ecological* Station
	Ecology Center
	Ecology Action
Earthquake	*Earthquake* Engineering Research Center
	Etc.
Transportation	Texas *Transportation* Institute
	Transportation Center
	Transportation Research Center
	Transportation Research Group
	Etc.

Flowchart

A depiction of the various steps involved in implementing this subsystem is included in Flowchart 1 (Fig. 10.4). This flowchart and the succeeding ones are outlines of a possible approach. The information manager implementing this system is encouraged to modify, amplify or simplify the processes/steps involved to meet actual needs and conditions. The flowcharts are comprehensive and should give those charged with organizing the system ample information and guidelines for setting-up an initial information sources bank. The flowcharts employ standard data processing symbols. Table 10.2 lists the symbols and their interpretations as used in the succeeding flowcharts. The information manager should become familiar with these symbols before attempting to "read" the flowcharts.

10.5 SUBSYSTEM 2: INITIAL CONTACTS AND PROFILING OF INFORMATION SOURCES

Subsystem Objectives:

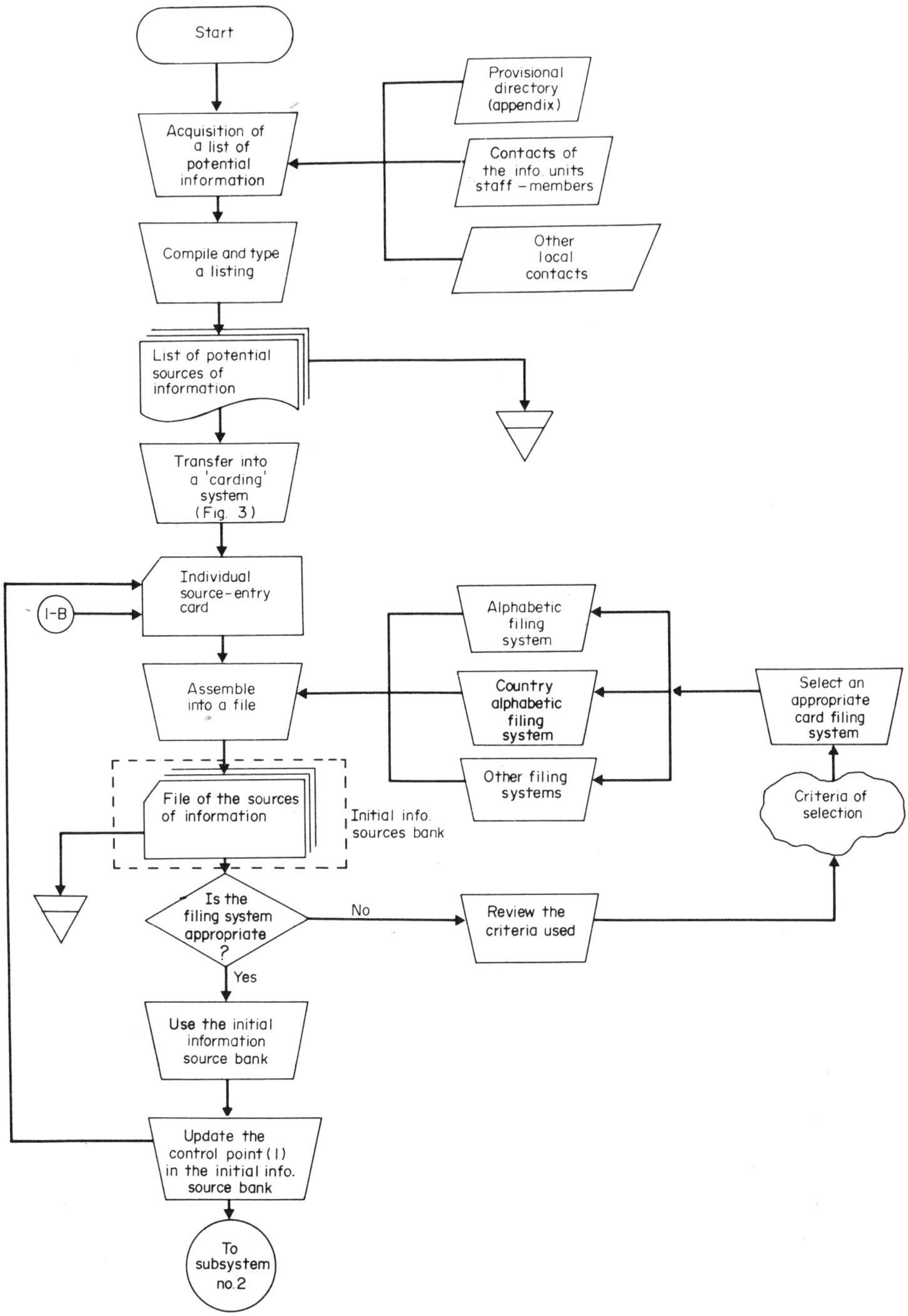

Fig. 10.4. Flowchart No.1 : Subsystem No. 1
Process of Setting-up an Initial Information
Sources Bank

TABLE 10.2

Flowcharting symbols and their interpretations*

Symbols

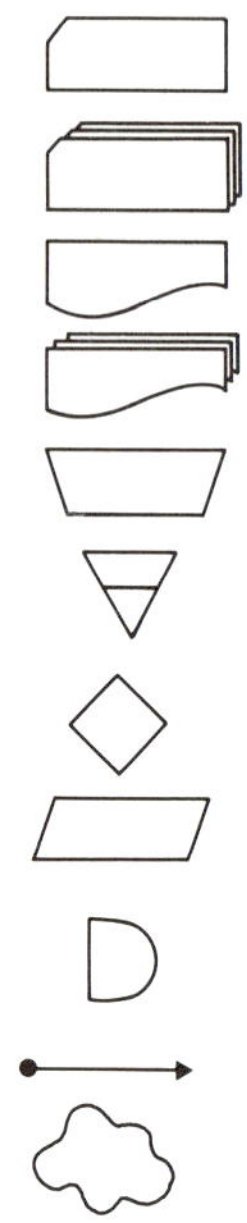

Interpretations

Entry card / filing card

Dock of filing or entry card

Document

File of documents

Manual operation

File

Decision point

Input/output symbol (Manual)

Delay

Sequence of flow

Outside the systems boundary

Based on the "Summary of flowchart symbols" from U.S.A. X3.5–1960 Flowchart Symbols for Information Processing.

After the completion of the operations involved in this subsystem, the information manager will be able to:

(a) Design or redesign a questionnaire for profiling potential institutional sources of information;

(b) Establish initial information exchange contacts with the potential sources of information;

(c) Gather experiences necessary in profiling sources of information;

(d) Revise or adapt the subsystem's structure to accommodate unique local information needs and resources.

Subsystem Description:

Subsystem No. 2 is made up of six major activities, viz.,

1. DESIGN/REDESIGN A QUESTIONNAIRE FOR SURVEYING THE SPECIFIC INFORMATION EXCHANGE CAPABILITIES OF THE FILE OF POTENTIAL INFORMATION SOURCES GENERATED IN SUBSYSTEM 1. A sample questionnaire is included as an Appendix. This form can be redesigned or adapted depending on the needs of the information unit seeking linkage relationships with other institutions.

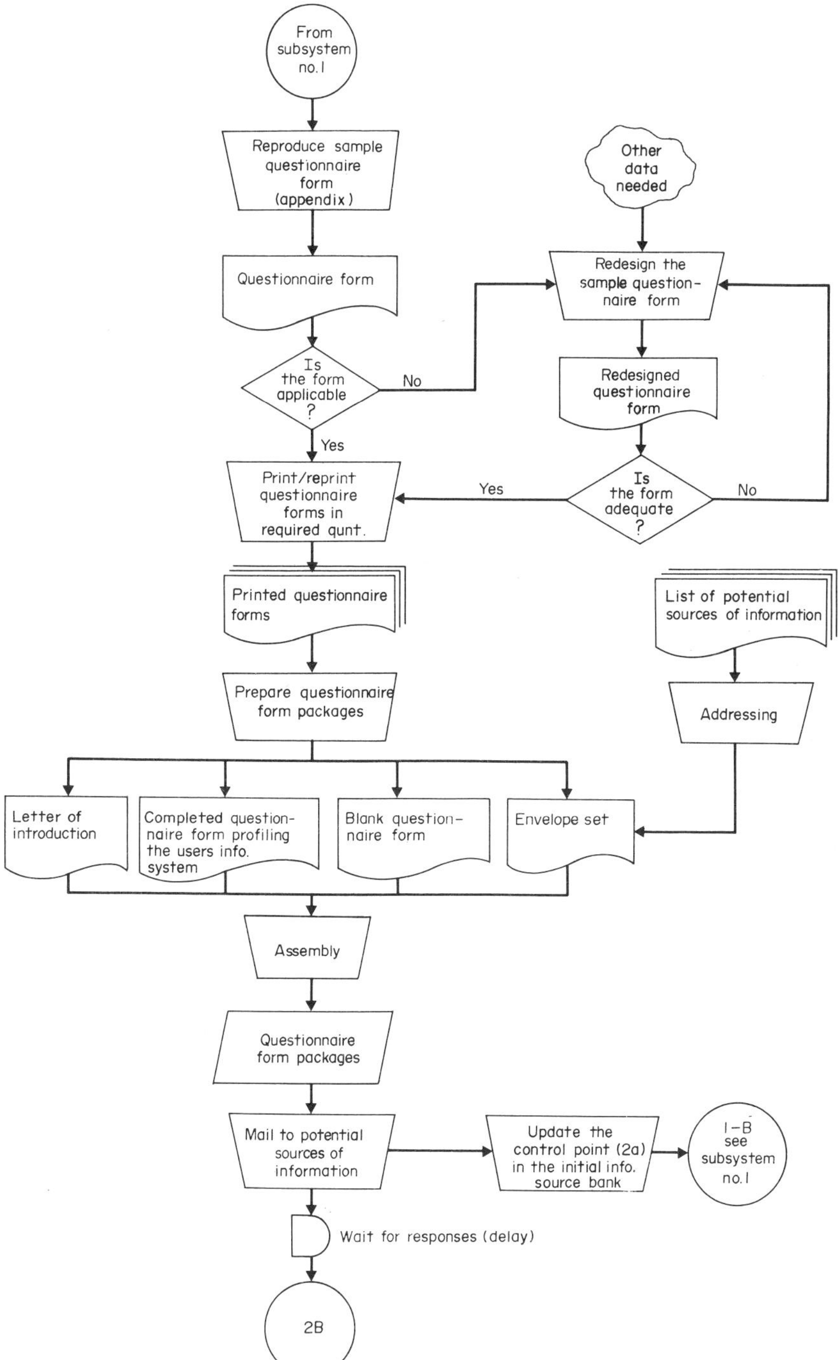

Fig. 10.5a Flowchart No.2a : Subsystem No. 2
Process of Profiling Institutional Sources of Information

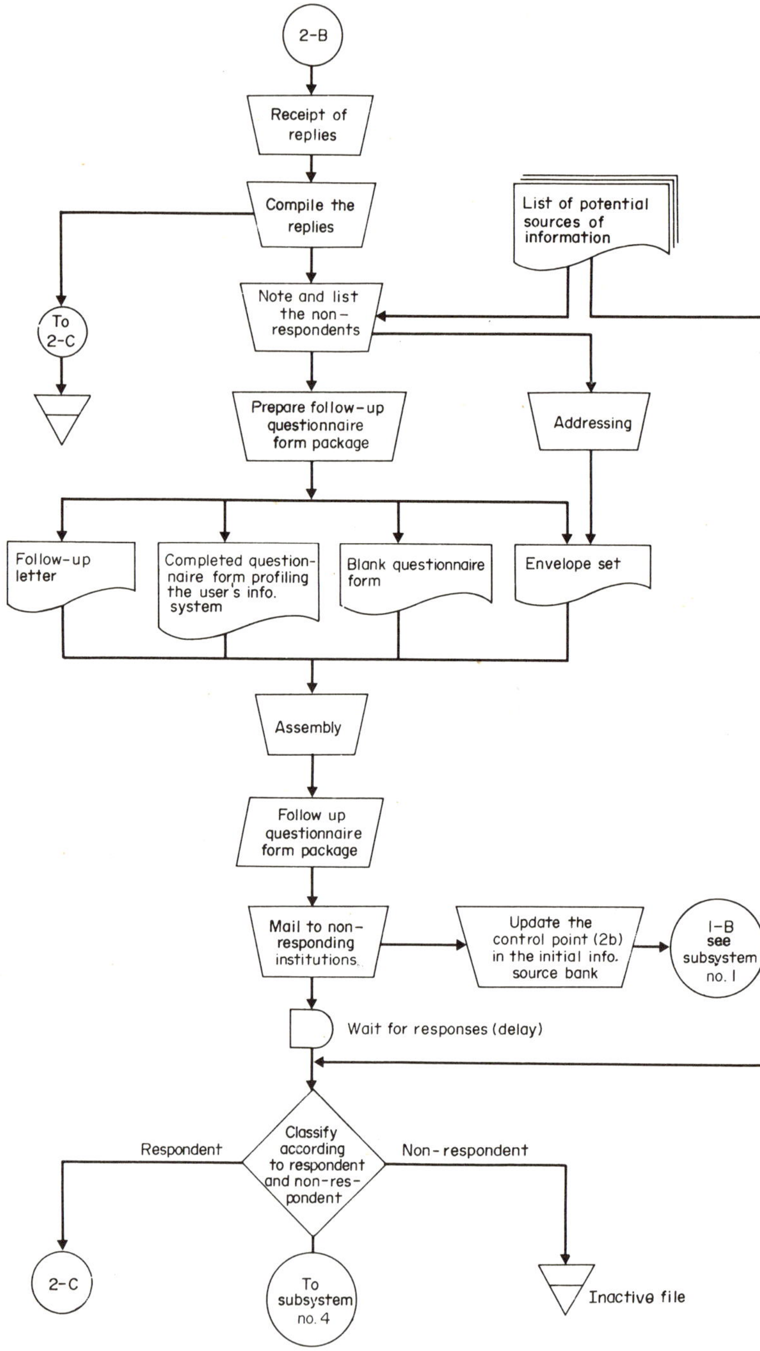

Fig. 10.5b. Flowchart No. 2b; Subsystem No. 2 (Cont.)

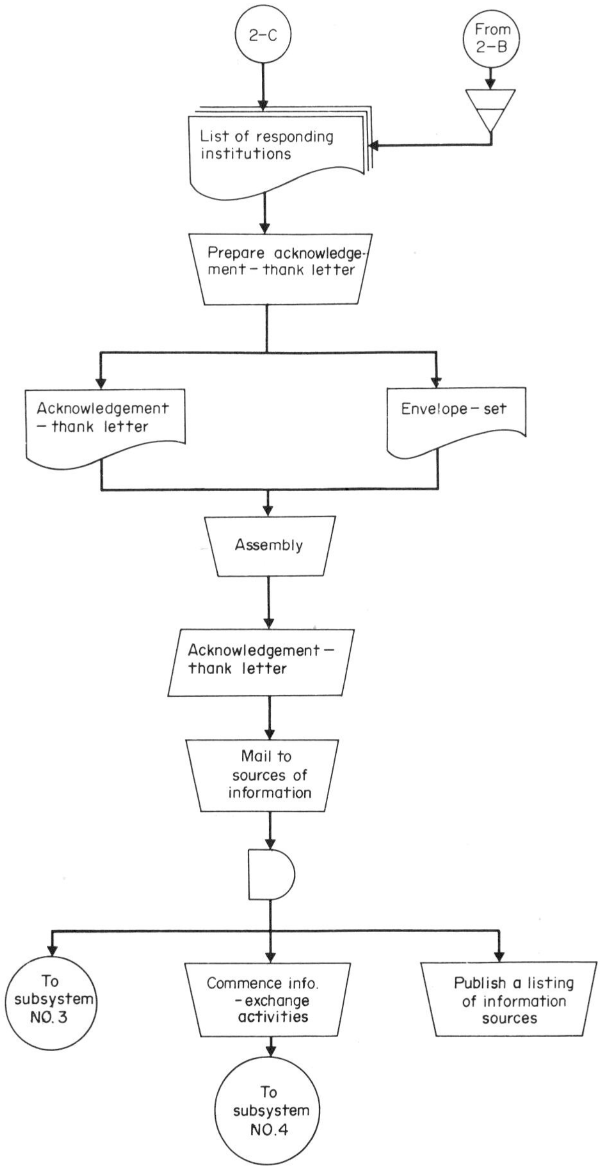

Fig. 10.5c. Flowchart No.2c : Subsystem No. 2 (Cont.)

2. PREPARATION OF A QUESTIONNAIRE PACKAGE. A questionnaire package sent to each potential information source on file will contain the following:

   (a) Letter of Introduction — This will introduce the linkage seeking institution to the potential information supplying institution. Structurally, the letter should contain a brief description of the initiating institution, rationale for the linkage activity and deadline for the submission of responses to the profiling questionnaire.

(b) Questionnaire form for the potential source of information to complete. The questionnaire should be short. Two double printed pages are ideal. The sample questionnaire included as an Appendix is adequate for constructing profiles of interest areas and information exchange capabilities as well as identifying additional information sources.

(c) A completed questionnaire describing the information exchange capabilities of the linkage initiating institution (user). This will not only serve as a guide for others in completing the blank questionnaire, but more importantly, the completed profile invites potential information suppliers to highlight information supplies and services which are compatible with the questioner's stated information interests and exchange capabilities.

3. SENDING THE QUESTIONNAIRE PACKAGES TO ALL/SELECTED POTENTIAL INFORMATION INSTITUTIONS ON FILE. Here decisions have to be made regarding the efficiency, reliability, and costs of the mailing channels to be used (see Flowchart 2a).

4. UPDATING THE CONTROL POINTS. Enter date questionnaire package is sent in control point 1-a on preliminary information source file cards (Fig.3). Enter receipt date of each completed questionnaire in control point No. 2-a.

5. SENDING FOLLOW-UP LETTER. If feasible, send a follow-up letter to institutions that did not respond initially, again enclosing the questionnaire package. Enter date follow-up questionnaire was sent in control point No. 1-b and date of response in control point No. 2-b on the entry card (see Flowchart 2b).

6. SENDING ACKNOWLEDGEMENT LETTER. Send a letter of acknowledgement and thanks to all responding institutions (see Flowchart 2c).

Flowchart

The flowcharts depicting the various steps and sequences of steps involved in the implementation of this subsystem are included as Flowcharts 2a, 2b and 2c.

10.6 SUBSYSTEM 3: EXPANDING THE INITIAL INFORMATION SOURCES BANK

Subsystem's Objectives

Upon the completion of the operations involved in this subsystem, the information manager will be able to:

(a) Acquire the necessary experiences in processing the information gathered from the information sources' profiling activities, detailed in Subsystem 2.

(b) Install a classification/indexing strategy in order to facilitate information retrieval from the information source bank.

(c) Revise or adapt the subsystem's structure to accommodate unique local information needs.

Subsystem Description

The initial information sources bank developed in Subsystem 1 offers only limited access to specific sources of information contained in the bank. The primary purpose of this subsystem is to introduce a more substantial range of entry paths into the information bank through the use of classification and cross indexing techniques which will facilitate retrieval of information stored. This subsystem is composed of four major activities, viz.,

1. Tabulation of Resources. Responses to the information sources profiling activity generated in Subsystem 2 are tabulated. This activity involves the aggregation of data for specified information parameters from responding sources of technological information. The information-exchange parameters specified in the questionnaire (Appendix) are tabulated in a summarized format using an expanded version of the initial entry card shown in Fig. 10.3. The expanded entry card is shown in Fig. 10.6 and contains the following index types:

INSTITUTION NAME, LOCATION and CONTACT: Name and complete address of institution (A); Country (B); and name/title of contact at institution (C). This file is arranged according to a country-alphabetic system.

INSTITUTION IDENTITY: Types of institutions involved in housing/human settlements work (D). This file is indexed in categories corresponding to the following institutional types:

(a) Governmental/Public Agency

(b) Educational Institution

(c) Research and Development Institute

(d) Professional Association

(e) Foundation

(f) Extension/Service-Organization

(g) Commercial Organization

(h) Industrial Organization

(i) Bank/Financial Organization

(j) International Agency

(k) (Other)

DISSEMINATION SERVICES: Information diffusion activities and services currently offered by institutions (E). This file is indexed according to the following types of diffusion instrumentalities:

(a) *Publications:* Publish books, monographs, journals, bulletins, newsletters, research/technical directories/reports, reprints.

(b) *Current Awareness:* Distribute periodic lists of your new information acquisitions. How often?

(c) *Bibliographies and Abstracts:* Compile indexes, abstracts, union lists, statistical abstracts of current holdings.

(d) *Inter-Library Loan:* Have reciprocal arrangements to lend library materials.

(e) *Literature Search:* Provide literature search services on special topics.

(f) *Technical Inquiry:* Answer technical inquiries in area of expertise.

(g) *Data Sharing:* Exchange processed data, surveys, preliminary analyses, computer data files.

(h) *Extension/Training:* Provide extension services and training for appropriate clientele.

(i) *Consultancy:* Expert/specialist contracting, seconding, sharing, personnel exchange.

INFORMATION EXCHANGE POLICIES: Describes the information exchange policies of institutions viz., each of the information dissemination activities/services described above (E) and is arranged according to the same categories (F).

INFORMATION EXCHANGE SETUP: Describes the scope of information exchange capabilities of institution as a function of both size of information base and extent of information personnel assigned to manage the information base. This file is divided into three categories (G):

Small-scale: 0-2 staff and/or less than 1,000 documents processed annually.

Medium-scale: 3-6 staff and/or 1,000-10,000 documents processed annually.

Large-scale: 7 or more staff and/or more than 10,000 documents processed annually.

AREAS OF CONCERN: Describes the areas of professional interest in housing/settlement technology and the availability of information in interest areas. A total of 29 areas are listed and defined. This file is created by producing an individual entry card copy for every interest area noted on each entry card and sorting them under the 29 areas.

1. *Housing Documentation/Statistics:* Project reports, statistics and survey, case studies, feasibility studies and evaluation for low-income housing.

2. *Housing Research/Types:* Demonstration/experimental houses, model houses, expansible houses, demountable houses, emergency shelters, disaster housing, mobile homes.

3. *Indigenous/Rural Housing:* Rural and indigenous housing solutions, rural improvements, rural community facilities.

4. *Co-operative Housing:* Management organization, economics/financing, case studies.

5. *Self-Help Construction:* Self-help, aided self-help, mutual help, public participation, settlement improvement, organization.

6. *Construction Materials:* Materials research, composition, properties, uses of construction materials, both indigenous and new, effects of wind, earthquake, fire, etc. on materials, low-cost applications.

7. *Construction Methods and Management:* Housing construction industry, industrialized and prefab building, management, factor mix (labor/capital intensive), structural engineering, on site construction methods, prefabrication, owner built housing, construction manuals, construction management.

8. *Architectural Design:* Design factors and concepts for low-income groups, recreational and community facilities, traditional architecture, aesthetics, physical/site relationships.

9. *Climatological Aspects:* Climatic factors in low-income housing, influence of climate on health, environment, building materials, comfort factors, current and indigenous solutions.

10. *Housing Policy and Legal Aspects:* Sites/services, government subsidies, rent control, land use, land reform, building codes, zoning, para legal aspects, standards, national housing policies.

11. *Housing Management:* Maintenance, modernization and conversion, repair, internal rules, tenant relations, contractual agreements, training housing managers, housing management, for low-income people.

12. *Demographic Aspects:* Population profiles, population research, migration, density-growth, family size, income and effect on housing.

13. *Social and Health Aspects:* Social and health aspects of human settlements, ways, methods and concepts for their improvement, impact and case studies.

14. *Economic Aspects:* Economic development, housing finance, feasibility studies, economics of housing and housing analysis, base studies, economic conditions, development forecasting, planning, policy, research, employment/income generation.

15. *Financing Methods:* Savings and loans, direct loans, property improvements loans, construction loans, amortization, mortgage and other methods/ arrangements used to finance housing for/by low-income groups.

16. *Ecological Aspects:* Pollution (water, air, and land), ecological management and environment, as related to low-income human settlements.

17. *Energy:* Conventional and alternative, renewable and nonrenewable sources of energy (direct solar, wind, tidal, bio-mass, geo-thermal, marine-thermal, fossil, atomic, etc.) conversion processes, conservation, recycling, and applications, especially as appropriate to the needs of low-income people.

18. *Sanitary Engineering:* Waste and sewage disposal water supply, drainage, and waste recycling, utilization as related to low-income settlements.

19. *Urbanization:* Growth of cities, rural-urban migration, urban sprawl, especially as these trends relate to low-income people.

20. *Regional Planning:* Macro/micro level planning, state/provincial planning, planning legislation, land use controls, industrial and commercial (location) policies and planning, planning for new and remote settlements, preservation of historical sites.

21. *Rural Planning:* Planning specifically related to the improvement of human settlements in rural areas and their integration into the fabric of the region, province/state, and nation.

22. *City Planning:* City planning development growth, and management, metropolitan area planning, master plan studies and surveys, metropolitan/city level legislation, zoning, land use control, municipal government, new towns, garden cities, satellite cities, model cities, planned communities.

23. *Community/Neighborhood Planning:* Community/neighborhood level planning, urban renewal, slum clearance, and surveys. Planning of site residential area planning, community centers and planning legislation.

24. *Education/Community Development:* Community awareness, technical assistance, education, literacy, community development, home economics, community nutrition, community improvement projects, case and impact studies, surveys.

25. *Squatters and Marginal Settlements:* Slums, squatters, resettlement schemes, slum clearance/improvement, or slum development.

26. *Land Reclamation:* Desert control, swamp, fill or drainage, terracing, and other techniques for human settlements.

27. *Appropriate Technology:* Philosophy, concepts, and theories of development and use of appropriate technologies.

28. *Agricultural Development:* Irrigation, agro-industry, agronomy, extension and training, production, farm management, for low-income settlements.

29. *Traffic and Transportation:* Traffic engineering/lay-out, and materials that affect housing and settlement of low-income people.

2. Indexing System. Decisions on the types of information classification/indexing systems to be used in the expanded version of the initial information sources bank. The criteria for selecting appropriate classification/indexing systems will be similar to those described in Subsystem 1. Transferring the tabulated entry cards into discrete files. The number of files will correspond to the number or types of classification/indexing systems to be used.

3. Control Point Update. Update the control point (enter control point No.3) in the initial information sources bank when sorting the completed deck of cards.

Flowchart

The flowchart depicting the various steps and sequences of steps involved in the implementation of this subsystem is included as Flowchart 3 (Fig. 10.7).

10.7 SUBSYSTEM 4: GENERATING ADDITIONAL INFORMATION SOURCES

Subsystem's Objectives

Upon the completion of the operations involved in this subsystem, the users will be able to continuously generate a listing of additional sources of information originating from the institutions currently participating in the information exchange network.

Subsystem Description:

This subsystem in general, tries to capture other sources of information, which were overlooked during the installation of the initial information sources bank. It also expands the final information sources bank (Subsystem 3), making the users' information systems adaptable to the dynamics of the interaction between the demand and supply of technical resource information.

This subsystem uses the lists of new information sources identified by the responding institutions in the profiling strategy (see Subsystem 2). The new sources are contacted and profiled using the steps detailed in subsystem 2. The procedure is repeated as additional sources are identified. Theoretically, a "saturation point"

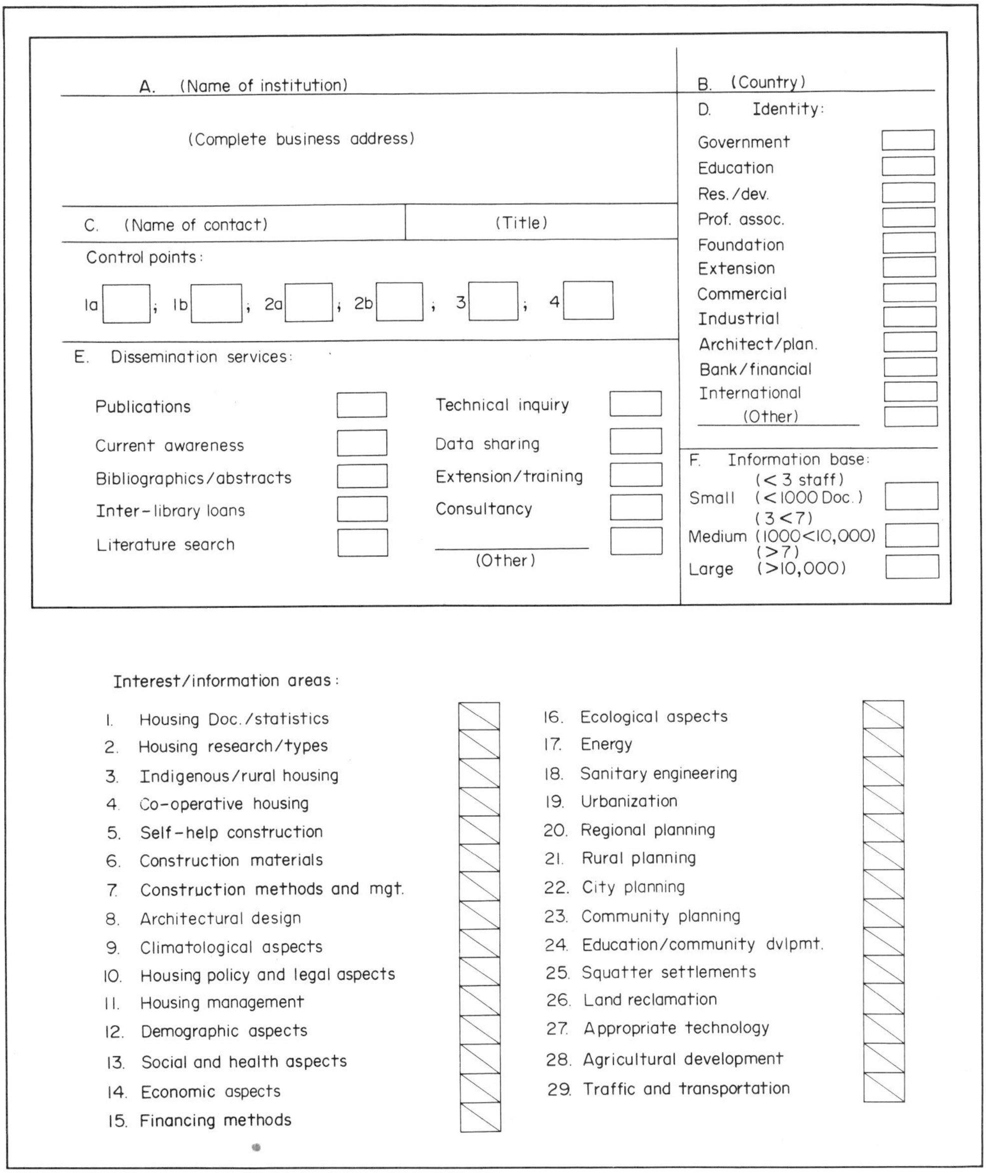
A. (Name of institution) | B. (Country)

(Complete business address)

D. Identity:
Government
Education
Res./dev.
Prof. assoc.
Foundation
Extension
Commercial
Industrial
Architect/plan.
Bank/financial
International
(Other)

C. (Name of contact) | (Title)

Control points:

1a ; 1b ; 2a ; 2b ; 3 ; 4

E. Dissemination services:

Publications
Current awareness
Bibliographics/abstracts
Inter-library loans
Literature search

Technical inquiry
Data sharing
Extension/training
Consultancy
(Other)

F. Information base:
Small (< 3 staff) (< 1000 Doc.)
Medium (3<7) (1000<10,000)
Large (>7) (>10,000)

Interest/information areas:

1. Housing Doc./statistics
2. Housing research/types
3. Indigenous/rural housing
4. Co-operative housing
5. Self-help construction
6. Construction materials
7. Construction methods and mgt.
8. Architectural design
9. Climatological aspects
10. Housing policy and legal aspects
11. Housing management
12. Demographic aspects
13. Social and health aspects
14. Economic aspects
15. Financing methods
16. Ecological aspects
17. Energy
18. Sanitary engineering
19. Urbanization
20. Regional planning
21. Rural planning
22. City planning
23. Community planning
24. Education/community dvlpmt.
25. Squatter settlements
26. Land reclamation
27. Appropriate technology
28. Agricultural development
29. Traffic and transportation

Fig. 10.6. Sample Information Sources Bank Entry Card

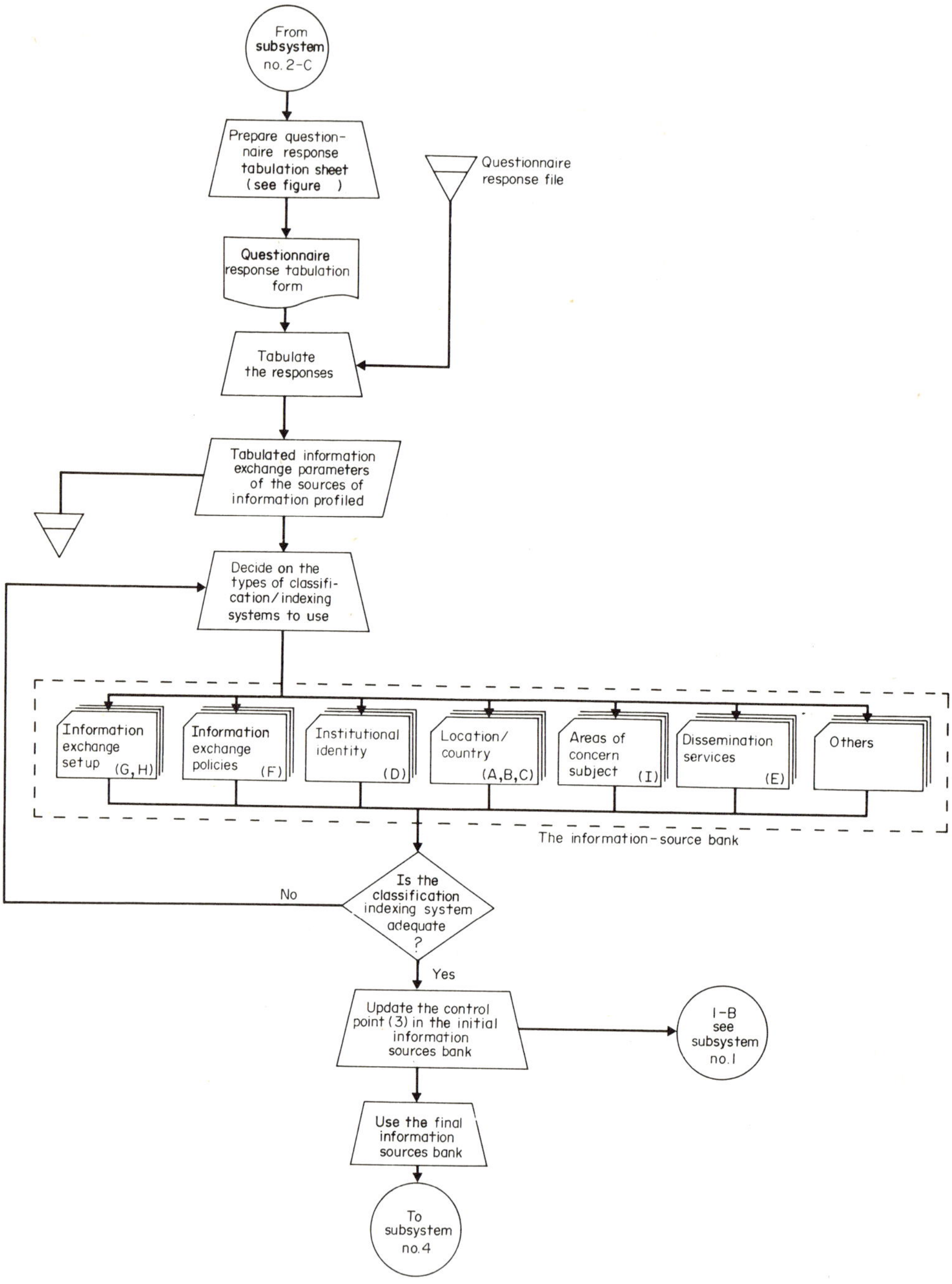

Fig. 10.7. Flowchart No.3 : Subsystem No. 3
Process of Expanding the Initial Information-Sources Bank

of contacts with information sources will be attained. This is the point wherein the participating institutions can no longer recommend sources of information not already contacted.

Throughout the series of information exchange transactions, every effort should be made to generate more sources of information from the participating institutions. This might require periodic requerying of all participating institutions.

Flowchart

The flowchart depicting the various steps and the sequences of steps involved in the implementation of this subsystem is included in Flowchart 4 (Fig. 10.8).

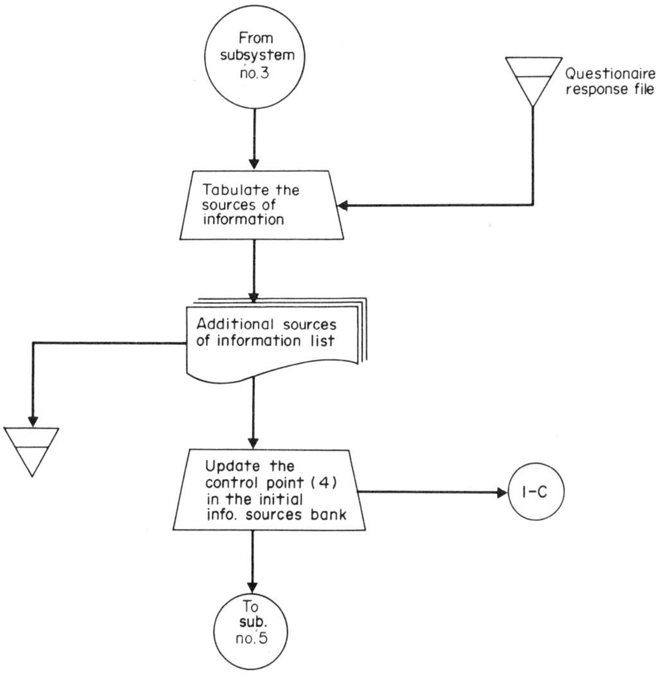

Fig. 10.8. Flowchart No.4 : Subsystem No. 4
Process of Generating More Sources of Information

10.8 SUBSYSTEM 5 : INITIATING INFORMATION EXCHANGE

Subsystem Objectives:

After completion of the operations involved in this subsystem, the information manager will be able to:

(a) Identify institutional sources of information which are compatible in areas of interest and information exchange capabilities.

(b) Establish information exchange arrangements with these institutions.

(c) Monitor the maturation of these exchange arrangements.

Subsystem Description:

Subsystem 5 is made of three activities, viz.,

1. Evaluation of Interest Capability. Based on profiling response to list of topics or areas of interest (Questionnaire Part G) the initiating institutes identified information sources compatible with its own interests. A high level of mutuality in subject interest warrants development of formalized information exchange arrangements.

2. Evaluation of Exchange Compatibility. The hierarchy of institutionalized information exchange arrangements described earlier (Table I and Questionnaire Part E and F) is applied to evaluate the compatibility between the initiating institution and selected information sources. The key here is *quid pro quo* matching — where the initiating institution seeks exchange arrangements only at the level which it can reciprocate. Three levels of exchange are identified, requiring increased mutual commitment to pass from one level to the next. The levels are:

   LEVEL I

   (a) *Correspondence exchange.* One or more two way exchanges of letters, inquiries, etc.

   (b) *Profile/questionnaire exchange.* The completion and return of a formal questionnaire profiling information sources.

   LEVEL II

   (c) *Publications exchange.* One or more exchanges of print and/or multi-media documents.

   (d) *Current awareness list exchange.* Periodic (at least two) exchange of lists of new information acquisition titles.

   (e) *Bibliography/abstract exchange.* Periodic exchange of bibliographies, indexes and/or abstracts of currently held document titles.

   (f) *Inter-library loan/exchanges.* At least one exchange of library loan materials.

   LEVEL III

   (g) *Literature search service.* At least one exchange of literature search services on a specified topic.

   (h) *Technical inquiry service.* The response to at least one technical inquiry by each exchange partner.

   (i) *Data sharing service.* At least one instance where primary data (raw or compiled), surveys and/or computer data files are exchanged or shared.

   (j) *Extension/training.* The exchange of curriculum and/or personnel needed to inaugurate or strengthen existing extension or training programs.

   (k) *Consultancy.* The exchange or secondment of at least one staff personnel as the contractury for professional service.

3. Link Information and Users. Use the information source bank to link supplier information system with initiating institution users and their clients.

10.9 SUMMARY

Linkage establishment is an important stage in the development of information networks. It is at this stage that a small information system will be able to:

(a) Initiate a "contact" with other sources of information in a given field of human interest.

(b) Investigate their system's compatibility with other information systems.

(c) Identify "common" information transfer capabilities which could be utilized as initial channels of information exchange.

(d) Decide on the need for the development linkages with other information networks.

The strategy proposed in this paper is but one approach to linkage development. The linkage development strategies presented here have been substantially simplified in order to present the processes involved in a much more manageable fashion. The user is advised to adapt the implementation steps outlined here, to his own environment, his resources and its capabilities.

If by some good fortune several institutions in developing countries take up the operationalization of these linkage formation strategies there will emerge the possibility of cross-linking/aggregating these mini-networks into a more globalized information exchange structure. It would seem appropriate that if such a condition came to pass, one or more international agency mandated to supported information exchange might sponsor this stage of development. At the very least, such an agency would be assured that these user-initiated mini-networks were developed out of the self interest of localized information users. Such a partnership might stand as the first user-initiated global information system among a plethora of faltering supplier induced systems.

Appendix

SOURCES OF HOUSING AND HUMAN SETTLEMENTS INFORMATION

INSTITUTIONAL PROFILE QUESTIONNAIRE

A. Name of Institution:

Complete Business Address: B. Country:

C. Contact: Position:

D. INSTITUTIONAL IDENTITY (Check-off your type of institution)

Governmental/Public	☐	Commercial	☐
Educational	☐	Industrial	☐
Research and Development	☐	Architect/Planning	☐
Professional Association	☐	Bank/Financial	☐
Foundation	☐	International	☐
Extension/Service	☐	____________	☐

DESCRIBE BRIEFLY THE OBJECTIVES OF YOUR INSTITUTION/ORGANIZATION.

E. DISSEMINATION SERVICES (Check the activities/services your institution currently offers)

Publications	☐	Publish books, monographs, journals, bulletins, newsletters. research/technical directories, reprints. (UNDERLINE TYPES)
Current Awareness	☐	Distribute periodic lists of your new information acquisitions. How often?________________.
Bibliographies and Abstracts	☐	Compile indexes, abstracts, union lists, statistical abstracts of current holdings.
Inter-Library Loan	☐	Have reciprocal arrangements to lend library materials.
Literature Search	☐	Provide literature search services on special topics.
Technical Inquiry	☐	Answer technical inquiries in area of expertise.
Data Sharing	☐	Exchange processed data, surveys, preliminary analyses, computer data files.
Extension/Training	☐	Provide extension services and training for appropriate clientele.
Consultancy	☐	Export/specialist contracting, seconding, sharing, personnel exchange.

(Please send us examples of your publications and/or your publications list/directory).

INFORMATION EXCHANGE POLICIES (Please describe briefly your information dissemination policies vis. dissemination activities/services in (E) or send written policies/procedures).

F. INFORMATION STAFF (Personnel Assigned to information management duties):

		NUMBER
1.	Professional Librarians/ Documentalists	☐
2.	Library Technicians/ Abstractors/Cataloguers	☐
3.	Area/Subject Specialists	☐
4.	Editors/Publications Specialists	☐
5.	Graphic/Media Specialists	☐
6.	System/Computer Specialists	☐
7.	Translators	☐
8.	Administrators	☐
9.		☐
10.		☐

INFORMATION BASE/ACQUISITIONS (Number of Housing/Human Settlements titles only):

		THIS YEAR	THIS YEAR
1.	Books/Monographs	☐	☐
2.	Serials/Periodicals	☐	☐
3.	Technical/Research Reports	☐	☐
4.	Country/Regional Reports	☐	☐
5.	Microfiche/Forms	☐	☐
6.	Photos/Slides/ Films	☐	☐
7.	Maps/Prints/ Graphics	☐	☐
8.	Vertical Files/ Fujitive	☐	☐
9.	Pamphlets/Reprints	☐	☐
10.	Machine Readable Data	☐	☐

G. Please check-off topics of institutional interest with special reference to housing/settlements for low-income people ⧅✓ and if you collect information/ data on topic ⧅✓

	TOPIC		DEFINITION and SCOPE of TOPIC
1.	Housing Documentation/ Statistics	⧅	Project reports, statistics and survey, case studies, feasibility studies and evaluation for low-income housing.
2.	Housing Research/Types	⧅	Demonstration/experimental houses, model houses, expansible houses, demountable houses, emergency shelters, disaster housing, mobile homes.
3.	Indigenous/Rural Housing	⧅	Rural and indigenous housing solutions, rural improvements, rural community facilities.
4.	Co-operative Housing	⧅	Management organization, economics/financing, case studies.
5.	Self-Help Construction	⧅	Self-help, aided self-help, mutual help, public participation, settlement improvement, organization.
6.	Construction Materials	⧅	Materials research, composition, properties, uses of construction materials, both indigenous and new, effects of wind, earthquake, fire, etc. on materials, low-cost applications.
7.	Construction Methods and Management	⧅	Housing construction industry, industrialized and prefab building, management, factor mix (labor/capital intensive), structural engineering, on site construction methods, prefabrication, owner built housing, construction manuals, construction management.

8.	Architectural Design	Design factors and concepts for low-income groups, recreational and community facilities, traditional architecture, aesthetics, physical/site relationships.
9.	Climatological Aspects	Climatic factors in low-income housing, influence of climate on health, environment, building materials, comfort factors, current and indigenous solutions.
10.	Housing Policy and Legal Aspects	Sites/services, government subsidies, rent control, land use, land reform, building codes, zoning, para legal aspects, standards, national housing policies.
11.	Housing Management	Maintenance, modernization and conversion, repair, internal rules, tenant relations, contractual agreements, training housing managers, housing management, for low-income people.
12.	Demographic Aspects	Population profiles, population research, migration, density-growth, family size, income and effect on housing.
13.	Social and Health Aspects	Social and health aspects of human settlements, ways, methods and concepts for their improvement, impact and case studies.
14.	Economic Aspects	Economic development, housing finance, feasibility studies, economics of housing and housing analysis, base studies, economic conditions, development forecasting, planning, policy, research, employment/income generation.
15.	Financing Methods	Savings and loans, direct loans, property improvements loans, construction loans, amortization, mortgage and other methods/arrangements used to finance housing for/by low-income groups.
16.	Ecological Aspects	Pollution (water, air, and land), ecological management and environment, as related to low-income human settlements.
17.	Energy	Conventional and alternative, renewable and non-renewable sources of energy (direct solar, wind, tidal, bio-mass, geo-thermal, marine-thermal, fossil, atomic, etc.) conversion processes, conservation, recycling, and applications, especially as appropriate to the needs of low-income people.
18.	Sanitary Engineering	Waste and sewage disposal water supply, drainage, and waste recycling, utilization as related to low-income settlements.
19.	Urbanization	Growth of cities, rural-urban migration, urban sprawl, especially as these trends relate to low-income people.
20.	Regional Planning	Macro/micro level planning, state/provincial planning, planning legislation, land use controls, industrial and commercial (location) policies and planning, planning for new and remote settlements, preservation of historical sites.

21.	Rural Planning	⧅	Planning specifically related to the improvement of human settlements in rural areas and their integration into the fabric of the region, province/state, and nation.
22.	City Planning	⧅	City planning development growth, and management, metropolitan area planning, master plan studies and surveys, metropolitan/city level legislation, zoning, land use control, municipal government, new towns, garden cities, satellite cities, model cities, planned communities.
23.	Community/ Neighborhood Planning	⧅	Community/neighborhood level planning, urban renewal, slum clearance, and surveys. Planning of site residential area planning, community centers and planning legislation.
24.	Education/Community Development	⧅	Community awareness, technical assistance, education, literacy, community development, home economics, community nutrition, community improvement projects, case and impact studies, surveys.
25.	Squatters and Marginal Settlements	⧅	Slums, squatters, resettlement schemes, slum clearance/improvement, or slum development.
26.	Land Reclamation	⧅	Desert control, swamp, fill or drainage, terracing, and other techniques for human settlements.
27.	Appropriate Technology	⧅	Philosophy, concepts, and theories of the development and use of appropriate technologies.
28.	Agricultural Development	⧅	Irrigation, agro-industry, agronomy, extension and training, production, farm management, for low-income settlements.
29.	Traffic and Transportation	⧅	Traffic engineering/lay-out, and materials that affect housing and settlement of low-income people.

J. SOURCES OF INFORMATION

Please provide us with a list of your *most valuable sources* of information in the areas of housing/human settlements

1. Name of Institution: Country:
   Business Address:

   Contact: Title:

2. Name of Institution: Country:
   Business Address:

   Contact: Title:

3. Name of Institution: Country:
   Business Address:

   Contact: Title:

4. Name of Institution: Country:
Business Address:

Contact: Title:

5. Name of Institution: Country:
Business Address:

Contact: Title:

6. Name of Institution: Country:
Business Address:

Contact: Title:

7. Name of Institution: Country:
Business Address:

Contact: Title:

8. Name of Institution: Country:
Business Address:

Contact: Title:

9. Name of Institution: Country:
Business Address:

Contact: Title:

10. Name of Institution: Country:
Business Address:

Contact: Title:

(Please use additional sheets, if necessary)

Index